A. Hein

Processing of SAR Data

Springer-Verlag Berlin Heidelberg GmbH

Engineering ONLINE LIBRARY

http://www.springer.de/engine/

Achim Hein

Processing of SAR Data

Fundamentals, Signal Processing, Interferometry

With 176 Figures, 32 in colour

 Springer

Dr. Achim Hein
Fürther Str. 212
D - 90429 Nürnberg
Achim.Hein@dr-hein.com

ISBN 978-3-642-05710-6   ISBN 978-3-662-09457-0 (eBook)
DOI 10.1007/978-3-662-09457-0

Library of Congress Cataloging-in-Publication-Data

Hein, Achim.
Processing of SAR data : fundamentals, signal processing, interferometry / Achim Hein.
p. cm. -- (Signals and communication technology)

1. Synthetic aperture radar. I. Title. II Series.
TK6592.S95.H45 2004
621.3848--dc22

http://www.springer.de

© Springer-Verlag Berlin Heidelberg 2004
Originally published by Springer-Verlag Berlin Heidelberg in 2004
Softcover reprint of the hardcover 1st edition 2004

Cover design: design & production, Heidelberg
Printed on acid free paper   62/3020/M - 5 4 3 2 1 0

Preface

The work on hand has been developed in the project sector 2: optimal signal processing, sensor data fusion, remote sensing – SAR, of the Center of Sensor Systems ZESS, as well as in the Institute for Data Processing of the University of Siegen.

I would like to thank all those people who directly or indirectly helped to make this work possible.

Special thanks to Prof. Dr.-Ing. habil. O. Loffeld whose exceedingly dedicated encouragement and precious discussions considerably contributed to the realisation of this work.

Furthermore I would like to thank the staff of ZESS and the Institute for Data Processing for their cooperation and their kindly support.

Thank you very much also to the following people of project sector 2 who contributed with their commitment to the success of this work: Ch. Arndt, D. Balzer, F. Klaus, R. Krämer, A. Stadermann, S. Oerder, F. Schneider, S. Knedlik, H. Nies, I. Stöver, A. Kunz, J. Hegerfeld, Ch. Gessner, E. Steinmann, Th. Bischoff.

I would like to thank Melanie Setz, Robert Setz and Michael Thiele for the assistance in reprocessing and translating this publication.

A special thank you also to my partner Ricarda who in many ways had been affected directly by the realisation of this work.

Processing of SAR data with special emphasis in interferometry

Summary

This work describes in a comprehensive manner new methods and algorithms concerning the interferometric processing of SAR data. It provides a general description with strong emphasis on system and signal theory. In this context a study is presented which derives a new, non-linear, two-dimensional, locally independent SAR-processing algorithm. It shows excellent phase consistency. A comparison is presented of different SAR-focussing algorithms by making use of a uniform starting point, the two-dimensional point target spectrum. With regard to sub-pixel accuracy, a method is introduced for the co-registration process. Based on the geometrical effects of the side-looking geometry, computationally efficient algorithms are developed to correct foreshortening, shadowing and layover. Based on a method of automatically generating reference heights, the outfinal SAR data product is geo-coded.

Zusammenfassung

In der vorliegenden Arbeit werden umfassend bestehende und neuartige Verfahren und Algorithmen zur interferometrischen Verarbeitung von SAR-Daten beschrieben. Es erfolgt eine durchgängige system- und signaltheoretische Ableitung der interferometrischen SAR-Signalverarbeitung. Hierbei werden – ausgehend von einem gemeinsamen Ausgangspunkt, dem zweidimensionalen Punktzielspektrum – verschiedene SAR-Prozessoren abgeleitet, gegenübergestellt und ein neuartiger zweidimensionaler, nichtlinearer, ortsvarianter Prozessierungsalgorithmus entwickelt, der zu den Präzisionsprozessoren zu zählen ist. Im weiteren Verlauf wird ein Verfahren zur subpixelgenauen Co-Registrierung erläutert und Korrekturverfahren zu geometrisch aufnahmebedingten Aufzeichnungsfehlern wie Layover, Shadowing und Foreshortening vorgestellt. Die komplexen SAR-Daten werden in den folgenden Schritten geocodiert, wobei zur Höhenreferenzierung insbesondere ein Verfahren zur automatischen Höhenpaßpunktgenerierung vorgestellt wird.

Summary

Ever since the first satellite equipped with a remote sensing system of synthetic aperture (SAR sensor: Seasat) was put into the orbit (1978), the use of SAR images has become an inherent part of numerous scientific disciplines. Apart from imaging areas for mapping purposes, SAR scenes have already been used in the fields of geology, agriculture and forestry for classifying and inventory controlling of cultivation areas and forests as well as determination of soil humidity. By means of the SAR interferometry (SAR-IF) a new discipline in remote sensing has quickly been established over the last few years in a way that those interferometric generated SAR images such as digital terrain models or digital motion models of volcanoes have become regular components within the product range of SAR remote sensing results.

In the work on hand a deduction of the entire interferometric processing of SAR data is done. The objective target is a system-theoretical description of the working steps necessary within the field of SAR interferometry starting with the received raw data to the description of geometric substructures to the generation of interferograms. All essential problems within the SAR interferometry are solved. Apart from the SAR processing also the adjustment of different geometrically caused decorrelation effects is carried out. A modern method of SAR processing is deduced and after a discussion on the already existing well-known methods a generalised derivation of all methods is introduced. The geometric decorrelation effects are egalised according to their fundamental description by means of both novel methods of orthometric projection and adjustment of inevitable geometrical effects as well as improved approaches concerning registration problems and also novel solutions in generating height pass points.

All in the course of this study developed methods and algorithms are basically orientated on the description of an interferometry processor. This processor must unify all complex work steps from the raw data tapes of different sources to be read to the two-dimensional non-linear distance-dependent scene focussing to the geo-referencing digital terrain model. In the Center for Sensor Systems, project area 2: optimal signal processing, sensor data fusion, remote sensing – SAR (where this work has been designed), the biggest parts of the whole interferometric conceptual problem could be solved. In other words, general problems are defined, theoretically compiled, different approaches realised and final solutions implemented. Future developments will extend the fields of SAR applications at a progressive rate. Apart from multidimensional mapping, particularly the methods for classifying and using of remote sensing data within a complex geo-information system will become more and more important. The work on hand defines the necessary interferometric basics and reveals the required steps of signal processing.

Table of Contents

1 Introduction

▪ 1.1 Applications of SAR sensors

Radar has been employed for electronic observing since World War II and ever since has experienced a substantial technical boost. Conventional radar mainly is used for both locating an object and measuring its distance and speed. The progressing technical development as well as experiences over many years soon provided the opportunity to construct devices supplying planar images of areas, for example the Synthetic Aperture Radar (SAR). In 1953, Carl A. Wiley for the first time published the basic principles of a radar with synthetic aperture, SAR (Synthetic Aperture Radar) as a patent specification with the title "Pulsed Doppler Radar and Means". SAR is an active, imaging method based on microwaves which is used on mobile platforms such as aeroplanes or satellites. It makes possible pictographic illustration (similar to maps) of areas with different backscattering characteristics of electromagnetic waves and compared with conventional radar systems offers fundamental improvements in geometric resolution. Reflective characteristics of individual objects are rendered with grey tones or colours.

Among other things SAR differs from other known mapping methods by its active operational mode. In other words, the sensor is equipped with its own energy source for lighting the area. In opposition to passive methods deployed in the optical and short-wave infrared sector that image an area by means of reflected sunlight, SAR allows unrestricted service irrespective of solar radiation and daytime. Furthermore it is possible to choose the transmitted wave length in such a way that the attenuation of electromagnetic waves caused by the atmosphere can be disregarded. Thus, the SAR sensor can be operated under almost any weather conditions and is able to even observe an area deep in clouds.

In 1978, the USA brought the first satellite equipped with a SAR sensor for picturing the earth's surface into the orbit. Since July 1991, ERS-1 ("European Remote Sensing Satellite") has been in a polar orbit around the earth supplying amongst other things also SAR images of the entire earth's surface with a resolution of about 10 m. This SAR sensor is operating with a wave length of 5.6 cm (C-band).

In Image 1.1 is taken a SAR image taken by ERS-1 whereas Image 1.2 is an infrared photo taken by the satellite Landsat which is equipped with sensors operating wavelengths from 0.5 µm to 1.1 µm [BUC94]. Both images were taken on 9th August 1991 in intervals of 42 minutes showing the Irish coastline close to Waterford.

Image 1.1: ERS-1 image

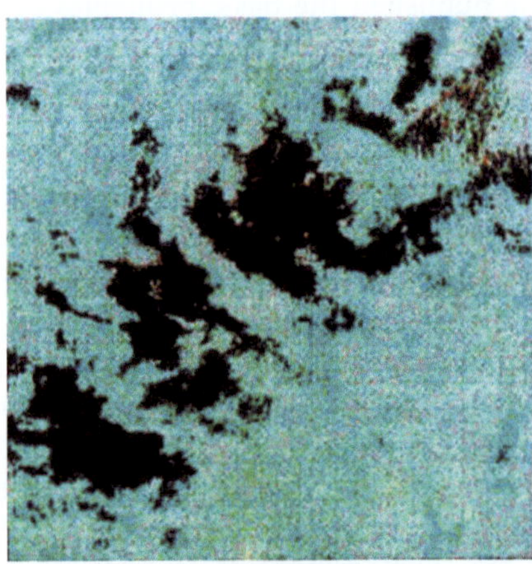

Image 1.2: Optical image

At this time the area was covered with clouds so that the optical sensors of Landsat for the most part failed to observe the surface. The SAR image, however, is not affected by the clouds. Only 10% of all European images taken by optical satellites are not affected by clouds which means that only every six months useful images can be taken.

With the SAR sensors of the European Space Agency (ESA) it is possible to take meaningful images every day.

Particularly these characteristics in conjunction with the SAR principle so far have been the only way to map the surface of the venus [BUC94]. As the venus is always covered with clouds it is not possible to take high resolution images for mapping of its surface by only using optical or infrared sensors. In 1983, the radar satellites Venera 15 and 16 of the former Soviet Union for the first time took SAR images of the venus' surface with a resolution of about 1 km.

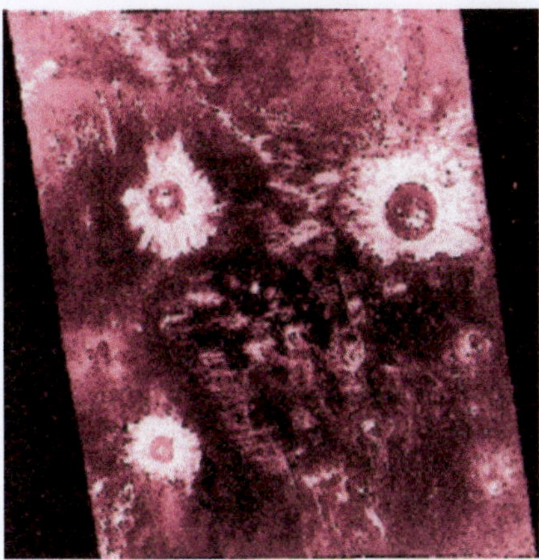

Image 1.3: Venus image

Since 1989 the American radar satellite Magellan has delivered SAR images of the venus with a resolution of approximately 100 m. Image 1.3 reveals an area of the venus' surface. Easily visible are the three large craters with diameters from 37 km up to 50 km. The edge and emissions of the craters as well as the surrounding areas consist of fairly rough material explaining the high radar backscattering coefficient which results in the bright coloured marks.

In the meantime SAR images have become an integral part in numerous scientific disciplines. Apart from imaging areas for mapping purposes, SAR images have already been used in the fields of geology, agriculture and forestry for classifying and inventory controlling of cultivation areas and forests as well as determination of soil dampness (Image 1.4).

Image 1.4 a: ERS scene Flevoland

Image 1.4 b: Humidity-classified ERS scene Flevoland (DLR)

By imaging oceanic and polar areas, oceanography and climatic research are provided with important information on sea currents, depths, waves, formation and back-formation of glaciers and ice. In the range of environment protection for example oil slicks polluting the oceans can be detected. Particularly the intensively researched fields of various SAR applications have become wider and wider over the last years.

Further applications of SAR image processing can be used to represent three-dimensional images. Here one can estimate an altitude information using the interferogram of two SAR images by means of a special processing of stored phase information of the raw data (height estimation) [LOF94] (Image 1.5: *Interferometric SAR image of the volcano Aetna*). With the aid of SAR interferometry (SARIF) a new discipline in remote sensing could be established so quickly over the last few years that in the meantime interferometric generated SAR images such as Image 1.5 or 1.6 have been added to the common product range of SAR remote sensing results.

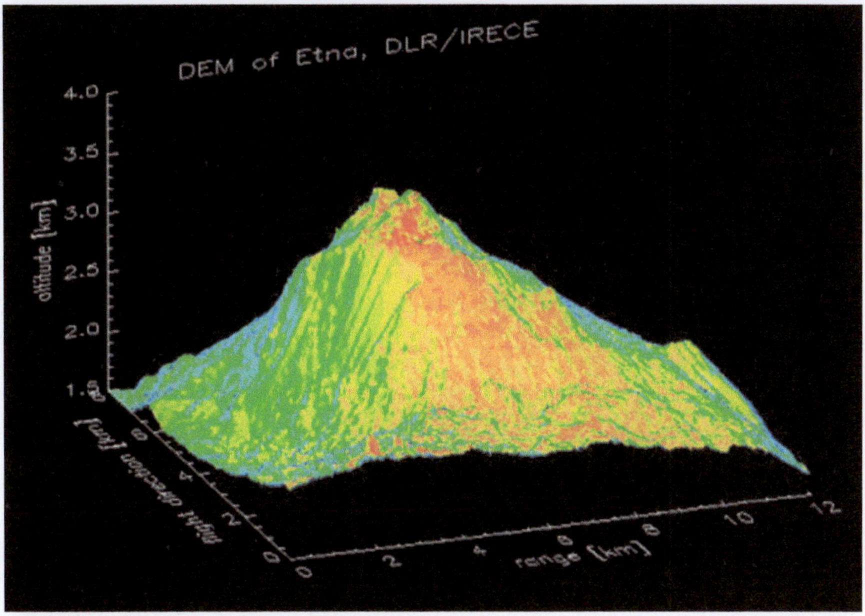

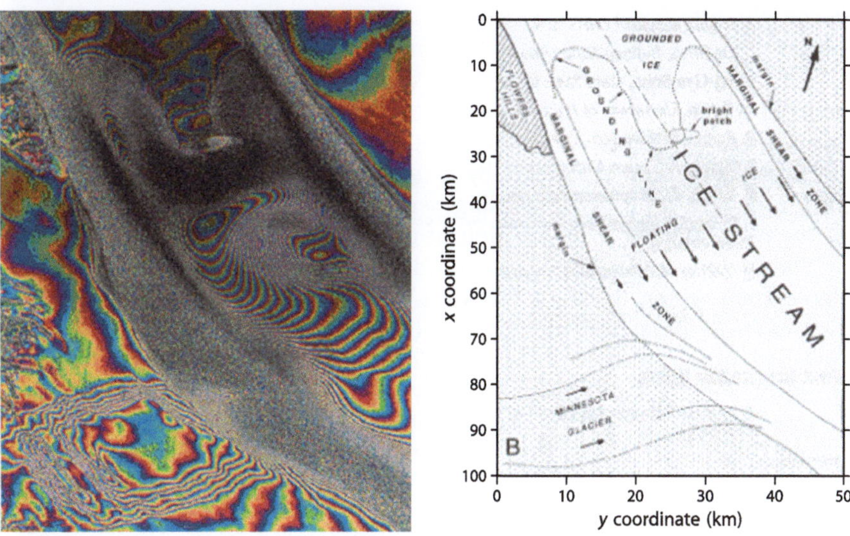

Image 1.5: Interferometrical SAR scene of aetna (DLR)

Image 1.6: Velocity of the Minnesota Glaciers
[http://www.asf.alaska.edu/step/abstracts/abstract15.html]

▪ 1.2 Problem and analysis

The SAR of ERS-1 transmits microwave pulses with a pulse repetition frequency (PRF) of about 1.680 Hz of linear-vertical polarisation (vertical-vertical) in a frequency range of 5.29 GHz (C-band) and stores the received echo signals coherently. During the processing of the stored raw data at a later date, the coherency of the radar receiving signals is used to create an aperture with a distance-independent azimuth resolution to synthesise a substantially bigger aperture (factor 600). Therefore is used the flight movement of the carrier system, the knowledge of the respective position of the antenna and the afterwards completed integration out of the relatively small antenna into the SAR processor. The conventional SAR as any runtime-measurement is at a disadvantage as two point-targets with exactly the same distance to the sensor can not be separated spatially since their echoes reach the receiving antenna at the same time.

The interferometic SAR (Image 1.7) uses a second sensor position (sensor 2) spatially separated from the first one (sensor 1) to dissolve the just mentioned ambiguity and to enable a three-dimensional representation of images. Thus the recording geometry is similar to well known stereometric methods and implicates that equal point-targets have a varying range distance difference to the diverse sensor positions. Whereas stereometric analysis systems try to determine the parallax by measuring the respective range, the interferometric SAR takes advantage of the coherency of the signal and the characteristics of the monochromatic electromagnetic wave. For these used waves the signal phase changes due to the range which is covered twice, means the distance to the interesting illuminated object. Measuring the difference of these appearing phase shifts by means of two receiving antennas separated by a so-called baseline (baseline = distance vector between the two recording sensors) allows an exact determination of the range distance at a fraction of the used wave length λ.

There are two fundamental methods to implement an interferometric radar system which differ in the kind of the baseline configuration. In the first case the baseline is determined by the distance between two antennas installed side by side to the carrier. The observed area is illuminated temporarily, simultaneously and spatially-parallely to be recorded. According to this there is only one overflight with two antennas necessary to produce a interferometric image. That is why for this kind of overflight normally the term "Single Pass Interferometry" is referred to. Dornier e.g. possesses an interferometric SAR system "DO-SAR", which works within the C-band and can be installed on SAR carriers such as the aeroplanes DO-228 or TransAl.

In the second case the SAR carrier carries along only one antenna and thus has to overfly the same area twice to obtain two images – 'Multi Pass Interferometry'. This method comes into operation mostly with satellite-based systems which for financial and also weight reasons are equipped with only one SAR antenna, e.g. the ERS-1/2 satellites.

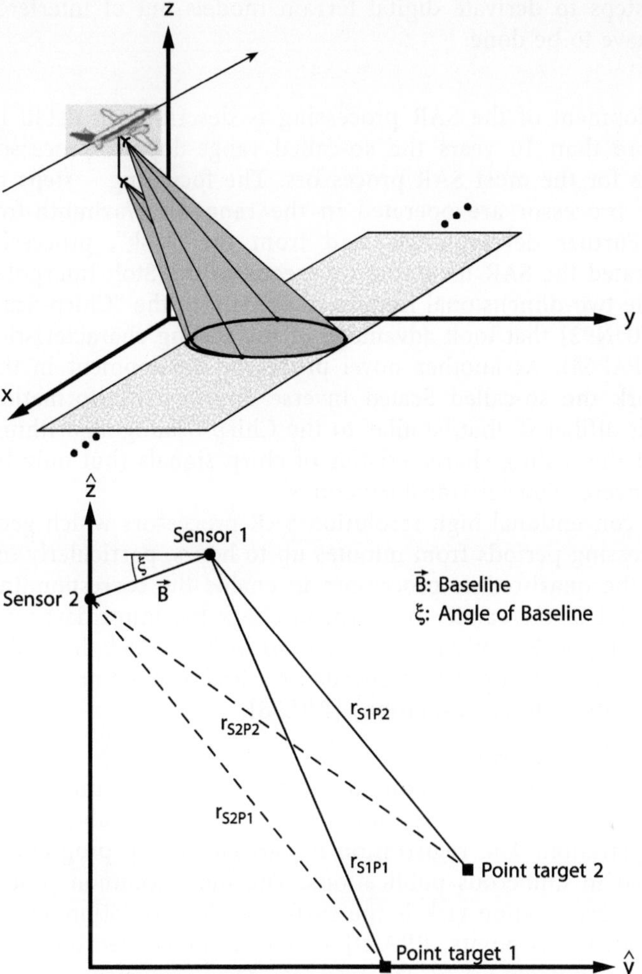

Image 1.7: Interferometrical geometry

■ 1.3 Short overview in interferometry

The principle of interferometry has been known for a long time [GRA74], whereas the method of satellite-based interferometry firstly was described by Li and Goldstein [LI90]. First studies took place with SEASAT data. But only with ERS-1 a quasi operational satellite could be started whose SAR data have made possible a global interferometric application. Particularly the exact repeatorbits and the accurate orbit determinations resulted in a substantial improvement of the SARIF development (also called InSAR: Interferometric SAR).

The following steps to derivate digital terrain models out of interferometric SAR data have to be done:

■ *SAR processing*
The historic development of the SAR processing is described in detail in [BAM92]. For more than 10 years the so-called range-doppler-processor had been the basis for the most SAR processors. The focussing – steps of the range-doppler processor are operated in the range-time/azimuth-frequency domain. Further developments lead from the "ω–k", processor [ROC89] that operated the SAR focussing by means of the Stolt-Interpolation [STO78] in the two-dimensional frequency domain to the "Chirp-Scaling algorithms" [RUN92] that took advantage of the scaling characteristics of chirp signals [PAP68]. As another novel processor development in the course of this work the so-called Scaled Inverse Fourier Transformation (SIFT) processor is affiliated, that, similar to the Chirp Scaling algorithms, takes advantage of the scaling characteristics of chirp signals (but only by using during the inverse Fourier transformations).

Apart from the conventional high resolution SAR processors which generally feature processing periods from minutes up to hours, particularly interesting are also the quasi-online processors to enable the corresponding operator or even pilot of the carrier platform to check the immediate possibilities for an overflight. For missions carried out by the NASA the development of fast processors to operation speeds of a few seconds are meanwhile satisfactory (with reduced resolution) [BRE98].

■ *Coregistration of the data pair*
A precondition for generating interferograms (phase difference images) is the exact superposition of two corresponding SAR scenes with sub-pixelaccuracy, the co-registration. The registration is carried out by progressive steps and described in numerous publications. The most common procedure for solving the registration task is the surface-wide correlation analysis of before generated pass points [PRA90] which can be carried out both in the time/distance-domain [GEU95] and in the frequency domain [ZEB93]. Also different ways of registration have been chosen, e.g. an interactive formulation on optimisation of irregularity function [LIN92] respectively a difference phase analysis of several interferograms to determine registration errors [SCHR98].

The method presented here benefits from the corresponding advantages of well-known strategies, however automatises the sub-pixel exact registration by means of an information-theoretical approach to a new choice of pass points [KNE98].

■ *Filter operations to reduce noise*
In order to improve the quality of the interference pattern geometrically induced signal decorrelations must be suppressed. For that purpose a series of studies were carried out which are noted as follows [PRA89] [JUS94].

■ *Evaluation of the interferogram*
The interferogram consists of a phase image and an amplitude image. The phase difference between two images is evaluated by a pixel-wise multiplication of the complex reference image with the complex conjugate image partner.

■ *Determination of coherence*
To determine the coherence resp. similarity of the two images the complex correlation coefficient is evaluated. Different publications describe the decorrelation effect between two acquisitions [ZEB92]. By means of the correlation coefficient it is possible to collect changes on the ground qualitatively (temporally during a period between taking the images and dimensionally in the range of the wavelength). To create calibrated coherency maps the geometric effects on the signal correlation must be eliminated [GEU94]. The evaluations of the coherence are not part of this work.

■ *Unwrapping of the ambiguity in measuring phases*
The different methods of the phase "Unwrapping" are demonstrated in [KRä98] and will not be dealt with here.

■ *Deduction of the digital terrain model and geometric correction procedures*
The principle of the deduction of digital terrain models on the interferometric, optical way had been proved by [GRA74] for the first time. [ZEB86] extended the method to the digital processing of aeroplane InSAR data and [HAR93] successfully generated by means of satellite data digital terrain models of different test areas in Europe. In [LOF] the theoretical and geometric basics for the SAR interferometry are affiliated. The effects of the geometrical deformations (layover, shadowing, foreshortening) are explained by [MEI89] and corrected with the aid of interferometric SAR data of several overflights in different directions [GEL96] or by stereo images [BOL98]. The novel method presented here uses geometric a priori knowledge, conventional interpolation algorithms and data fusion techniques to detect and remove geometric deformation effects [HEI98].

■ 1.4 Objective and methodology

The task of two-dimensional SAR signal processing consists of collecting the images reflected by the earth's surface via SAR-carrier and to process them the way that a two-dimensional brightness depiction is created. This so-called focussing process basically consists of a two-dimensional, distance-dependent optimal filter operation taking advantage of the temporally coherence of the form the to-be-mapped planet's surface reflected microwave signals are used to obtain the necessarily high resolution.

The SAR data processing mainly consists of two steps, the azimuth compression (processing in flight direction) and the range compression (processing in range direction). The range compression can be carried out smoothly as the for correlating of transmitter and receiver necessary parameters are known. The implementation of the azimuth compression is much more difficult due to the influence of many unknown factors such as the dependence on data of the geometric position of the carrier and last but not least the distance-dependent deformation of the two-dimensional image. That is why it has to be carried out in more worksteps.

To carry out three-dimensional, interferometric processing of SAR data, apart from the actual SAR image processing more, worksteps are needed which are described in Image 1.8 *Interferometric processing chain.*

In Image 1.8 all necessary steps to create a digital terrain model are described. After the data acquisition all SAR raw data are processed by a phase preservance SAR processor; the complex SAR images are available then (SLC – single look complex). The two images of a chosen area taken by a satellite during two orbital circulations are blurred and shifted against each other as the orbital circulations of the satellite are not exactly equal.

In Image 1.9: *Simplified geometric arrangement at the two-pass-interferometry* the appearing effects are indicated. Apart from the necessary corrections of the geometrical effects they themselves are basically desired since different side looking perspectives (a base line) are a pre-condition for interferometric SAR data processing.

In order to use the corresponding points of each scene whilst generating an interferogram both images must be geometrically superposed. In other words the coordinates of the corresponding points on an image one have to be transformed in a way that they match the coordinates of the corresponding point on the other image. The matching procedure, the so-called coregistration, has to be carried out absolutely thoroughly to avoid phase errors. After that the interferogram can be obtained. It makes sense to eliminate with proper filters the spectral shifting of the system transferfunctions and the scenes (caused by different recording positions) which could result in phase noise [JUS94].

After further methods to increase the phase accuracy and to remove the part of the phases which results in the recording geometry (side looking geometry) a method to estimate the absolute phase out of the relative phase (which is ambiguous due to its co-domain from $-\pi$ to π) is added, the phase "Unwrapping". Out of this absolute phase altitude information now can be obtained.

This height information, however, is created in consequence of the recording geometry (the so-called slant range) which means they are not in a rectangular coordinate system but in a side looking geometry – the slant-range-geometry. Image 1.10 *Slant/ground-recording geometry* prescribes this context.

The respectively prevailing surface geometry can cause errors on the SAR image such as shadowing effects which must be considered and corrected whilst adjusting the slant- on the groundgeometry [HEI98].

After converting the non-ambiguous phase in the terrain altitude in Cartesian coordinates a digital terrain model is available (DHM). To utilise the

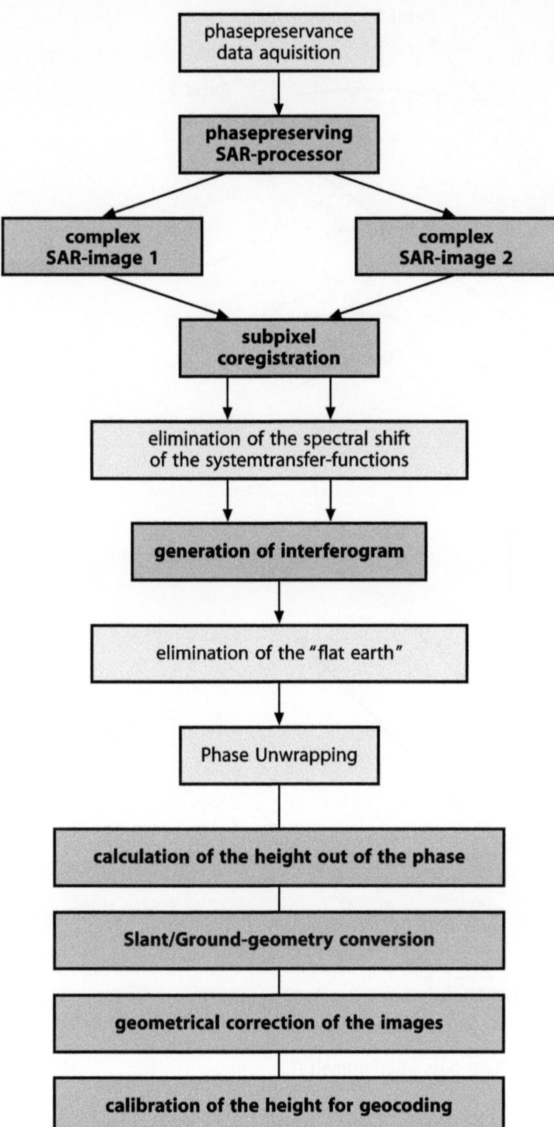

Image 1.8: Interferometrical processing chain *geometry* prescribes this context

DHM for further applications it has to be referenced to the corresponding earth's surface. This step is called geocoding resp. geo-referencing [SCHW95]. Not only the cartographic conversions of the coordinate system must be calibrated but particularly the acquired height values of the DHM. Here usually corner cubes are placed in the interesting area that is measured by GPS.

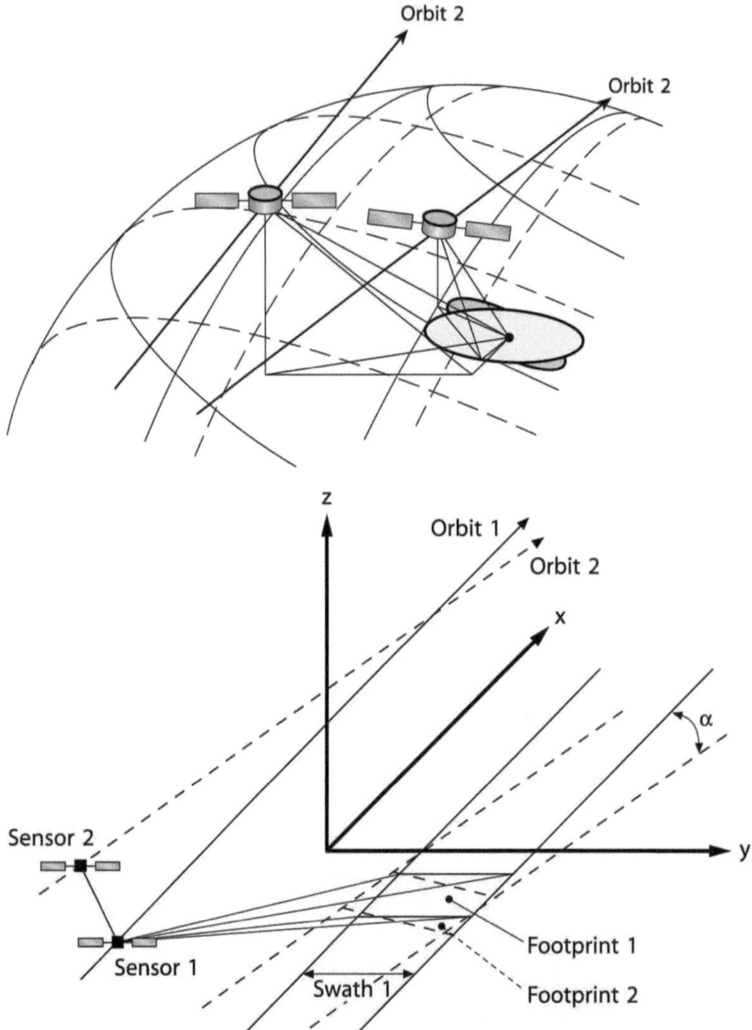

Image 1.9: Simplified geometry of the "two-pass interferometry"

In the later interferogram those corner cubes can be tracked and the whole height model corresponding to the reference matrix can be calibrated [DO-SAR]. This method requires the placement of corner cubes in the area to be observed.

As a novel method to generate the calibrating points an algorithm will be presented which replaces the corner cubes. Only the corresponding sensor positions for detecting the absolute height position during the two necessary overflights are compulsory.

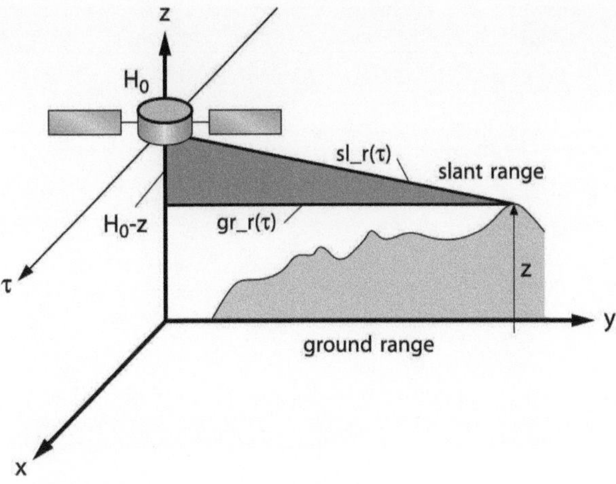

Image 1.10: Slant/ground geometry

Methodology

The main focus whilst realising the single worksteps is the continuous system theoretical description of the worksteps commencing with the SAR sensor signal to image processing and finally to the generation of interferograms. The image processing creates a standardised starting point for the deduction of all different processing algorithms. The interferometric part

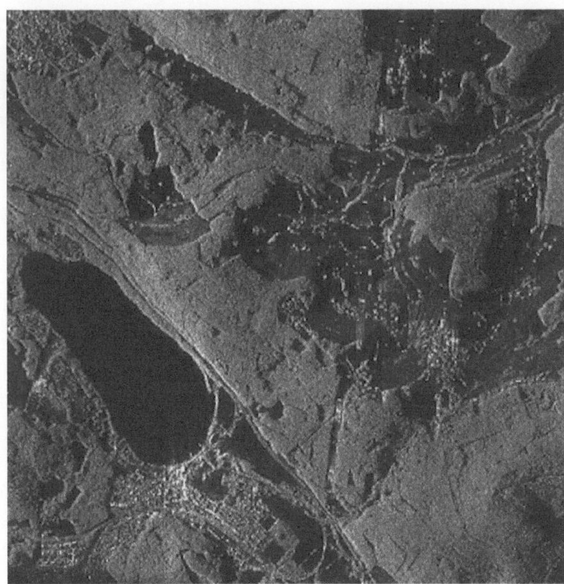

Image 1.11: DO-SAR scene Titisee (Schwarzwald), Germany

Image 1.12: SAR-raw data Flevoland ERS-1

Image 1.13: SAR scene ERS-1

deals particularly with the novel ways on how to solve different problems such as image processing, co-registration, layover and shadowing correction and the generation of height pass points.

In chapter 2 of this work explanations on basic principles concerning signal theoretical considerations and geometric effects can be found.

SAR processing demands a description of the SAR system as a dynamic system to deduce considerable characteristics such as limitation in resolutions.

In chapter 3 the SAR system is accordingly regarded dynamically and the hence expected two-dimensional system description deduced from. Starting from a point-target description, the transforming functions are explained in different two-dimensional time domains and frequency domains.

This description allows to lead the different existing descriptions in literature into one another and thus creating a common point of origin for derivations in the future.

In reference to chapter 3 a novel two-dimensional, distance-dependent image processing algorithm is deduced in the following chapter 4 which bases on the so-called scaling techniques. For the first time a global connection to other well-known processors by using the standardised derivations described in chapter 3 is indicated and discussed.

Chapter 5 describes different new ways of realisation within the interferometric processing chain which are required to generate digital terrain models.

As test data both ERS-1/ERS-2 raw data (Images 1.4, 1.12, 1.13, 4.19) and images taken by the airborne system DO-SAR (Image 1.11: SAR scene Titisee) had been used.

2 SAR basics

This chapter contains derivations of several basic relations of radars with synthetic aperture which are used for more differentiated explanations within the following chapters.

2.1 Introduction

Corresponding to the two dimensions of an SAR image, it is differentiated between the azimuth direction (or flight direction) and the crosswise distance direction, the range direction. Primarily, it is important to achieve a high spatial resolution in both the azimuth direction and the range direction as well as a high geometrical similarity and a high contrast of the SAR-image. In contrast to the conventional method of a radar with real aperture, particularly the requirement of a high spatial resolution is decisive for the concept of a radar with synthetic aperture. Both methods determine the resolution in range direction by means of the bandwidth of the transmitted impulse. However, regarding the radar with real aperture, the resolution in azimuth is limited by the directional radio pattern respectively the full width at half maximum of the antenna's main lobe. The full width at half maximum can be reduced by extending the aperture of the antenna in azimuth direction. However, to gain an azimuth resolution of 5 m with a wave length of 5.65 cm at a distance of 800 km to the object to be mapped, an antenna consisting of thousands of components with a length of about 8 km would be needed (see chapter 3.2.2.1). Of course, neither an aircraft nor a satellite would be in a position to transport such an antenna. That is why C. A. Wiley broke new grounds by inventing an antenna on a computer, called the "antenna with synthetic aperture". Instead of constructing an antenna out of thousands of components only one single element is installed on the platform of the SAR carrier. The travel speed of the platform in forward direction respectively the relative movement between the radar's antenna and the illuminated object plays a major role in the creation of the substantially longer, effective antenna. If it is known when the elementary antenna is at a certain time at a certain place, it can henceforth be assembled virtually. By coherent processing, in other words by integration of the back-scattering signal in consideration of value and phase, an antenna pattern can be created which is equivalent to that one of a real antenna whose maximum length is only limited by the lighting duration of a point target.

A pre-condition for synthesizing the aperture or the azimuth compression is the exact knowledge of the phase history of the backscattering signal. The phase history is a result of the chronological process of the slant range between a point target and the radar scanner during the fly-by which can be calculated by taking the speed above ground and the minimum slant range respectively the running time of a radar impulse into account. In chapter 3 *SAR-signal processing* those dynamic system descriptions are discussed at length.

First of all, let us have a look at the radar equation, particularly in consideration of the synthetic aperture.

2.1.1 The radar equation

2.1.1.1 Radar equation for radar with real aperture (unfocussing systems)

The radar equation describes the physical correlation between transmitted and received output as a function of the system parameters of a radar-unit and the propagation features of the electromagnetic wave. It serves as an estimation for the achievable range in consideration of a predetermined signal-noise ratio (SNR) at the receiver's input. With the help of that a system tuning can be done as all device-specific parameters except for the backscattering coefficient of the object can be influenced on the part of the developer.

Assisted by that the system can be optimized, because all system-specific parameters, except the radar back-scattering-coefficient of the object, can be affected by the developer. Image 2.1 *Measurement geometry for radar equations*, illustrates the measurement geometry which is needed for deducing radar equations.

A radar-transmitter radiates via an antenna the power P_S. If this power is radiated consistently in all directions we will get the emittance $D_s(r)$ on a sphere surface with the radius r:

$$D_s(r) = \frac{P_S}{4\pi \cdot r^2} \tag{2.1}$$

The term $1/(4\pi r^2)$ of the geometric dilution is equivalent to the electromagnetic energy during the spreading. Normally, in the radar technology multiplex antennas with predefined levelling characteristics are used. The proportions of the maximum emittance of such a focussing antenna to the emittance of an isotropic (spherically emitting) antenna is called the antenna gain. Thus the emittance D_0 of the object in the distance R_0 is retrieved:

$$D_0(R_0) = \frac{P_S \cdot G}{4\pi \cdot R_0^2} \tag{2.2}$$

Multiplying the emittance D_0 on the object with its radar backscattering cross-section σ results in the power P_{Rausch} which is scattered evenly consistent in all directions by the object:

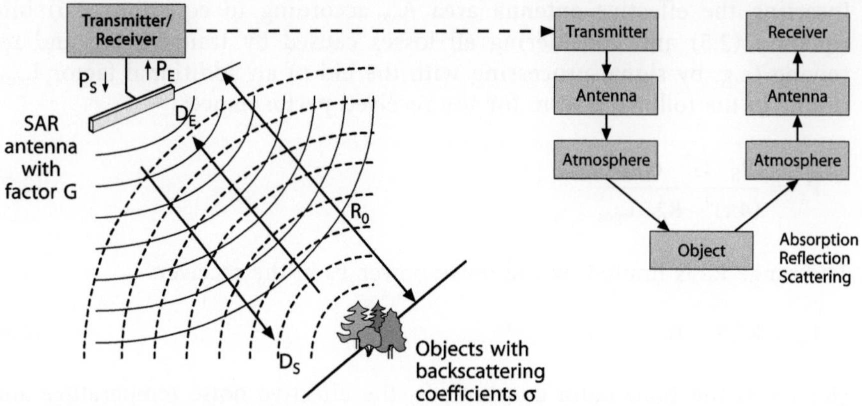

Image 2.1: Measurement geometry

$$P_{Rausch} = D_0(R) \cdot \sigma \tag{2.3}$$

The radar backscattering coefficient describes the scattering characteristics of an object.

Thus, the scattering is assumed as incoherent, that means the scattering points illuminated that way by an antenna are statistically distributed. Objects whose dimensions are large compared with the illuminated wave length λ, the value of σ is determined substantially by the geometric form of the object. Possibilities of calculating different backscattering coefficients are explained in different literature [SKO70].

The power spectrum D_e on the receiver in the distance R_0, considering the above mentioned, is:

$$D_e = \frac{P_R}{4\pi \cdot R_0^2} = \frac{P_S \cdot G \cdot \sigma}{(4\pi)^2 \cdot R_0^4} \tag{2.4}$$

Here, the scattering decay again is taken into account by the geometrical thinning. Usually, antennas possess an effective area, called effective antenna area A_w, the received performance is described with as follows:

$$P_e = D_e \cdot A_w = \frac{P_S \cdot G \cdot \sigma \cdot A_w}{(4\pi)^2 \cdot R_0^4} \tag{2.5}$$

The correlation between the effective antenna area A_w and it's antenna gain results from:

$$A_w = \frac{\lambda^2 \cdot G}{4\pi} \tag{2.6}$$

Inserting the effective antenna area A_w, according to equation (2.6) into equation (2.5) and considering all losses caused by transmitting and receiving (e.g. by signal-processing with the aid of an additional factor L_{ges}) results in the following term for the received performance:

$$P_e = \frac{P_S \cdot G^2 \cdot \sigma \cdot \lambda^2}{(4\pi)^3 \cdot R_0^4 \cdot L_{ges}} \tag{2.7}$$

The range R_0 is limited by the noise power P_r in the receiver.

$$P_r = k \cdot \vartheta_0 \cdot B \tag{2.8}$$

Here, k is the Boltzmann constant, ϑ_0 the effective noise temperature and B the bandwidth of the receiver. For an ideal, signal-adapted receiver applies the following with a rectangular transmitting impulse of the duration t_{send}:

$$B \cdot t_{send} = 1 \tag{2.9}$$

The maximum range $R_{0\,max}$ is given when the receiving power P_e decreases to the minimally detectable power $P_{e\,min}$. The minimum SNR (Signal to Noise Ratio) can therefore be defined as:

$$SNR_{min} = \frac{P_{e\,min}}{P_{Rausch}} \tag{2.10}$$

With equation (2.10) the maximum range $R_{0\,max}$ is:

$$R_{0\,max} = \sqrt[4]{\frac{P_S \cdot G^2 \cdot \lambda^2 \cdot \sigma \cdot t_{send}}{(4\pi)^3 \cdot k \cdot \vartheta_0 \cdot (SNR)_{min} \cdot L_{ges}}} \tag{2.11}$$

The equation (2.11) is valid on the condition that the remote field term does not change, a homogeneous scattering of the electromagnetical takes place and no parasitic ways have to be considered.

2.1.1.2 Radar term for radars with synthetic aperture-focussing systems

For the radar with synthetic aperture follows a coherency integration of the echo signals during the illumination time T of an object. Thus the receiving power increases and thus the SNR increases proportionally to the illumination time T.

To derivate the radar equation for the SAR let us again have a look at the correlation concerning the radar equation for the radar with real aperture, see eq. (2.11):

$$R_{0\,max}^4 = \frac{P_S \cdot G^2 \cdot \lambda^2 \cdot \sigma \cdot t_{send}}{(4\pi)^3 \cdot k \cdot \vartheta_0 \cdot SNR_{min} \cdot L_{ges}}$$

For the SAR, the number N_{send} of the coherently integrated echo signals during an illumination time T (T is proportional to S) and with the puls-repeating frequency PRF is:

$$N_{send} = T \cdot PRF = \frac{x_{repeat}}{v_a} \cdot PRF \tag{2.12}$$

The further consideration is done with the aid of the derivation in chapter 3.2.2. Geometrical resolutions in azimuth direction.

The transmission-sequence duration is defined by the covered distance x_{repeat} between the transmitted pulses. Φ_{HB} as the half-life angle and R_0 is the shortest distance between the SAR-sensor and the point target:

$$x_{repeat} = \Phi_{HB} \cdot R_0 = 0.88 \cdot \frac{\lambda}{D_x} \cdot R_0 \tag{2.13}$$

A further result was the possible limitation of the azimuth resolution of the radar with synthetic aperture which is:

$$\Delta x = \frac{D_x}{2} \tag{2.14}$$

Inserting equation (2.14) in equation (2.13) and then equation (2.13) in equation (2.12) delivers the following term:

$$N_{send} = \frac{x_{repeat}}{v_a} \cdot PRF = 0.88 \cdot \frac{\lambda \cdot R_0}{2 \cdot \Delta x \cdot v_a} \cdot PRF \tag{2.15}$$

The radar equation valid for a radar with synthetic aperture results in the radar equation of a radar with real aperture, whereto the received power is multiplied by the number N_{send} of the received echo signals. The maximum radar range is larger on a radar with synthetic aperture after signal processing than at a radar with real aperture. The range is calculated with equation (2.11) and equation (2.15):

$$R_{0\,max\,syn}^4 = \frac{P_S \cdot G^2 \cdot \lambda^2 \cdot \sigma \cdot t_{send}}{(4\pi)^3 \cdot k \cdot \vartheta_0 \cdot SNR_{min} \cdot L_{ges}} \cdot N_{send} \tag{2.16}$$

$$R_{0\,max\,syn}^3 = 0.44 \cdot \frac{P_S \cdot G^2 \cdot \lambda^3 \cdot \sigma \cdot t_{send} \cdot PRF}{(4\pi)^3 \cdot k \cdot \vartheta_0 \cdot SNR_{min} \cdot L_{ges} \cdot \Delta x \cdot v_a} \tag{2.17}$$

The radar equation for the radar with synthetic aperture is effective only after the signal processing as the processed azimuth resolution Δx is also part of the equation. In contrast to the conventional radar equation for non-focussed systems, the slant-range distance-dependency for the synthetic array enters the radar equation cubically. This substantial advantage is described by means of the comparing consideration of the signal/noise-distances.

2.1.1.3 Signal/noise distance

The signal/noise distance is defined as the relation of the received power to the noise power. For a synthetic radar, the result is:

$$\mathrm{SNR_{syn}} = \frac{P_e}{P_r} = 0.44 \cdot \frac{P_S \cdot G^2 \cdot \lambda^3 \cdot \sigma \cdot \mathrm{PRF} \cdot t_{send}}{(4\pi)^3 \cdot k \cdot \vartheta_0 \cdot L_{ges} \cdot \Delta x \cdot v_a \cdot R_{0\ max_{syn}}^3} \tag{2.18}$$

For a radar with real aperture results:

$$\mathrm{SNR_{real}} = \frac{P_e}{P_r} = \frac{P_S \cdot G^2 \cdot \lambda^3 \cdot \sigma \cdot t_{send}}{(4\pi)^3 \cdot k \cdot \vartheta_0 \cdot L_{ges} \cdot R_{0\ max}^4} \tag{2.19}$$

The calculating comparison of both methods results in (ERS-1 parameters presupposed):

$$\mathrm{SNR_{syn}} \cong 10\ \mathbf{dB}$$

$$\mathrm{SNR_{real}} \cong -80\ \mathbf{dB}$$

The radar with synthetic aperture is about by 90 dB more sensitive than a radar with real aperture – this is a factor of more than 30 000.

Used parameters [ERS92]:

$$P_S = 4800\ \mathrm{W}$$
$$\lambda = 5.6\ \mathrm{cm\ C\text{-}band}$$
$$\vartheta = 550\ \mathrm{K}$$
$$B = 15.55\ \mathrm{MHz}$$
$$t_{send} = 37\ 10^{-6}\ \mathrm{s}$$
$$R_0 = 827\ 10^3\ \mathrm{m}$$
$$G = 3160 = 35\ \mathrm{dB}$$
$$L = 0.5 = -3\ \mathrm{dB}$$
$$k = 1.38\ 10^{-23}\ \mathrm{J/Grad}$$
$$\Delta x = 5\ \mathrm{m}$$
$$v_a = 7040\ \mathrm{m/s}$$

Whereas the derivations of the radar equation allow a consideration of the corresponding performance ratio of different radar systems, the system-theoretical description of the SAR-system requires in advance a definition of the used transmitting signal. Besides the signal descriptions in the time domain also definitions to describe the frequency domain will be provided.

2.1.2 Transmitted and received signal

When signal-processing, preferred signals are those with a quadratic phase progression. Signals of that kind ("Chirps") are also qualified for SAR-processing as they feature a very narrow autocorrelation function and thus a high range resolution as well as a high signal/noise distance.

2.1.2.1 Description in bandpass domain and equivalent lowpass domain

The signal transmitted by the SAR carrier has the following form:

$$s(t) = \text{rect}\left(\frac{t}{T}\right) \cdot \cos\left(2\pi f_0 t + \pi k_r t^2\right) \tag{2.20}$$

and looks like illustrated in Image 2.7.

The so-called chirp is a linear frequency-modulated signal featuring a quadratic phase progression:

$$s(t) = \text{rect}\left(\frac{t}{T}\right) \cdot \cos(\psi(t)) \tag{2.21}$$

with $\text{rect}\left(\frac{t}{T}\right)$ as amplitude modulation and $\cos(\psi(t))$ as phase modulation.

The quadratic term (signal with quadratic phase) is called momentary phase:

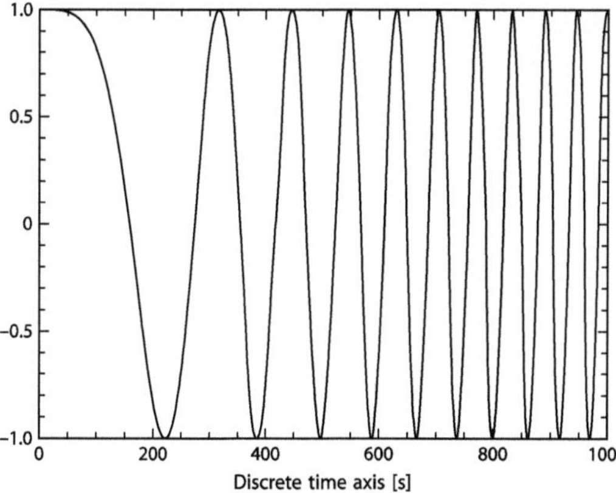

Image 2.7: Chirp

$$\psi(t) = 2 \cdot \pi \cdot f_0 \cdot t + \pi \cdot k_r \cdot t^2 \tag{2.22}$$

The factor k_r is called sweep rate. It characterises the frequency of the periodic changes or the phase-modulation velocity of the chirp signal. The current frequency is defined by the linear term (linear frequency-modulated signal):

$$f_0(t) = \frac{1}{2 \cdot \pi} \cdot \frac{d\psi(t)}{dt} \tag{2.23}$$

$$f_0(t) = f_0 + k_r \cdot t \tag{2.24}$$

In [LÜK92] the possibilities are revealed how to describe a bandpass signal by a so-called equivalent lowpass signal that bandpass system theory and the bandpass signal-processing are simplified without loosing information. In the following, all equivalent lowpass signals are marked with a lower placed "T". The bandpass signals are shown without further indices. According to the theory of the equivalent lowpass signals, the transmitting signal (here still in the bandpass domain) is described as follows:

$$s(t) = \text{rect}\left(\frac{t}{T}\right) \cdot \cos\left(2\pi f_0 t + \pi k_r t^2\right) = \text{Re}\left\{s_T(t) \cdot e^{j2\pi f_0 t}\right\} \tag{2.25}$$

The equivalent "transmitting=" lowpass signal now results from equation (2.25) to:

$$s(t) = \text{rect}\left(\frac{t}{T}\right) \cdot \cos(\psi(t)) = \text{Re}\left\{\text{rect}\left(\frac{t}{T}\right) \cdot e^{j\psi(t)}\right\} \tag{2.26}$$

$$s(t) = \text{Re}\left\{\text{rect}\left(\frac{t}{T}\right) \cdot e^{j2\pi\left(f_0 t + \frac{k_r}{2} \cdot t^2\right)}\right\} \tag{2.27}$$

$$s(t) = \text{Re}\left\{\text{rect}\left(\frac{t}{T}\right) \cdot e^{j\pi k_r t^2} \cdot e^{j2\pi f_0 t}\right\} \tag{2.28}$$

$$s(t) = \text{Re}\left\{s_T(t) \cdot e^{j2\pi f_0 t}\right\} \tag{2.29}$$

$$s_T(t) = \text{rect}\left(\frac{t}{T}\right) \cdot e^{j\pi k_r t^2} \tag{2.30}$$

Summing up we can describe the used real bandpass signal in SAR-signal processing

$$s(t) = \text{rect}\left(\frac{t}{T}\right) \cdot \cos(2\pi f_0 t + \pi k_r t^2)$$

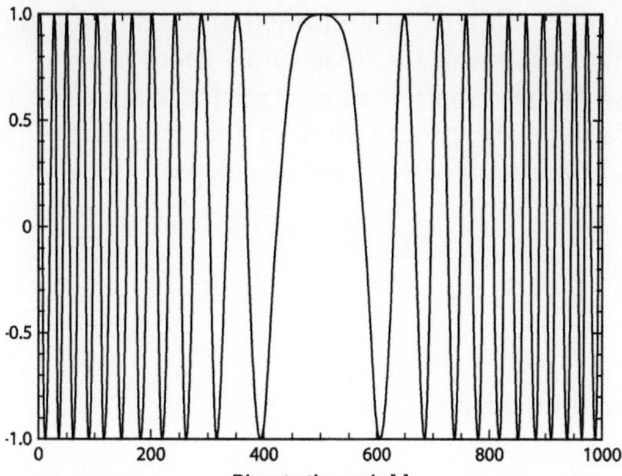

Image 2.7 a: Equivalent lowpass chirp

by a correspondingly equivalent lowpass signal:

$$s_T(t) = \text{rect}\left(\frac{t}{T}\right) \cdot e^{j\pi k_r t^2}$$

From equation (2.24) for the bandpass domain:

$$f_0(t) = f_0 + k_r \cdot t$$

$$f_0(t) = k_r \cdot t \qquad \text{(equivalent lowpass domain)}$$

can be retrieved that such a linear frequency-modulated signal with a bandwidth B_r phase modulated by the carrier frequency f_0. The bandwidth is determined by the sweep rate and the corresponding transmitting-pulse length T as a cutout of the endless timeline:

$$B_r = k_r \cdot T \qquad (2.31)$$

The appearing maximum frequencies are calculated as:

$$f_{max} = f_0 + \frac{B_r}{2} \; ; \; f_{min} = f_0 - \frac{B_r}{2}$$

2.1.2.2 Fourier transformation of the chirp

As substantial correlations and derivations in the SAR-signal-processing, in particular as a common initial point for reflections of the SAR interferometry are often watched and calculated in the frequency domain, in the following the corresponding Fourier transformation are derivated. There will be introduced not only different methods of the spectral calculation but also the methods of the approximative calculation of time-limited and also infinitely chirp which are necessary when explaining the SAR-signal theory.

We are looking for a Fourier transformation of a term that enables us to describe a signal with quadratic phase progression in the frequency domain. Here is differentiated between temporally limited chirps:

$$F(f) = \int_{-\infty}^{+\infty} f(t) \cdot e^{-j2\pi ft} \cdot dt = \int_{-\infty}^{+\infty} \text{rect}\left(\frac{t}{T}\right) \cdot e^{j\pi k_r t^2} \cdot e^{-j2\pi ft} \cdot dt$$

and temporally infinitely expanding chirps:

$$F(f) = \int_{-\infty}^{+\infty} f(t) \cdot e^{-j2\pi ft} \cdot dt = \int_{-\infty}^{+\infty} e^{j\pi k_r t^2} \cdot e^{-j2\pi ft} \cdot dt$$

Temporally infinitely chirps can exactly be described in the frequency domain and serve later as an initial point for the approximative calculation of temporally limited chirps.

2.1.2.2.1 Temporally infinite chirp signal

The calculation of the spectrum is performed by solving the Fourier integral. A temporally infinite chirp signal is described as:

$$S_\infty(t) = e^{j\pi k_r t^2} \tag{2.32}$$

The spectrum results in solving the Fourier integral:

$$S_\infty(f) = \int_{-\infty}^{+\infty} e^{j\pi k_r t^2} \cdot e^{-j2\pi ft} \cdot dt = \int_{-\infty}^{+\infty} e^{j\pi k_r (t^2 - 2ft/k_r)} dt \tag{2.33}$$

With quadratic completion of the phase

$$\left(t - \frac{f}{k_r}\right)^2 = t^2 - \frac{2ft}{k_r} + \frac{f^2}{k_r^2} \tag{2.34}$$

follows:

$$S_\infty(f) = e^{-j\pi f^2/k_r} \cdot \int_{-\infty}^{+\infty} e^{j\pi k_r (t - f/k_r)^2} \cdot dt \tag{2.35}$$

Substituting

$$x = \sqrt{\pi k_r} \cdot \left(t - \frac{f}{k_r} \right) \tag{2.36}$$

results in:

$$S_\infty(f) = \frac{e^{-j\pi f^2/k_r}}{\sqrt{\pi k_r}} \cdot \int\limits_{-\infty}^{+\infty} e^{jx^2} \cdot dx \tag{2.37}$$

With

$$e^{jx^2} = \cos(x^2) + j \cdot \sin(x^2) \tag{2.38}$$

and

$$\int\limits_{-\infty}^{+\infty} \sin(x^2) \cdot dx = \int\limits_{-\infty}^{+\infty} \cos(x^2) \cdot dx = \sqrt{\frac{\pi}{2}} \tag{2.39}$$

results in the spectrum to:

$$S_\infty(f) = \frac{e^{-j\pi f^2/k_r}}{\sqrt{\pi k_r}} \cdot \left[\sqrt{\pi/2} + j \cdot \sqrt{\pi/2} \right] = \frac{1}{\sqrt{k_r}} \cdot e^{-j\pi(f^2/k_r - 1/4)} \tag{2.40}$$

For the spectrum of the infinite chirp:

$$\boxed{S_\infty(f) = \frac{1}{\sqrt{k_r}} \cdot e^{j\frac{\pi}{4}} \cdot e^{-j\pi \cdot \frac{f^2}{k_r}} = \frac{1}{\sqrt{k_r}} \cdot e^{-j\pi \left(\frac{f^2}{k_r} - \frac{1}{4} \right)}} \tag{2.41}$$

As correspondence, the following relation can be specified:

$$S_\infty(t) = e^{j\pi \cdot k_r \cdot t^2} \quad \circ\!\!-\!\!\bullet \quad S_\infty(f) = \frac{1}{\sqrt{k_r}} \cdot e^{j\frac{\pi}{4}} \cdot e^{-j\pi \cdot \frac{f^2}{k_r}} \tag{2.42}$$

The particular correspondences for a temporally infinitely expanding chirp are derivated in the time domain and the frequency domain and described in equation (2.42). A realistic reflection, however, demands an additional derivation of chirp signals not featuring an infinite expansion, which in other words are generated temporally limited.

2.1.2.2.2 Temporally limited chirp signal

As a transmitting signal chirps are not available that are temporally infinitely expanded (therefore a infinite bandwidth has to be available). Indeed, signals have to be used that feature a finite pulse duration T. In opposite to temporally infinite signals with quadratic phase an exact, whole solution for the Fourier integral is not yet and has to be calculated approximatively:

$$F(f) = \int_{-\infty}^{+\infty} f(t) \cdot e^{-j2\pi ft} \cdot dt = \int_{-\infty}^{+\infty} rect\left(\frac{t}{T}\right) \cdot e^{j\pi k_r t^2} \cdot e^{-j2\pi ft} \cdot dt$$

Calculation of the spectrum with the help of the "Fresnel integrals".
A temporally limited chirp is described as:

$$S_{T_n}(t) = rect\left(\frac{t}{T}\right) \cdot e^{j\pi k_r t^2} \tag{2.43}$$

The corresponding Fourier integral is:

$$S_{T_n}(f) = \int_{-T/2}^{T/2} e^{j\pi k_r\left(t^2 - \frac{2tf}{k_r}\right)} \cdot dt = e^{-j\pi \frac{f^2}{k_r}} \cdot \int_{-T/2}^{T/2} e^{j\pi k_r\left(t - \frac{f}{k_r}\right)^2} \cdot dt \tag{2.44}$$

With the substitution

$$x = \sqrt{\pi k_r} \cdot \left(t - \frac{f}{k_r}\right) \tag{2.45}$$

the result is:

$$S_{T_n}(f) = \frac{e^{-j\pi f^2/k_r}}{\sqrt{r}} \cdot \left[\int_0^{\sqrt{\pi k_r}\left(\frac{T}{2} - \frac{f}{k_r}\right)} e^{jx^2} \cdot dx - \int_0^{\sqrt{\pi k_r}\left(-\frac{T}{2} - \frac{f}{k_r}\right)} e^{jx^2} \cdot dx\right] \tag{2.46}$$

$$S_{T_n}(f) = \frac{e^{-j\pi f^2/k_r}}{\sqrt{2k_r}} \cdot \frac{2}{\sqrt{2\pi}} \left[\int_0^{\sqrt{\pi k_r}\left(\frac{T}{2} - \frac{f}{k_r}\right)} e^{jx^2} \cdot dx - \int_0^{\sqrt{\pi k_r}\left(-\frac{T}{2} - \frac{f}{k_r}\right)} e^{jx^2} \cdot dx\right] \tag{2.47}$$

With the Fresnel integral

$$K(x) = C(x) + j \cdot S(x) \tag{2.48}$$

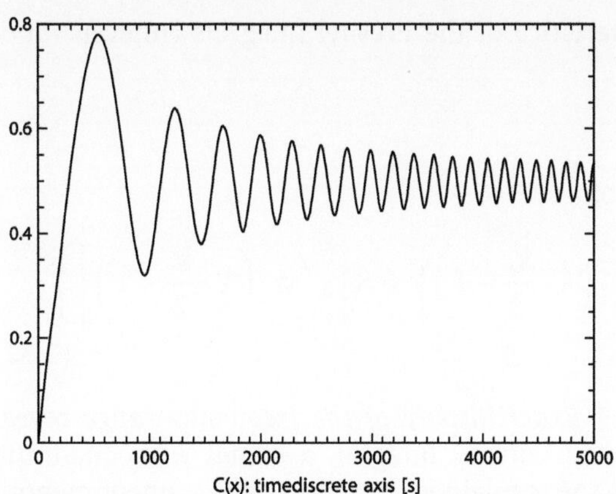

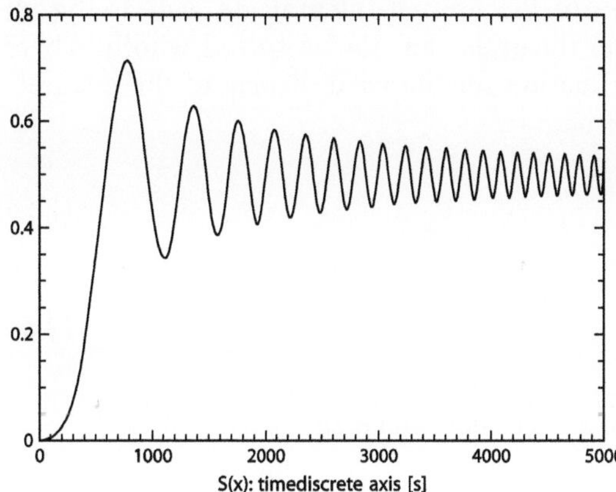

Image 2.8: Fresnel integrals

with respect to [LI93]

$$S(x) = \frac{2}{\sqrt{2\pi}} \int\limits_0^x \sin(t^2) \cdot dt \tag{2.49}$$

and

$$C(x) = \frac{2}{\sqrt{2\pi}} \int\limits_0^x \cos(t^2) \cdot dt \tag{2.50}$$

results:

$$S_{T_n}(f) = \frac{e^{-j\pi f^2/k_r}}{\sqrt{2k_r}} \left\{ K\left[\sqrt{\pi k_r}\left(\frac{T}{2} - \frac{f}{k_r} \right) \right] - K\left[\sqrt{\pi k_r}\left(-\frac{T}{2} - \frac{f}{k_r} \right) \right] \right\} \tag{2.51}$$

Due to the symmetric characteristics of the Fresnel integrals (uneven function)

$$K(-f) = -K(f)$$

equation (2.51) is simplified to:

$$S_{T_n}(f) = \frac{e^{-j\pi f^2/k_r}}{\sqrt{2k_r}}\left\{K\left[\sqrt{\pi/k_r}\left(T\cdot\frac{k_r}{2}-f\right)\right] + K\left[\sqrt{\pi k_r}\left(T\cdot\frac{k_r}{2}+f\right)\right]\right\} \tag{2.52}$$

Equation (2.52) and Image 2.9 *Exact display of the frequency range* reveal the exact solution (value) of the Fourier integral, a signal with quadratic phase trend. The ripples can be explained by the Gibbs's phenomenom and by the decaying oscillation of the Fresnel integral. As well as the initially described task the Fresnel integrals can not be solved wholly. Therefore we have to find an approximation for the used display of the integral:

Looking at the function $x \to \infty$:

$$\left|\sqrt{\frac{\pi}{k_r}}\left(\frac{Tk_r}{2}-f\right)\right| \gg 0 \tag{2.53}$$

$$\left|\sqrt{\frac{\pi}{k_r}}\left(\frac{Tk_r}{2}+f\right)\right| \gg 0 \tag{2.54}$$

becomes because of (see Image 2.8 *Fresnel integral*):

$$\lim_{x\to\infty} S(x) = \lim_{x\to\infty}\frac{2}{\sqrt{2\pi}}\int_0^x \sin(t^2)\cdot dt = \lim_{x\to\infty} C(x) = \frac{1}{2}$$

$$\lim_{x\to\infty} K(x) = \frac{1}{2}+j\frac{1}{2} = \frac{1}{\sqrt{2}}e^{j\pi/4} \tag{2.55}$$

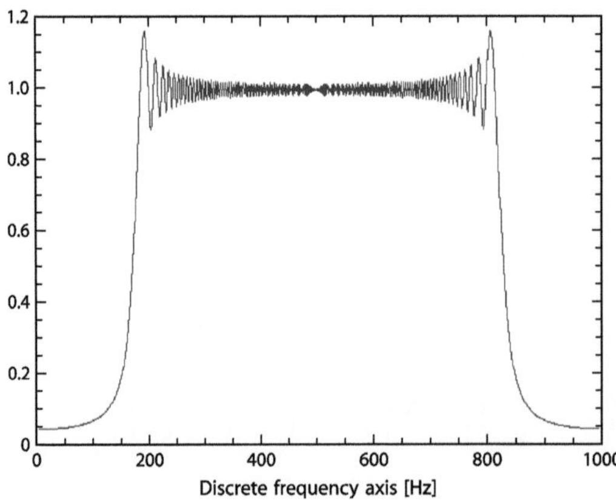

Image 2.9: 'Exact' display of the frequency range (absolute)

the spectrum of the finite chirp in a first step to that one of the infinite chirp:

$$S_{T_{\cap k \to \infty}}(f) \approx \frac{1}{\sqrt{k_r}} \cdot e^{-j\pi\left(\frac{f^2}{k_r} - \frac{1}{4}\right)} = S_{T_\infty}(f) \tag{2.56}$$

Further considering of the factors $Tk_r/2$ in the equations (2.54) and (2.55) let us realise the same as an ax-symmetric spectral displacement in the frequency direction. The resulting "shifting" signal can be interpreted as a rectangular function in the range:

$$-\frac{T \cdot k_r}{2} \leq f \leq \frac{T \cdot k_r}{2}$$

(see Image 2.10 *Approximative solution*)

The spectrum is calculated as:

$$S_{T_\cap}(f) = \frac{1}{\sqrt{k_r}} \cdot e^{-j\pi\left(\frac{f^2}{k_r} - \frac{1}{4}\right)} \cdot \text{rect}\left(\frac{f}{T \cdot k_r}\right) \tag{2.57}$$

Further it can be shown [LOF] that a displacement of the rectangular envelope of the chirp in the time domain by $-t_0$ is equivalent to a displacement in the frequency domain by $-f_0$ with $|f_0| = k_r \cdot |t_0|$.

If a finite chirp is regarded as a "windowed" chirp in the time domain, the result is the Fourier transformation (FT) of the finite chirp as FT of the infinite chirp "windowed" with the equivalent window function in the frequency domain.

As "transformation pair" results:

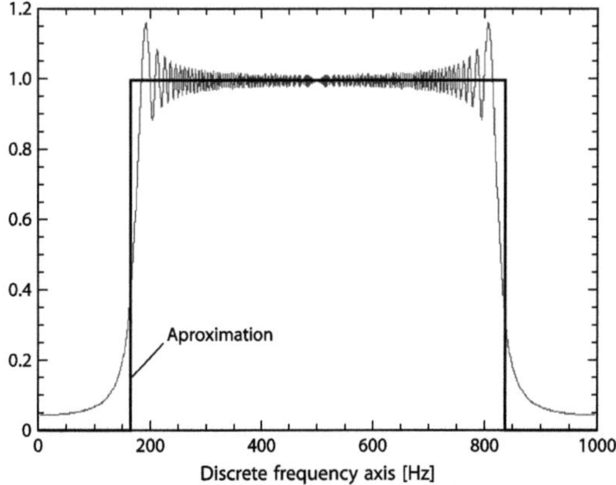

Image 2.10: Approximative solution

$$S_{T_\cap}(t) = \text{rect}\left(\frac{t - t_0}{T}\right) \cdot e^{j\pi k_r t^2} \tag{2.58}$$

$$\boxed{S_{T_\cap}(f) \approx \frac{1}{\sqrt{k_r}} \cdot e^{-j\pi(f^2/k_r - 1/4)} \cdot \text{rect}\left(\frac{f - f_0}{T \cdot k_r}\right)} \qquad \text{with} \quad f_0 = k_r \cdot t_0 \tag{2.59}$$

or generally with w(t) as window function:

$$S(t) = w(t) \cdot e^{j\pi k_r t^2} \tag{2.60}$$

$$S(f) \approx W(f) \cdot \mathscr{F}\left\{e^{j\pi k_r t^2}\right\} \tag{2.61}$$

whereas the function W(f) represents the "equivalant" window function in the frequency domain and not the Fourier transformation of w(t).

Calculation of the spectrum with the help of the stationary-phase principle

A further method to solve not wholly solvable integrals, in particular of signals with strong oscillating functions, is the so-called method of the stationary phase.

First approximation methods towards the solution of this integral using the method of stationary phase were published by Papoulis [PAP68]. Basically, the idea is:

the integral over the range where the function is oscillating rapidly tends towards zero. In comparison the integration over entire periods can be regarded. Here, the positive and negative parts of the function eliminate each other;

the main parts in the Fourier occur when the sweep rate of the oscillation is minimal. This condition is fullfilled in the point of the stationary phase. The integral delivers a significant value where the derivation is zero at the spot τ^*.

Using the results from chapter 2.1.2.2.1 *Temporally infinite chirp-pulse*, the equation is calculated as follows:

$$S_T(f) = \mathscr{F}\left\{\text{rect}\left(\frac{t}{T}\right) \cdot e^{j\pi k_r t^2}\right\}$$

$$S_T(f) = T \cdot \text{si}(\pi f T) \cdot \frac{1}{\sqrt{k_r}} \cdot e^{-j\pi\left(\frac{f^2}{k_r} - \frac{1}{4}\right)}$$

$$S_T(f) = \frac{T}{\sqrt{k_r}} \cdot e^{j\frac{\pi}{4}} \cdot \int_{-\infty}^{\infty} e^{-j\pi\frac{\tau^2}{k_r}} \cdot \text{si}(\pi T(f - \tau)) \cdot d\tau \tag{2.62}$$

The resulting convolution integral can not be solved wholly but can be considered as an explaining illustration. The convolution operation is made

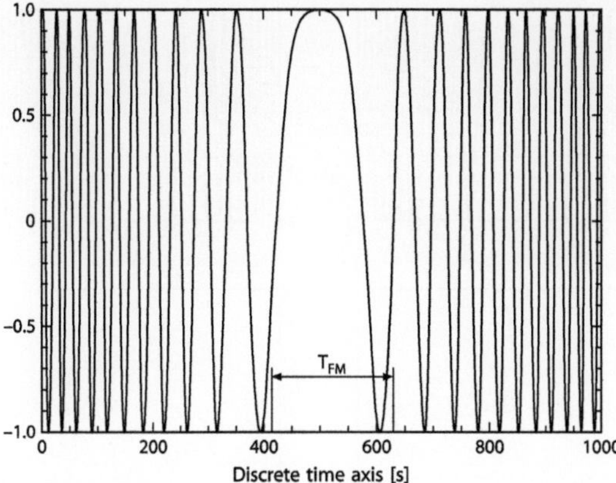

Image 2.11: Oscillating function

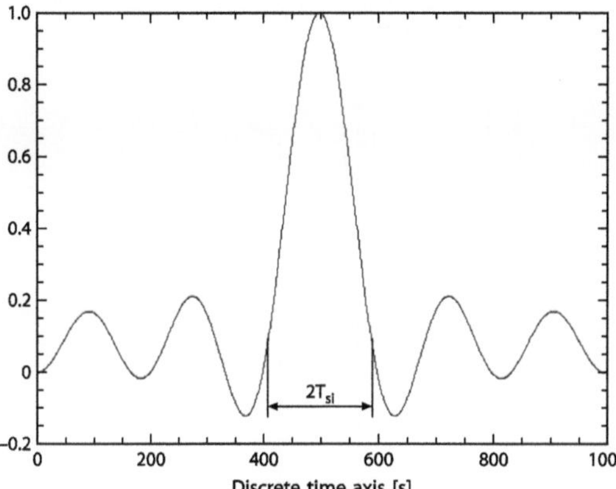

Image 2.12: si-function

up of a succession of displaced superposition, multiplication and integration. It is known that integrals result in zero via periodical, average-free functions, integrated over a certain period.

The two terms in the equation (2.62) are presented as follows:

In a train of thoughts, shifting the si-function steadily (means sin(x)/x) along the strongly oscillating function (Image 2.13 *Stationary phase*), the respective value of the integral differs significantly from zero only when the windowed si-function is not stretched over one complete period of the oscillating function. The integration e.g. would result close to zero if it was integrated over the range $T_{FM} = 2T_{si}$ (to get an exact solution it would have to be integrated a bit more).

In Image 2.13 the si-function is shifted from left to the right side over the chirp, multiply the respective functions and are integrated to clarify the idea of the resulting multiplication (no convolution integral calculation). The largest part of the convolution integral obviously is contributed by (see Image 2.13) that part of the oscillating function that varies in the slowliest way or in other words whose phase variation is minimal (stationary point).

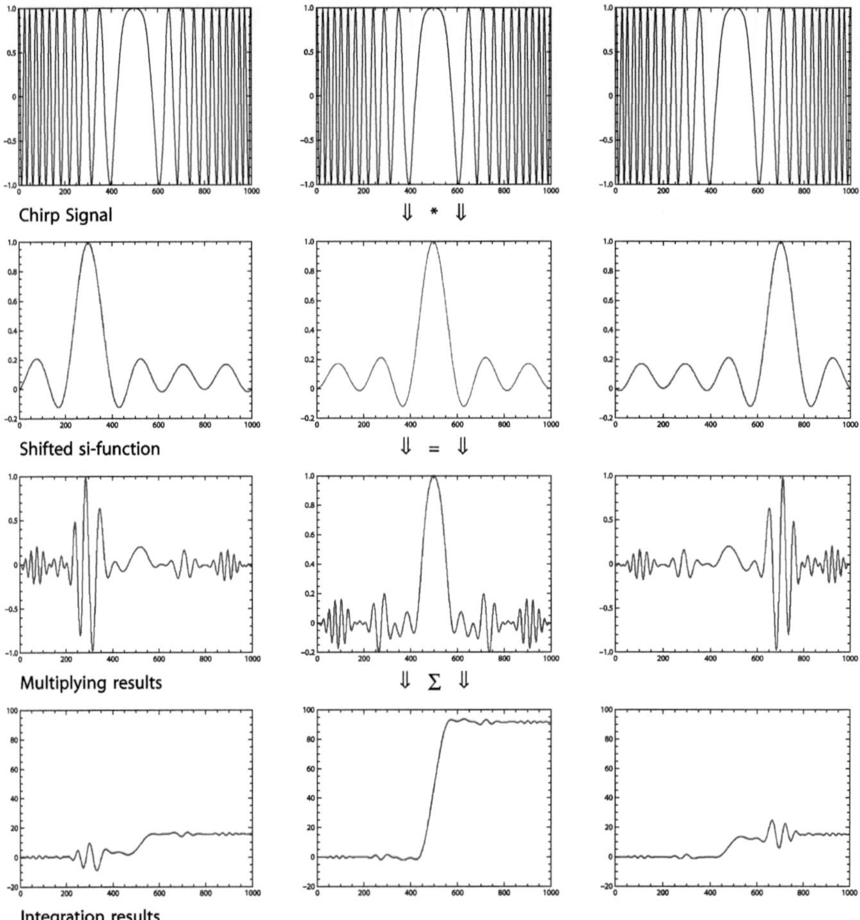

Image 2.13: Stationary phase, time-discrete ranges [s]

We calculate:

$$
\begin{aligned}
S_{T_\cap}(f) &= \int_{-\infty}^{\infty} \text{rect}\left(\frac{t}{T}\right) \cdot e^{j\pi \cdot k_r \cdot t^2} \cdot e^{-j2\pi f \cdot t} dt \\
&= \int_{-\infty}^{\infty} \text{rect}\left(\frac{t}{T}\right) \cdot e^{j\pi \cdot [k_r \cdot t^2 - 2f \cdot t]} dt \\
&= \int_{-\infty}^{\infty} \text{rect}\left(\frac{t}{T}\right) \cdot e^{j\Phi(t)}
\end{aligned}
\tag{2.63}
$$

with a quickly variating phase term:

$$\Phi(t) = \pi \cdot [k_r \cdot t^2 - 2f \cdot t] \tag{2.64}$$

The momentaneous frequency calculates to

$$f_0(t) = \frac{1}{2\pi} \frac{d}{dt} \Phi(t) = \frac{1}{2} \cdot [2k_r \cdot t - 2f] \tag{2.65}$$

Please note:
 the integral over the ranges the phase oscillates quickly, disappears;
 the integral only delivers a contribution on that spot the phase does not vary, in other words: $f_0(\tilde{t}) = 0$.

There is valid:

$$f_0(\tilde{t}) = 0 \qquad \tilde{t} = \text{stationary point}$$

and

$$\left. \frac{d^2\Phi(t)}{dt^2} \right|_{t=\tilde{t}} \overset{!}{\neq} 0 \qquad \text{phase minimum or maximum}$$

Search for the stationary point:

$$\left. \frac{d\Phi}{dt} \right|_{\tilde{t}} = \pi[2k_r \cdot \tilde{t} - 2f] \overset{!}{=} 0 \tag{2.66}$$

$$\tilde{t} = \frac{f}{k_r} : \text{stationary point} \quad \tilde{t} \in \left[\frac{-T}{2}, \frac{T}{2} \right]$$

$$\frac{d^2\Phi(t)}{dt^2} = \pi \cdot 2k_r \begin{cases} > 0 & \text{phase minimum} \\ < 0 & \text{phase maximum} \end{cases}$$

$$\frac{d^3\Phi(t)}{dt^3} = 0! \qquad \text{higher derivatives disappear}$$

We develop the phase according to Taylor:

$$\Phi(t) = \Phi(\tilde{t}) + \dot{\Phi}(\tilde{t}) \cdot [t - \tilde{t}] + \frac{\ddot{\Phi}(\tilde{t})}{2} [t - \tilde{t}]^2 \tag{2.67}$$

$$\Phi(t) = \Phi(\tilde{t}) + \frac{\ddot{\Phi}(\tilde{t})}{2} [t - \tilde{t}]^2 \tag{2.68}$$

with

$$\dot{\Phi}(\tilde{t}) \cdot [t - \tilde{t}] = 0 \quad \text{because} \quad \dot{\Phi}(\tilde{t}) = 0 \tag{2.69}$$

and simplify the integral equation (2.63):

$$S_{T_n}(f) = \int\limits_{-\infty}^{\infty} \text{rect}\left(\frac{\tilde{t}}{T}\right) \cdot e^{j[\Phi(\tilde{t}) + \frac{1}{2}\ddot{\Phi}(\tilde{t})[t-\tilde{t}]^2]} dt = e^{j\Phi(\tilde{t})} \cdot \int\limits_{-\infty}^{\infty} \text{rect}\left(\frac{\tilde{t}}{T}\right) \cdot e^{j\frac{1}{2}\ddot{\Phi}(\tilde{t})[t-\tilde{t}]^2} dt \tag{2.70}$$

$$S_{T_n}(f) \cong e^{j\Phi(\tilde{t})} \cdot \text{rect}\left(\frac{\tilde{t}}{T}\right) \cdot \int\limits_{-\infty}^{\infty} e^{j\frac{1}{2}\ddot{\Phi}(\tilde{t})[t-\tilde{t}]^2} dt \tag{2.71}$$

with the substitution of $t - \tilde{t} = u$ equation (2.71) changes to:

$$S_{T_n}(f) = e^{j\Phi(\tilde{t})} \cdot \text{rect}\left(\frac{\tilde{t}}{T}\right) \cdot \int\limits_{-\infty}^{\infty} e^{j\frac{1}{2}\ddot{\Phi}(\tilde{t})\cdot u^2} du \tag{2.72}$$

with:

$$\frac{1}{2}\ddot{\Phi}(\tilde{t}) \cdot u^2 = v^2 \Rightarrow v = \frac{1}{\sqrt{2}} \cdot \sqrt{\ddot{\Phi}(\tilde{t})} \cdot u \Rightarrow du = \frac{dv\sqrt{2}}{\sqrt{\ddot{\Phi}(\tilde{t})}}$$

the integral forms

$$S_{T_n}(f) \cong e^{j\Phi(\tilde{t})} \cdot \text{rect}\left(\frac{\tilde{t}}{T}\right) \cdot \int\limits_{-\infty}^{\infty} e^{jv^2} dv \cdot \frac{\sqrt{2}}{\sqrt{\ddot{\Phi}(\tilde{t})}} \tag{2.73}$$

Note the Fresnel integral:

$$\int\limits_{-\infty}^{\infty} e^{jv^2} dv = \sqrt{\frac{\pi}{2}} + j\sqrt{\frac{\pi}{2}} \tag{2.74}$$

The spectrum can now be rewritten as follows:

$$S_{T_n}(f) \cong e^{j\Phi(\tilde{t})} \cdot \text{rect}\left(\frac{\tilde{t}}{T}\right) \cdot \sqrt{\pi} \cdot [1 + j] \cdot \frac{1}{\sqrt{\ddot{\Phi}(\tilde{t})}} \tag{2.75}$$

Inserting the values

$$\ddot{\Phi}(\tilde{t}) = \pi \cdot 2k_r \qquad \tilde{t} = \frac{f}{k_r}$$

$$\Phi(\tilde{t}) = \pi \cdot [k_r \cdot \tilde{t}^2 - 2f \cdot \tilde{t}]$$

$$\Phi(\tilde{t}) = \pi \cdot \left[k_r \cdot \frac{f^2}{k_r^2} - 2f \cdot \frac{f}{k_r} \right] = -\pi \cdot \frac{f^2}{k_r}$$

delivers:

$$S_{T_\cap}(f) \cong \mathrm{rect}\left(\frac{f}{k_r \cdot T}\right) \cdot e^{-j\pi \cdot \frac{f^2}{k_r}} \cdot [1 + j] \cdot \frac{1}{\sqrt{\pi} \cdot \sqrt{2}} \cdot \frac{\sqrt{\pi}}{\sqrt{k_r}} \tag{2.76}$$

$$\boxed{S_{T_\cap}(f) \cong \mathrm{rect}\left(\frac{f}{k_r \cdot T}\right) \cdot e^{-j\pi \cdot \frac{f^2}{k_r}} \cdot e^{j\frac{\pi}{4}} \cdot \frac{1}{\sqrt{k_r}}} \tag{2.77}$$

$$S_{T_\cap}(t) = \mathrm{rect}\left(\frac{t}{T}\right) \cdot e^{j\pi k_r \cdot t^2} \quad \circ\!\!-\!\!\bullet \quad S_{T_\cap}(f) = \frac{1}{\sqrt{k_r}} \cdot \mathrm{rect}\left(\frac{f}{B_r}\right) \cdot e^{-j\frac{\pi f^2}{k_r}} \cdot e^{j\frac{\pi}{4}}$$

$$\tag{2.78}$$

As a result we get the Fourier transformation pair as an approximation of the chirp spectrum according to the method of the stationary phase with the *bandwidth* $B_r = k_r \cdot T$.

$$S_{T_\cap}(t) = \mathrm{rect}\left(\frac{t}{T}\right) \cdot e^{j\pi k_r \cdot t^2} \quad \circ\!\!-\!\!\bullet \quad S_{T_\cap}(f) = \frac{1}{\sqrt{k_r}} \cdot \mathrm{rect}\left(\frac{f}{k_r \cdot T}\right) \cdot e^{-j\pi \cdot \frac{f^2}{k_r}} \cdot e^{j\frac{\pi}{4}}$$

$$\tag{2.79}$$

This identity is applicable e.g. in the range compression (see chapter 3.2 *Geometrical resolution/range compression*), where a matched filter in the frequency domain is used.

$$g_T(t) = \frac{1}{2} r_T(t)^* \cdot s_T(-t)^*$$

$$g_T(f) = \frac{1}{2} R_T \cdot S_T(f)^* = \frac{1}{2} R_T(f) \cdot H_T(f)^*$$

The ascertained spectrum then is used as a transfer function of the range-compression filter.

$$H_T(f) = S_T(f)^* = \frac{1}{\sqrt{k_r}} \cdot \mathrm{rect}\,\frac{f}{k_r \cdot T} \cdot e^{j\pi \frac{f}{k_r}} \cdot e^{-j\frac{\pi}{4}}$$

2.1.2.3 Autocorrelation function of a chirp signal

In particular the autocorrelation function of a chirp signal plays an important role in the system-theoretical explanation of the SAR-signal processing. Therefore it will be derivated whilst explaining the basics.

From a chirp with the form

$$S_T(t) = rect\left(\frac{t}{T}\right) \cdot e^{j\pi k_r t^2} \tag{2.80}$$

the autocorrelation function according to [LÜK92] can be expressed as follows:

$$\varphi_{SS_T}(t) = \frac{1}{2} \cdot s_T(-t)^* \cdot S_T(t) \tag{2.81}$$

$$\varphi_{SS_T}(t) = \frac{1}{2} \int_{-\infty}^{+\infty} rect\left(\frac{-\tau}{T}\right) \cdot e^{-j\pi k_r \tau^2} \cdot rect\left(\frac{t-\tau}{T}\right) \cdot e^{j\pi k_r(t-\tau)^2} \cdot d\tau \tag{2.82}$$

$$\varphi_{SS_T}(t) = \frac{1}{2} \cdot e^{j\pi k_r t^2} \cdot \int_{-\infty}^{+\infty} rect\left(\frac{\tau}{T}\right) \cdot rect\left(\frac{t-\tau}{T}\right) \cdot e^{-j2\pi k_r t\tau} \cdot d\tau \tag{2.83}$$

Exemplary, mainly to compare, is the calculation of an autocorrelation function of a real chirp in the bandpass domain:

$$S(t) = rect\left(\frac{t}{T}\right) \cdot \cos(2\pi \cdot f_0 \cdot t + \pi \cdot k_r \cdot t^2) \qquad \text{real chirp} \tag{2.84}$$

$$r(t) = a \cdot s(t - t_0) \qquad \text{delayed, received chirp} \tag{2.85}$$

$$h(t) = s(-t) \Rightarrow g(t) = a \cdot \varphi_{SS}(t - t_0) \tag{2.86}$$

The autocorrelation function we look for is calculated as follows:

$$\varphi_{SS}(t) = S(-t) \cdot S(t) = \int_{-\infty}^{\infty} rect\left(\frac{-\tau}{T}\right) \cdot \cos(\pi \cdot k_r \cdot \tau^2 - 2\pi \cdot f_0 \cdot \tau)$$

$$\cdot rect\left(\frac{t-\tau}{T}\right) \cdot \cos[2\pi \cdot f_0(t-\tau) + \pi \cdot k_r \cdot (t-\tau)^2]d\tau \tag{2.87}$$

This correlation function (see also equation (2.83)) is difficult to calculate in the bandpass domain! More easily to calculate and also to implement is the calculation of the correlation function in the equivalent lowpass domain. This is described as follows:

Processing of the correlation in the equivalent lowpass domain:

$$h_T(t) = S_T(-t)^* \tag{2.88}$$

$$g_T(t) = \frac{1}{2} \cdot r_T(t) \cdot h_T(t) = \varphi_{SS_T}(t - t_0) \cdot e^{-j2\pi \cdot f_0 \cdot t_0} \tag{2.89}$$

Calculating of the correlation product in the equivalent lowpass domain:

$$\begin{aligned}
S(t) &= \text{rect}\left(\frac{t}{T}\right) \cdot \cos(2\pi \cdot f_0 \cdot t + \pi \cdot k_r \cdot t^2) \\
&= \text{Re}\left\{e^{j2\pi \cdot f_0 \cdot t} \cdot e^{j\pi \cdot k_r \cdot t^2}\right\} = \text{Re}\left\{s_T(t) \cdot e^{j2\pi \cdot f_0 \cdot t}\right\}
\end{aligned} \tag{2.90}$$

$$S_T(t) = \text{rect}\left(\frac{t}{T}\right) \cdot e^{j\pi \cdot k_r \cdot t^2} \qquad \text{equivalent lowpass chirp} \tag{2.91}$$

Now we have a look at:

$$g_T(t) = \varphi_{SS_T}(t) = \frac{1}{2} \cdot S(-t)^* \cdot S_T(t) \tag{2.92}$$

$$\varphi_{SS_T}(t) = \frac{1}{2} \cdot \int_{-\infty}^{\infty} \text{rect}\left(\frac{-\tau}{T}\right) \cdot e^{-j\pi \cdot k_r \cdot \tau^2} \cdot \text{rect}\left(\frac{t - \tau}{T}\right) \cdot e^{j\pi \cdot k_r \cdot (t - \tau)^2} \, d\tau \tag{2.93}$$

$$\begin{aligned}
\varphi_{SS_T}(t) = \frac{1}{2} \cdot \int_{-\infty}^{\infty} &\text{rect}\left(\frac{\tau}{T}\right) \cdot e^{-j\pi \cdot k_r \cdot \tau^2} \\
&\cdot \text{rect}\left(\frac{t - \tau}{T}\right) \cdot e^{j\pi \cdot k_r \cdot t^2} \cdot e^{-j2\pi \cdot k_r \cdot t \cdot \tau} \cdot e^{j\pi \cdot k_r \cdot \tau^2} \, d\tau
\end{aligned} \tag{2.94}$$

$$\varphi_{SS_T} = \frac{1}{2} \cdot \int_{-\infty}^{\infty} \text{rect}\left(\frac{\tau}{T}\right) \cdot \text{rect}\left(\frac{t - \tau}{T}\right) \cdot e^{-j2\pi \cdot k_r \cdot t \cdot \tau} \cdot e^{j\pi \cdot k_r \cdot t^2} \, d\tau \tag{2.95}$$

$$\varphi_{SS_T}(t) = \frac{1}{2} \cdot e^{j\pi \cdot k_r \cdot t^2} \int_{-\infty}^{\infty} \text{rect}\left(\frac{\tau}{T}\right) \cdot \text{rect}\left(\frac{t - \tau}{T}\right) \cdot e^{-j2\pi \cdot k_r \cdot t \cdot \tau} \, d\tau \tag{2.96}$$

$$\text{rect}\left(\frac{t - \tau}{T}\right) = \text{rect}\left(\frac{\tau - t}{T}\right)$$

we define the corresponding limits of the interval (Image 2.14)

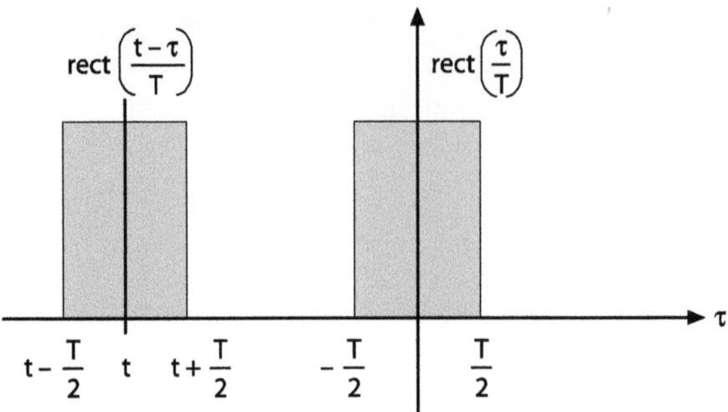

Image 2.14: Definition of interval limits

For further regarding of the autocorrelation function we have to consider the corresponding cases of the overlapping rectangular function:

1st case

$$t + \frac{T}{2} < -\frac{T}{2}$$

$$t < -T \quad \Rightarrow \quad \varphi_{SS_T}(t) = 0!$$

(2.97)

2nd case

$$t + \frac{T}{2} \geq -\frac{T}{2} \quad \text{and} \quad t + \frac{T}{2} < \frac{T}{2}$$

$$\Rightarrow \quad t \geq -T \quad \text{and} \quad t < 0$$

$$\Rightarrow \quad T \leq t < 0$$

(2.98)

here applies:

$$\varphi_{SS_T}(t) = \frac{1}{2} \cdot e^{j\pi \cdot k_r \cdot t^2} \cdot \int_{-\frac{T}{2}}^{t+\frac{T}{2}} e^{-j2\pi \cdot k_r \cdot t \cdot \tau} d\tau$$

(2.99)

$$\varphi_{SS_T}(t) = \frac{1}{2} \cdot \frac{e^{-j2\pi \cdot k_r \cdot t \cdot \tau}}{-j2\pi \cdot k_r \cdot t}\Bigg|_{-\frac{T}{2}}^{t+\frac{T}{2}} \cdot e^{j\pi \cdot k_r \cdot t^2}$$

(2.100)

$$\varphi_{SS_T}(t) = -\frac{1}{2j} \cdot \frac{1}{2\pi \cdot k_r \cdot t} \cdot \left[e^{-j2\pi \cdot k_r \cdot t \cdot \left(t+\frac{T}{2}\right)} - e^{j2\pi \cdot k_r \cdot \frac{T}{2} \cdot t} \right] \cdot e^{j\pi \cdot k_r \cdot t^2}$$

$$= -\frac{1}{2\pi \cdot k_r \cdot t} \cdot \frac{1}{2j} \cdot \left[e^{-j\pi \cdot k_r \cdot t^2} \cdot e^{-j2\pi \cdot k_r \cdot t \cdot \frac{T}{2}} - e^{j2\pi \cdot k_r \cdot t \cdot \frac{T}{2}} \cdot e^{j\pi \cdot k_r \cdot t^2} \right]$$

$$= \frac{1}{2\pi \cdot k_r \cdot t} \cdot \frac{1}{2j} \cdot 2j \sin(\pi \cdot k_r \cdot t \cdot (t+T))$$

$$= \frac{1}{2} \cdot \frac{t+T}{\pi \cdot k_r \cdot t \cdot (t+T)} \cdot \sin(\pi \cdot k_r \cdot t \cdot (t+T))$$

$$= \frac{1}{2} \cdot (t+T) \cdot \mathrm{si}(\pi \cdot k_r \cdot t \cdot (t+T))$$

$$\varphi_{SS_T}(t) = \frac{1}{2} \cdot (t+T) \cdot \mathrm{si}(\pi \cdot k_r \cdot t \cdot (t+T)) \tag{2.101}$$

3rd case

$$t + \frac{T}{2} \geq +\frac{T}{2} \quad \text{and} \quad t - \frac{T}{2} < \frac{T}{2} \tag{2.102}$$

$$\Rightarrow 0 \leq t < T$$

here applies:

$$\varphi_{SS_T}(t) = \frac{1}{2} \cdot \int_{t-\frac{T}{2}}^{\frac{T}{2}} e^{-j2\pi \cdot k_r \cdot t \cdot \tau} d\tau \cdot e^{+j\pi \cdot k_r \cdot t^2} \tag{2.103}$$

$$\varphi_{SS_T} = \frac{1}{2} \cdot \frac{e^{-j2\pi \cdot k_r \cdot t \cdot \tau}}{-j2\pi \cdot k_r \cdot t} \Bigg|_{t-\frac{T}{2}}^{\frac{T}{2}} \cdot e^{j\pi \cdot k_r \cdot t^2}$$

$$= -\frac{1}{2} \cdot \frac{1}{2\pi \cdot k_r \cdot t} \cdot \left[e^{-j2\pi \cdot k_r \cdot t \cdot \frac{T}{2}} - e^{-j2\pi \cdot k_r \cdot t \cdot \left(t-\frac{T}{2}\right)} \right] \cdot e^{j\pi \cdot k_r \cdot t^2} \tag{2.104}$$

$$\varphi_{SS_T}(t) = -\frac{1}{2\pi \cdot k_r \cdot t} \cdot \frac{1}{2j} \cdot \left[e^{-j\pi \cdot k_r \cdot t \cdot (t-T)} \cdot e^{-j\pi \cdot k_r \cdot t \cdot (t-T)} \right] \tag{2.105}$$

$$\varphi_{SS_T}(t) = -\frac{1}{2\pi \cdot k_r \cdot t} \cdot \sin(\pi \cdot k_r \cdot t \cdot (t-T))$$

$$= -\frac{1}{2} \cdot \frac{t-T}{\pi \cdot k_r \cdot t \cdot (t-T)} \cdot \sin(\pi \cdot k_r \cdot t \cdot (t-T)) \tag{2.106}$$

$$= \frac{1}{2} \cdot (T-t) \cdot \mathrm{si}(\pi \cdot k_r \cdot t \cdot (T-t))$$

$$\varphi_{SS_T}(t) = \frac{1}{2} \cdot (T-t) \cdot \mathrm{si}(\pi \cdot k_r \cdot t \cdot (T-t))$$

4th case

$$t \geq T \quad \Rightarrow \quad \Phi_{SS_T}(t) = 0 \tag{2.107}$$

Summary of case 2 and 3: $-T \leq t \leq T$

$$t < 0 \quad \text{2nd case:} \quad \varphi_{SS_T}(t) = \frac{1}{2} \cdot (t + T) \cdot si(\pi \cdot k_r \cdot t \cdot (t + T)) \tag{2.108}$$

$$t > 0 \quad \text{3rd case:} \quad \varphi_{SS_T}(t) = \frac{1}{2} \cdot (T - t) \cdot si(\pi \cdot k_r \cdot t \cdot (T - t)) \tag{2.109}$$

$$\varphi_{SS_T}(t) = \frac{1}{2} \cdot (T - |t|) \cdot si(\pi \cdot k_r \cdot t \cdot (T - |t|)) \tag{2.110}$$

$$\varphi_{SS_T}(t) = \frac{T}{2} \cdot \left(1 - \frac{|t|}{T}\right) \cdot si\left(\pi \cdot k_r \cdot T \cdot t \cdot \left(1 - \frac{|t|}{T}\right)\right)$$

$$\varphi_{SS_T}(t) = \frac{T}{2} \cdot \Lambda\left(\frac{t}{T}\right) \cdot si\left(\pi \cdot k_r \cdot T \cdot t \cdot \Lambda\left(\frac{t}{T}\right)\right)$$

$$\varphi_{SS_T}(t) = \frac{T}{2} \cdot \Lambda\left(\frac{t}{T}\right) \cdot si\left(\pi \cdot k_r \cdot T \cdot t \cdot \Lambda\left(\frac{t}{T}\right)\right) \tag{2.111}$$

and with $B_r = k_r \cdot T = f_\Delta \cong$ bandwidth of chirp

$$\varphi_{SS_T}(t) = \frac{T}{2} \cdot \Lambda\left(\frac{t}{T}\right) \cdot si\left(\pi \cdot B_r \cdot t \cdot \Lambda\left(\frac{t}{T}\right)\right) \tag{2.112}$$

The envelope $\Lambda(t/T)$ is determined by the width of the rectangular function.

The function $\varphi_{ss_T}(t)$ is very similar to a si-function. Therefore we determine a parameter of the si-function, the respective zeroes.

Determination of the 1st zero point for $B_r = k_r \cdot T \gg 1$

Close to the zero point is valid:

$$si\left(\pi \cdot k_r \cdot T \cdot t \cdot \Lambda\left(\frac{t}{T}\right)\right) \cong si(\pi \cdot B_r \cdot t) \tag{2.113}$$

$$si\left(\pi \cdot k_r \cdot T \cdot t \cdot \Lambda\left(\frac{t}{T}\right)\right) = 0 \quad \Longleftrightarrow \quad si(\pi \cdot B_r \cdot t) = 0 \tag{2.114}$$

$$t \cdot \pi \cdot B_r = \pm \pi \tag{2.115}$$

$$t = \pm \frac{1}{B_r} = \pm \frac{1}{k_r \cdot T} \tag{2.116}$$

The 1st zero point is also in the si-function at $\approx \dfrac{1}{B_r} = \dfrac{1}{\text{Bandwidth}}$

The autocorrelation function plays a determining role with respect to the resolution of a SAR-sensor. If the chirp bandwidth increased the zero points would shift in a way that the main lobe of the autocorrelation function would narrow and thus increase the resolution of the SAR sensor in range direction (see chapter 3.2.1 *Geometric resolution in range direction*)

$$B_r \to \infty \Rightarrow \frac{1}{B_r} \to 0$$

The autocorrelation function of the chirp in the bandpass domain:

$$\varphi_{SS}(t) = \mathrm{Re}\left\{ \varphi_{SS_T}(t) \cdot e^{j2\pi \cdot f_0 \cdot t} \right\} \tag{2.117}$$

$$\varphi_{SS}(t) = \frac{T}{2} \cdot \Lambda\left(\frac{t}{T}\right) \cdot \mathrm{si}\left(\pi \cdot k_r \cdot T \cdot t \cdot \Lambda\left(\frac{t}{T}\right)\right) \cdot \cos(2\pi \cdot f_0 \cdot t) \tag{2.118}$$

At this, $\mathrm{si}\left(\pi \cdot k_r \cdot T \cdot t \cdot \Lambda\left(\frac{t}{T}\right)\right)$ is the envelope of the cosine oscillation.

In Image 2.15 *Autocorrelation function of a chirp* the result of the autocorrelation referring equation (2.93) is displayed.

Please note the similarity with a si-function in the range of the maximum.

If the autocorrelation function in the frequency domain wants to be obtained, equation (2.81) has to be rewritten as follows:

$$\Phi_{SS_T}(f) = 1/2 \cdot S_T(f)^* \cdot S_T(f) \tag{2.119}$$

For $S_T(f)$ the derivated spectrum according to chapter 2.1.2.2.2 *Time-limited chirp signal* has to be used.

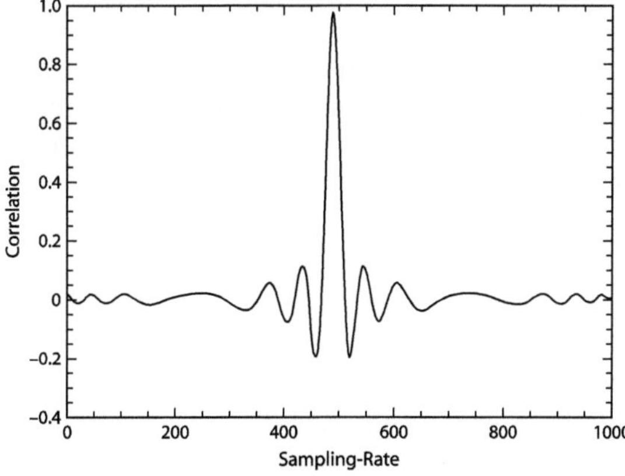

Image 2.15: Autocorrelation function of a chirp

$$S_{T_n}(f) = \frac{1}{\sqrt{k_r}} \cdot e^{-j\pi\left(\frac{f^2}{k_r}-\frac{1}{4}\right)} \cdot \mathrm{rect}\left(\frac{f}{T \cdot k_r}\right)$$

$$\Phi_{SS_T}(f) = \frac{1}{2 \cdot k_r} \cdot e^{-j\pi\left(\frac{f^2}{k_r}-\frac{1}{4}\right)} \cdot e^{j\pi\left(\frac{f^2}{k_r}-\frac{1}{4}\right)} \cdot \mathrm{rect}\left(\frac{f}{T \cdot k_r}\right) = \frac{1}{2 \cdot k_r} \cdot \mathrm{rect}\left(\frac{f}{T \cdot k_r}\right)$$

$$(2.120)$$

Now the approximate character of the frequency-domain display of the chirp signals can be recognised. If we calculate, as shown above, the autocorrelation function in the frequency domain using the conjugated complex multiplication of the chirp spectrum, we will get a rectangular function. A rectangular function is equivalent to a si-function in the time domain. In contrast, the calculation of the autocorrelation function in the time domain (eqs. (2.112), (2.118)) only results in an approximate si-function, but represents, as calculated through the convolution in the time domain, the exact solution of the autocorrelation function of a chirp.

According to the signal theory the description of an SAR system also demands the discussion of the geometrical influence effects that arise on the one hand due to the geometric positioning of the recording system on the other hand because of the own movements of the SAR-sensor.

Apart from the useful influence, these geometric effects also cause imaging errors which will be discussed further on.

2.2 SAR system

2.2.1 Recording geometry

2.2.1.1 Introduction

In principle, SAR systems require solutions for a classic task, the so-called inverse problem. As input-measuring value the backscattering coefficient of the respective illuminated surface is to our disposal. The backscattering coefficient $\sigma(x, y)$ gets falsified before recording of the received signal $r(t, \tau)$ by the 2-dimensional range variant transfer function $h(.,.)$ of the SAR sensor. Therefore the received and recorded signal (raw-data signal) contains the wanted information, but is overlayed with the transfer function of the sensor.

The task of SAR-processing is to process the raw-data in a way that the falsified backscattering information can be retrieved respectively made visible. The transferring function caused on one hand by the movement of the SAR sensor and on the other hand by the recording principle has to be egalised by inverse filtering. Presupposition is an optimised description of the sensor's movement and the principle of the synthetic aperture. The more exact the description is the more accurate the corresponding procession can be calculated. For this reason we will discuss in chapter 2 the geometrical basics in order to describe in chapter 3.1 *General dynamic system description* the recording geometry with the resulting signal consequences.

2.2.1.2 Geometry of sensor motion

2.2.1.2.1 Orbit

Orbit is called the particular line a sensor follows when overflying the object to be observed. The selection of the orbit depends on the kind of the results wanted. Different types of orbits are necessary for continually observing respectively mapping of a certain area. Important critera influencing the selection of an orbit for the SAR are:

minimizing of transmitting power: requires low orbital height
minimizing of gravitation-related interferences: high orbital height
continuing observing of an area (no SAR); geo-stationary orbit
measuring of gravitation-related interferences: low orbital height.

In SAR circular orbital tracks in opposite to the more commonly used elliptical orbits are chosen. Doing this, the carriage-control system of the satellite is not preliminarily engaged. In addition, the advantages of a low as well as a high orbital height can be utilised. Above all, signal-processing is simplified by maintaining the overflight height. In the following, the orbital geometry and its associated parameters are described.

In the coordinate system $K'(x', y', z')$ with the earth's center as origin, the circular orbit is in the $(x', y', 0)$-level (Image 2.18 *The sensors trajectory* and Image 2.17 *Local earth geometry*.

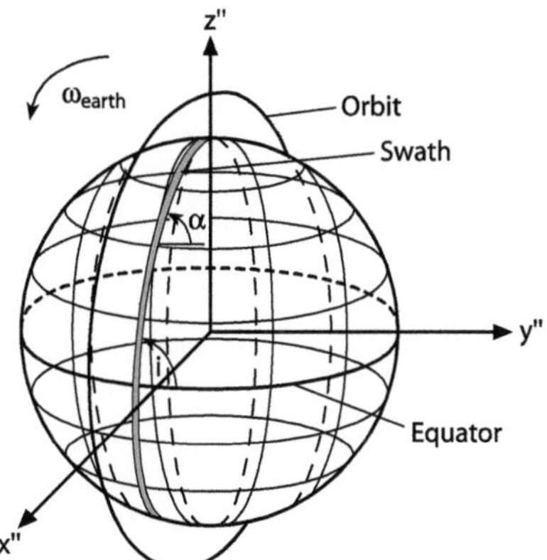

Image 2.16: Orbit

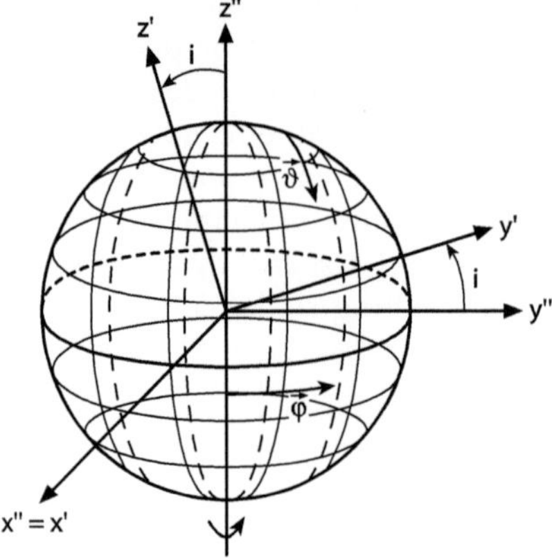

Image 2.17: Local earth geometry

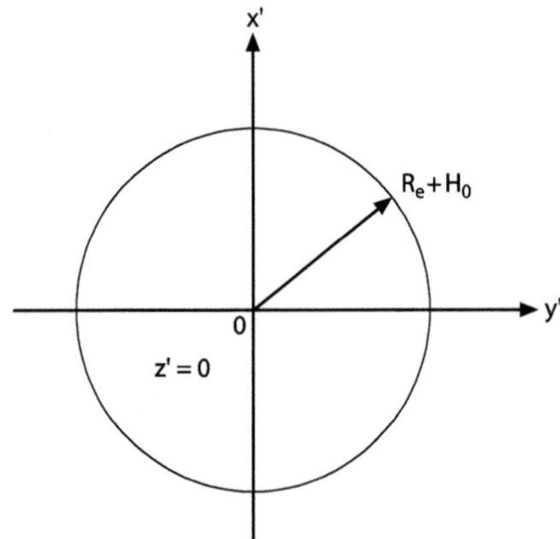

Image 2.18: Trajectory of sensor

In the moment $\tau = 0$ the satellite is in the position $(R_e + H_0, 0, 0)$.

At this, R_e is the earth's radius and H_0 the altitude of the satellite over ground [KAU].

We get the orbital equation of the coordinate system K':

$$x' = (R_e + H_0) \cdot \cos\left(2 \cdot \pi \cdot \frac{\tau}{T}\right) \tag{2.121}$$

$$y' = (R_e + H_0) \cdot \sin\left(2 \cdot \pi \cdot \frac{\tau}{T}\right) \tag{2.122}$$

The coordinate system K' is in opposite to the fixed earth-coordinate system $K''(x'', y'', z'')$ with its origin in the earth's center, is turned by the inclination angle i around the x''-axis ($=$x'-axis) (see Image 2.16 *Orbit*). The z''-axis is the rotation axis of the earth. The inclination angle is defined as angle between the orbital level $(x', y', 0)$ and the equatorial level $(x'', y'', 0)$. The conversion of the track equations is performed via the rotary matrix D:

$$D = \begin{bmatrix} 1 & 0 & 0 \\ 0 & \cos(i) & -\sin(i) \\ 0 & \sin(i) & \cos(i) \end{bmatrix} \tag{2.123}$$

The transformating instruction is

$$K'' = D \cdot K' \quad \text{with} \quad K'' = \begin{bmatrix} x'' \\ y'' \\ z'' \end{bmatrix} \quad \text{and} \quad K' = \begin{bmatrix} x' \\ y' \\ z' \end{bmatrix} \tag{2.124}$$

The new coordinate values are now:

$$x'' = (R_e + H_0) \cdot \cos\left(2 \cdot \pi \cdot \frac{\tau}{T}\right) = x' \tag{2.125}$$

$$y'' = (R_e + H_0) \cdot \sin\left(2 \cdot \pi \cdot \frac{\tau}{T}\right) \cdot \cos(i) \tag{2.126}$$

$$z'' = (R_e + H_0) \cdot \sin\left(2 \cdot \pi \cdot \frac{\tau}{T}\right) \cdot \sin(i) \tag{2.127}$$

Also the corresponding actual orbital positions of the SAR carrier in form of spherical coordinates are required. The general conversion formulas for further transformation of coordinates of cartesian coordinates are:

$$r'' = \sqrt{x''^2 + y''^2 + z''^2} \tag{2.128}$$

$$\vartheta'' = \arccos\left(\frac{z''}{\sqrt{x''^2 + y''^2 + z''^2}}\right) \tag{2.129}$$

$$\varphi'' = \arctan\left(\frac{y''}{x''}\right) \tag{2.130}$$

Regarding our special problem follows:

$$r'' = R_e + H_0 \tag{2.131}$$

$$\vartheta'' = \frac{\pi}{2} - \arcsin\left(\sin\left(2 \cdot \pi \cdot \frac{\tau}{T}\right) \cdot \sin(i)\right) \tag{2.132}$$

$$\varphi'' = \arctan\left(\tan\left(2 \cdot \pi \cdot \frac{\tau}{T} \cdot \cos(i)\right)\right) \tag{2.133}$$

The spherical coordinates are related directly in connection with the following equations with the latitude λ'' and the longitude ϕ'':

$$\lambda'' = \frac{\pi}{2} - \vartheta'' \tag{2.134}$$

$$\phi'' = \varphi'' + \phi''(\tau = 0) \tag{2.135}$$

$$\lambda'' = \arcsin\left(\sin\left(2 \cdot \pi \cdot \frac{\tau}{T}\right) \cdot \sin(i)\right) \tag{2.136}$$

$$\phi'' = \arctan\left(\tan\left(2 \cdot \pi \cdot \frac{\tau}{T}\right) \cdot \cos(i)\right) + \phi''(\tau = 0) \tag{2.137}$$

2.2.1.2.2 *The flight-trajectory angle α*

The flight-trajectory angle α describes the angle between the unit vector of the flight direction of the SAR-carrier and the corresponding unit vector of the earth's surface velocity of the latitude λ'' the SAR-carrier is located at the corresponding time.

At first we calculate the unit vector of the flight direction, in other words the unit-target vector of the orbital track by derivating the track equation:

$$\underline{e}_t = \frac{\underline{v}_s}{|\underline{v}_s|} \quad \text{with} \quad \underline{v}_s = \frac{\partial K''}{\partial \tau} \tag{2.138}$$

$$v_{sx} = (R_e + H_0) \cdot \frac{2\pi}{T} \cdot \left(-\sin\left(2 \cdot \pi \cdot \frac{\tau}{T}\right)\right) \tag{2.139}$$

$$v_{sy} = (R_e + H_0) \cdot \frac{2\pi}{T} \cdot \cos\left(2 \cdot \pi \cdot \frac{\tau}{T}\right) \cdot \cos(i) \tag{2.140}$$

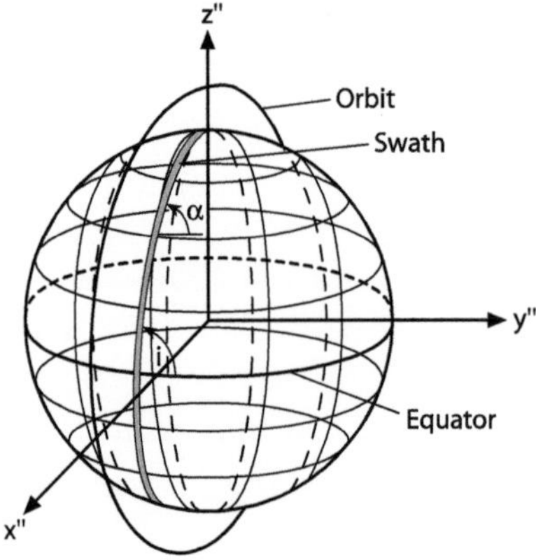

Image 2.19: Orbital track

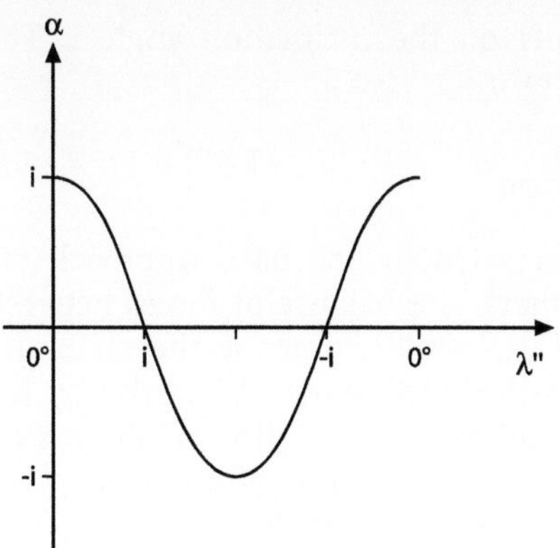

Image 2.20: Flight-trajectory angle

$$v_{sz} = (R_e + H_0) \cdot \frac{2\pi}{T} \cdot \cos\left(2 \cdot \pi \cdot \frac{\tau}{T}\right) \cdot \sin(i) \tag{2.141}$$

$$v_s = |\underline{v}_s| = \sqrt{v_x^2 + v_y^2 + v_z^2} = (R_e + H_0) \cdot \frac{2\pi}{T} \tag{2.142}$$

An object point on the earth's surface always moves vertically to the longitude. The unit vector's form of the earth's surface velocity is:

$$\underline{e}_b = \underline{e}_{\varphi''} = \begin{bmatrix} -\sin(\varphi'') \\ \cos(\varphi'') \\ 0 \end{bmatrix} \quad \text{with} \quad \varphi'' = \arctan\left(\tan\left(2\pi \cdot \frac{\tau}{T}\right) \cdot \cos(i)\right) \tag{2.143}$$

Now both unit vector's are known and the flight-track angle can be calculated with the scalar product:

$$\cos(\alpha) = \underline{e}_t \cdot \underline{e}_b \tag{2.144}$$

$$\alpha = \arccos\left[\begin{array}{l} \sin\left(2\pi \cdot \frac{\tau}{T}\right) \cdot \sin\left(\arctan\left(\tan\left(2\pi \cdot \frac{\tau}{T}\right) \cdot \cos(i)\right)\right) \\ + \cos\left(2\pi \cdot \frac{\tau}{T}\right) \cdot \cos\left(\arctan\left(\tan\left(2\pi \cdot \frac{\tau}{T}\right) \cdot \cos(i)\right)\right) \cdot \cos(i) \end{array} \right] \tag{2.145}$$

Image 2.20 *Flight-trajectory angle* describes a n example for the course of α. For $\tau = 0$ the satellite is over the equator. There the flight trajectory angle is equal to the inclination angle due to the definition. The most important vertices are:

$$\alpha = \pm i \quad \text{equator}$$

$$\alpha = 0° \quad \text{degree} \pm i$$

The course between the vertices depends on the inclination angle i. The wider i is the steeper will be the course of α.

2.2.1.2.3 Satellite velocity and circulation time

A circular orbit of a satellite can be derivated by the basic approach via force balance. The orbit is stable when there is a balance of forces between the gravitation force F_g that tries to pull the satellite down to the earth and the centrifugal force F_z that tries to push the satellite out of its orbit by the rotation velocity. This balance of forces allows a calculation of the necessary rotation velocity and the corresponding orbital time of a sensor.

The gravitation force is calculated according to Newton:

$$\underline{F}_g = -\gamma \cdot \frac{m_e \cdot m_s}{r''^2} \cdot \underline{e}_{r''} \tag{2.146}$$

The variables in equation (2.145) mean:
γ: gravitational constant ($6.67 \ 10^{-11} \ \mathrm{Nm^2 kg^{-2}}$)
m_e: mass of the earth ($5.98 \ 10^{24}$ kg)
m_s: mass of the satellite (2157.4 kg ERS1)
r'': $r'' = R_e + H_0$ distance between satellite and center of the earth ($R_e = 6.378140 \ 10^6$ m)
H_0: flying altitude of the satellite over ground (783444.1 m ERS1)

For low orbits of satellites equation (2.146) can be transferred into a calculation of potentials. The potential difference $\Delta\Phi$ between two point exists when the following integral is solved unequally zero (working integral).

$$\Delta\Phi = \Phi_2 - \Phi_1 = A_{1,2} \tag{2.147}$$

$$A_{1,2} = \int_{r_1}^{r_2} \underline{F}_g \cdot d\underline{r}'' = -\int_{r_1}^{r_2} F_g \cdot dr'' \quad \text{with} \quad \underline{F}_g \| dr'' \tag{2.148}$$

$$A_{1,2} = \gamma \cdot m_e \cdot m_s \cdot \left(\frac{1}{r_2} - \frac{1}{r_1} \right) \tag{2.149}$$

To obtain the potential of a point in the general distance r'' from the earth, the potential in the infinity is set to zero, and the border transient of r_2 to infinity:

$$\lim_{r_2 \to \infty} \frac{1}{r_2} = 0$$

The potential in a general distance r'' is formulated as:

$$\Phi(\underline{r}'') = -\gamma \cdot m_e \cdot m_s \cdot \frac{1}{\underline{r}''} \tag{2.150}$$

Thus, the potential in the correspondingly used orbit (height H_0) is:

$$\Phi(R_e + H_0) = -\gamma \cdot m_e \cdot m_s \cdot \frac{1}{R_e + H_0} = -\gamma \cdot m_e \cdot m_s \cdot \frac{1}{R_e} \cdot \frac{1}{\left(1 + \dfrac{H_0}{R_e}\right)}$$

$$(2.151)$$

With the approximation $H_0 \ll R_e$, equation (2.151) can be rewritten:

$$\Phi(R_e + H_0) = -\gamma \cdot m_e \cdot m_s \cdot \frac{1}{R_e} \cdot \left(1 - \frac{H_0}{R_e}\right) \tag{2.152}$$

The potential energy W between the orbiting satellite and the earth's surface is calculated by means of the equations (2.150) and (2.152):

$$W = \Phi(R_e) - \Phi(R_e + H_0) \tag{2.153}$$

$$W = -\gamma \cdot m_e \cdot m_s \cdot \frac{H_0}{R_e^2} \tag{2.154}$$

Comparing equation (2.154) with the conventional formulating of the potential energy of a body in the altitude h:

$$W = -m_{body} \cdot g \cdot h \tag{2.155}$$

we can describe the normal acceleration by comparing the coefficients:

$$g = \gamma \cdot \frac{m_e}{R_e^2} \tag{2.156}$$

To determine the orbital velocity and circulating time of a satellite we have to observe in addition the changing normal acceleration in the corresponding altitude:

$$g_H = g \cdot \left(1 + \frac{H_0}{R_e}\right)^{-2} \tag{2.157}$$

Inserting equation (2.157) in equation (2.56) and equation (2.146) for the gravitation force in altitude H_0 results in

$$F_g = |\underline{F}_g| = m_s \cdot g_H \cdot \frac{R_e^2}{r''^2} \tag{2.158}$$

The centrifugal force F_c of a corpus during a circular movement has to be regarded:

$$F_c = m_s \cdot r'' \cdot \omega_s^2 \quad \text{with} \quad \omega_s = \frac{v_s}{r''}$$

$$F_c = m_s \cdot \frac{v_s^2}{r''} \tag{2.159}$$

here

v_s: velocity of satellite
ω_s: angle velocity of satellite

For the determination of the satellite's velocity and time of circulation the balance of forces has to be taken into consideration:

$$F_g \overset{!}{=} F_c \tag{2.160}$$

Inserting equations (2.158) and (2.159) in equation (2.160) results in

$$v_s = \sqrt{\frac{g_H \cdot R_e^2}{r''}} = \sqrt{\frac{g_H \cdot R_e^2}{(R_e + H_0)}} \tag{2.161}$$

The respective time of circulation is calculated by means of the orbital course s and the satellite's velocity v_s:

$$T = \frac{s}{v_s} = \frac{2 \cdot \pi \cdot (R_e + H_0)}{v_s} \tag{2.162}$$

The above derivated calculation are considered to be approximatively. For an exact determination of the parameters additionally has to be considered the gravitation-related influences and its corresponding atmospheric impact. Generally, so-called side-looking systems are used as SAR-flying systems. Concerning side-looking systems, a SAR-carrier system illuminates the area to be observed by adjusting the antenna almost in a right angle to the flight direction. Thus, we define the corresponding geometry for further usage of the necessary parameters and use the usual approximations of the flight path and the illuminated surface, by name the linear presentation in the right-angled coordinate system.

2.2.1.3 Definition of the side-looking geometry

The radar antenna is mounted on a satellite or fixed alongside an aircraft and looks down in slant direction in order to be able to reproduce an area parallel to the flight direction. During the fly-by the point target on the ground is illuminated by the antenna's main lobe within the time range t_{send}. The duration of the illumination is the result of the velocity in forward direction v_a, the minimum slant distance R_0 to the time $\tau = \tau_0$ (Image 2.23) and the width of the antenna's main lobe. During this time, the point target is hit by thousands of pulses depending on the "pulse-repetition frequency", whose echoes are registered according to amount and phase and are coherently accumulated. This transaction can be compared with the construction of a linear array antenna alongside the flight path which is assembled out of the physically existing antenna.

The distance of the elements of this fictious array is the result of the velocity in forward direction of the SAR carrier and the pulse-repetition frequency.

In Image 2.21 *Synthetic array* the development of a synthetic array is illustrated. At any azimuth time τ_n the SAR sensor emits a transmitting pulse with the length of time $t_{send} = 37$ µs (the example applies for the parameters of the ERS-1 mission and are used in a qualitatively approximative way). Every transmitting pulse is repeated with a pulse-repetition frequency of $PRF = 1680$ Hz. This is equivalent to an azimuth distance of $x_{repeat} = 4.2$ m at a flight velocity of $v_a = 7000$ m/s. The antenna's footprint is calculated corresponding to the antenna theory in the shortest slant distance of the sensor to the target and can be approximatively expected as $\Delta x = 4$ km and $\Delta y = 50$ km. Please consider that the image is not true to scale!

During the recording of a complete SAR-scene, about 28 000 pulses ($= a$ swath length of about 120 km) are transmitted in the way mentioned. The reflected backscattered signals are recorded coherently.

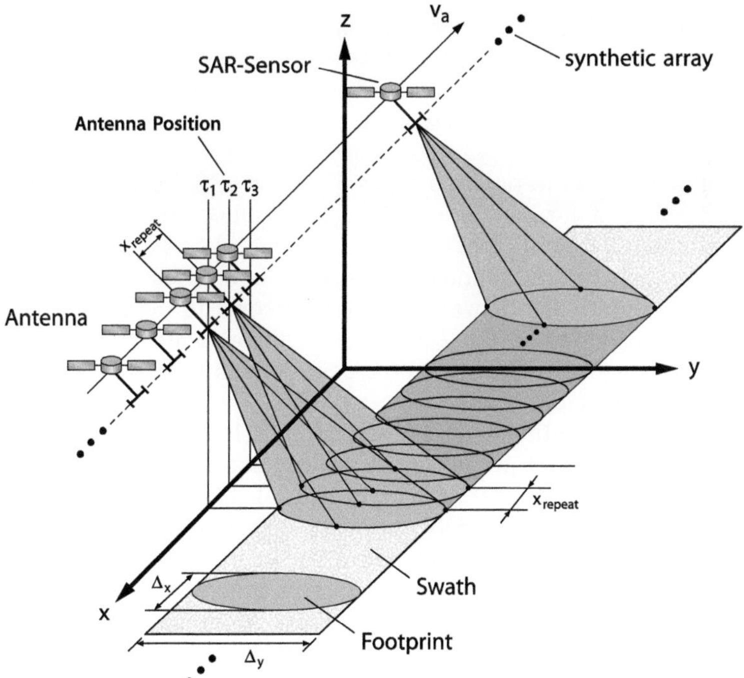

Image 2.21: Synthetic array (qualitative)

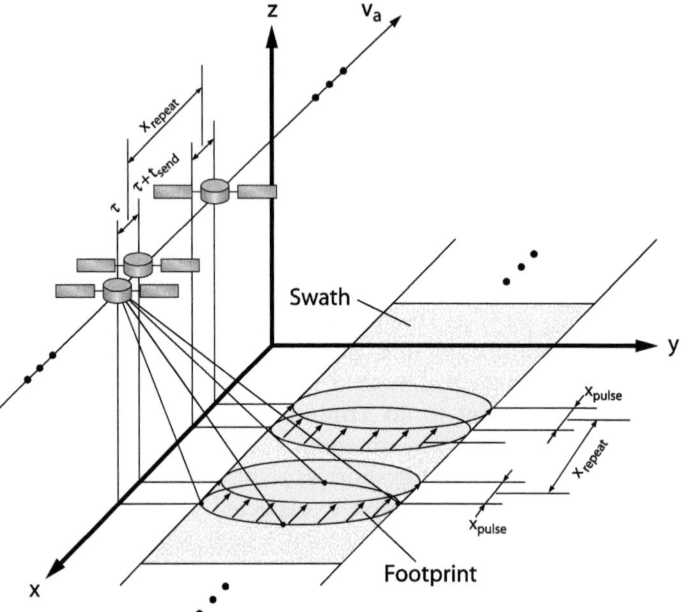

Image 2.22: SAR-footprint suspension (qualitative)

In Image 2.22 *SAR-footprint suspension* shows a distortion effect, which can be neglected when working with SAR-signal-processing algorithms as the corresponding phase history is processed correctly for each azimuth-pixel position (range-migration correction, see chapter 4 *Signal-processing algorithms*).

The transmitting pulse duration is $t_{send} = 37$ μs for ERS-1 satellites. Thus, the antenna's footprint for a duration of t_{send} is "pulled" over the surface in azimuth direction (azimuth length of the footprint: $\Delta x = 4$ km).

This duration can be determined as a surface coordinate $x_{pulse} = 0.3$ m and is negligible due to the above mentioned reasons. In addition to signal-theoretical considerations the exact system theoretical description of the illustrated sensor movement allows the development of a corresponding useful processing-algorythm.

When a point target on the earth is observed, then Image 2.23 *SAR-geometry* illustrates a simple SAR geometry which does not consider neither the curvature of the earth nor the circular flight path of the satellite.

Besides the description of the SAR geometry displayed in Image 2.23 as a relative geometry from the illuminating SAR sensor to the illuminated surface, further geometric effects resulting from movements of the SAR sensor itself have to be taken into consideration. Movements of the SAR sensor itself can occur on aircraft-based systems by turbulences or path corrections during the flight respectively the data recording. The mentioned platform movements can strongly influence the processing parame-

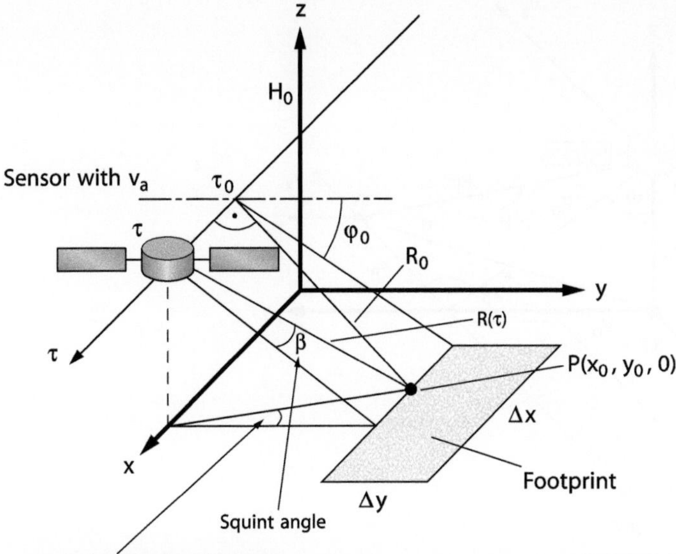

Image 2.23: SAR geometry

ters and have to be compensated [MOR92]. Therefore, the movement of the SAR-carrier platform is also defined in this work in the so-called *local sensor geometry*.

2.2.1.4 Local sensor geometry – sensoric movements

The local sensor geometry helps to describe the effects due to local sensoric movements basically. The carrier circulates around the earth at the altitude H_0. As the altitude H_0 is much smaller than the earth's radius R_0 in this first step we consider the earth's surface as an even surface.

$R(\tau)$: slant-range distance
$r(\tau)$: ground-range distance
v_x: along-track-component of earth rotation
v_y: across-track component or earth rotation
β: squint angle
R_0: shortest slant distance

Point P describes an impact point of the wave transmitted by the carrier. Point P is displaced by the instable position of the SAR carrier. The drifting movements occur in addition to the flight movements and the earth's rotation around the 3-body axis of the SAR carrier yaw, nick, roll. The temporal progression of the drifting movements can not be predicted. To achieve the required preciseness for signal processing, the position of the SAR carrier must be stabilized in a way that the determined critical angle and the drifting velocity are not exceeded.

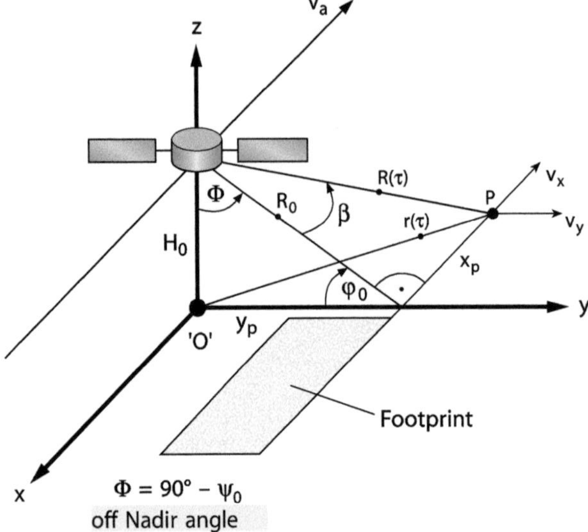

Image 2.24: Sensoric geometry

In the following we describe the effects of the different motion axes as well as the effects of the earth's rotation. Initial point is the sensoric geometry shown in Image 2.24. We describe the movement of the point P depending on the different local carrier angles and with regard to the correspondingly changing relative velocity between the carrier and the illuminated area on the earth's surface.

The aim is the description of the two components of the velocity (along track-flight direction and across track-viewing direction) of a point observed by the sensor depending on the corresponding local sensor movements and the earth's motion. The relative velocity is required to determine the doppler displacement which is necessary for the SAR-image processing.

2.2.1.4.1 *Yaw/gier angle*

The yaw/gier angle describes the rotation angle of the SAR carrier around its yaw axis. Point P moves over the earth's surface on a circular arc around the center of the coordinate system.

Please note: The yaw angle is a local carrier-motion angle and therefore as shown in Image 2.25 only to be measured on the carrier.

To describe the effects of the carrier's movement on the doppler-frequency, we have a look at the corresponding motion-angle projection on the earth's surface.

For SAR-satellites, the yaw angle ψ has only very small values. The movement is assumed as linear along the x-axis.

From Image 2.26 *Yaw angle* we learn:

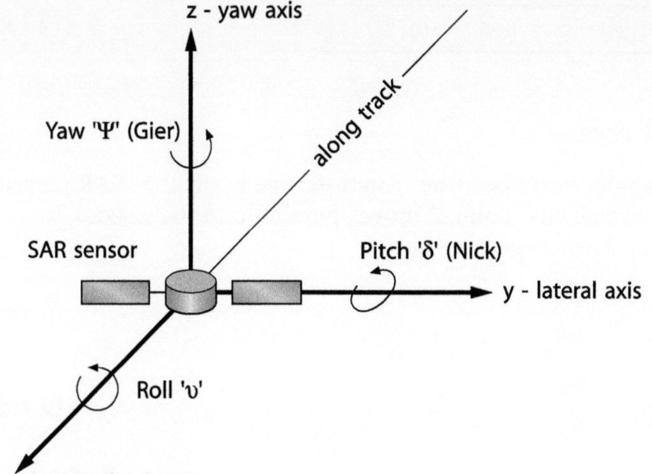

Image 2.25: Local sensoric geometry

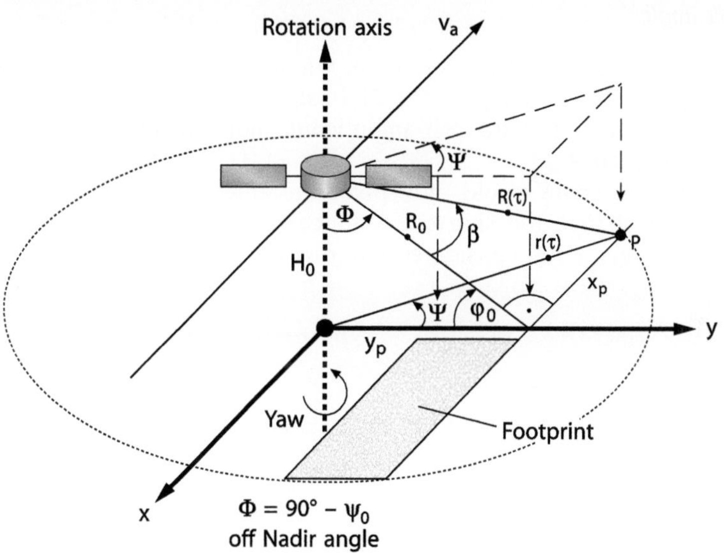

Image 2.26: Yaw angle

$$\tan(\psi) = \frac{x_p}{y_p} \tag{2.163}$$

$$\sin(\Phi) = \cos(\varphi_0) = \frac{y_p}{R_0} \tag{2.164}$$

$$\boxed{x_p = \tan(\psi)y_p = \tan(\psi)\cos(\varphi_0) \cdot R_0 = \sin(\psi)\cos(\varphi_0) \cdot r(\tau)} \tag{2.165}$$

$$\boxed{y_p = \tan(\psi)\sin(\Phi) \cdot R_0 = \cos(\psi)\sin(\Phi) \cdot r(\tau)} \qquad (2.166)$$

2.2.1.4.2 Pitch/nick angle

The pitch/nick angle describes the rotation angle of the SAR carrier around its lateral axis. Thus, point P moves parallelly to the x-axis.

From Image 2.27 *Pitch angle* we learn:

$$\tan(\delta) = \frac{x_p}{H_0} \qquad (2.167)$$

$$\cos(\Phi) = \sin(\varphi_0) = \frac{H_0}{R_0} \qquad (2.168)$$

$$x_p = \tan(\delta) \cdot H_0 = \tan(\delta)\sin(\varphi_0) \cdot R_0 = \tan(\delta)\cos(\Phi) \cdot R_0 \qquad (2.169)$$

2.2.1.4.3 Roll angle

The rotation of the SAR carrier around its longitudinal axis is called roll motion. During the data acquisition the ERS-satellites are in a nominal position $\vartheta' = 0°$, a changing of the roll angle would not lead to an influence of the doppler frequency.

An occurring roll angle would only pivot the antenna's taper in range direction but not change the shortest distance to the footprint. For that reasons, the roll angle is not considered any further.

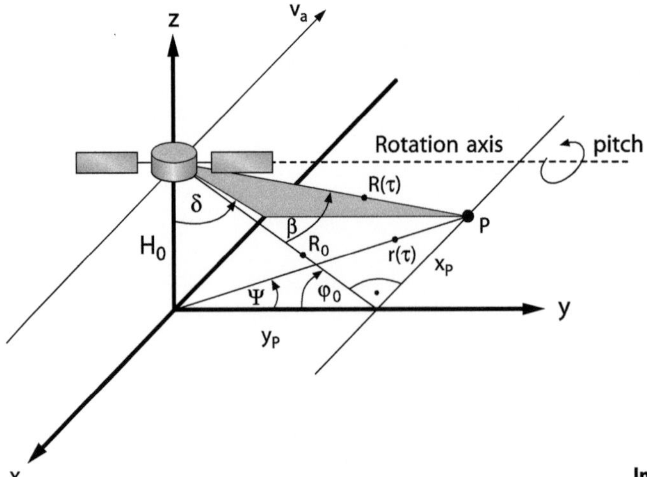

Image 2.27: Pitch angle

2.2.1.4.4 Summary of the positioning

Having considered the yaw/gier and nick angle revealed that the angles have only an influence on the along-track parameter x_p. For this reason, both effects are superposed additively.

The resulting rectangular distances are adding as follows:

$$x_p = R_0[\cos(\varphi_0)\tan(\psi) + \sin(\varphi_0)\tan(\delta)] \tag{2.170}$$

$$x_p = H_0[\cot(\varphi_0)\tan(\psi) + \tan(\delta)] \tag{2.171}$$

$$y_p = \frac{H_0}{\tan(\varphi_0)} = \tan(\Phi) \cdot H_0 = R_0\cos(\varphi_0) \tag{2.172}$$

$$x_p = R_0[\tan(\psi)\sin(\Phi) + \tan(\delta)\cos(\Phi)] \tag{2.173}$$

$$\boxed{x_p = r(\tau)[\sin(\psi)\sin(\Phi) + \tan(\delta)\cos(\Phi)]} \tag{2.173a}$$

$$\boxed{x_p = H_0[\tan(\psi)\tan(\Phi) + \tan(\delta)]} \quad \text{with} \quad R_0 = \frac{H_0}{\cos(\Phi)} \tag{2.174}$$

Equation (2.174) describes the along-track motion of a point P due to the local sensor-motion angle yaw and nick and with the help of the "off-Nadir"-angle.

2.2.1.4.5 Rotation of the earth

Moving point targets cause an additional doppler displacement by their velocity. If we observe fixed targets on the earth's surface we have to consider

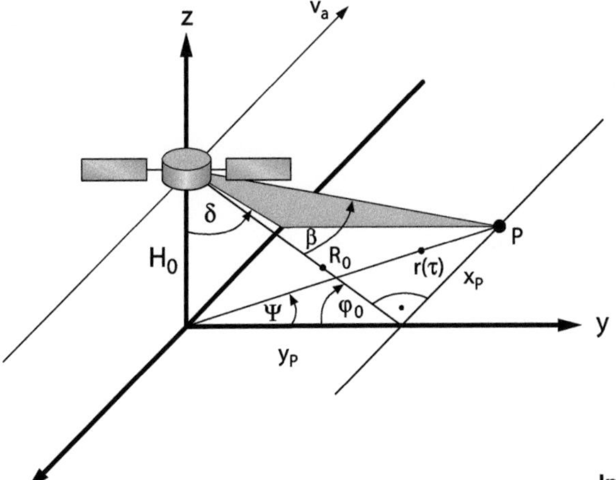

Image 2.28: Pitch and yaw angle

the earth's rotation as the source of the movement. While circulation around the earth, the target point P moves on the earth's surface with the earth's rotation velocity v_e. This velocity only depends on the latitude the observed point is located on.

$$v_e = R_e \cdot \omega_e \cdot \cos(\lambda_e) \qquad (2.175)$$

$$\omega_e = \frac{2\pi}{T_p} \qquad (2.176)$$

with
ω_e: circular velocity of the earth
λ_e: earth's radius
R_e: latitude of point P

The observating point is not on the same latitude as the SAR-carrier due to the angle of reflected beam and the instabilities of the position.

We introduce now the local earth's coordinate system $K_e(x_e,y_e,z_e)$.

Image 2.30 *Local earth's coordinates* shows the displacement K_e in opposite to the orbital coordinate system $K(x,y,z)$ and the flight-path angle α in the z- resp. z_e-axis.

The transformation of the coordinates can be performed as in chapter 2.2.3.2.1:

$$D = \begin{bmatrix} \cos(\alpha) & -\sin(\alpha) & 0 \\ \sin(\alpha) & \cos(\alpha) & 0 \\ 0 & 0 & 1 \end{bmatrix} \qquad (2.177)$$

$$\underline{K}_e = D \cdot \underline{K} \qquad (2.178)$$

Image 2.29: Earth

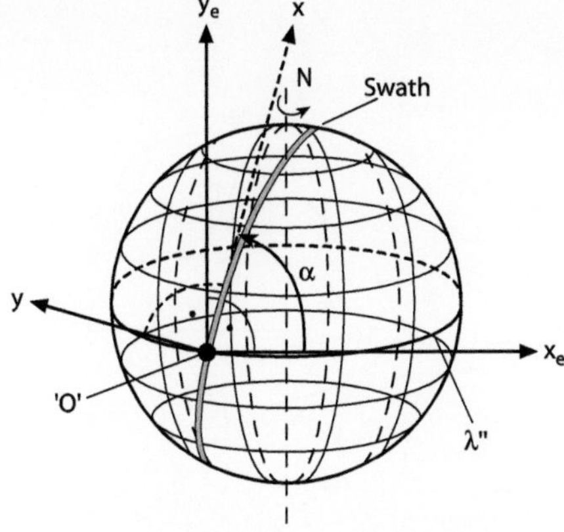

Image 2.30: Local earth

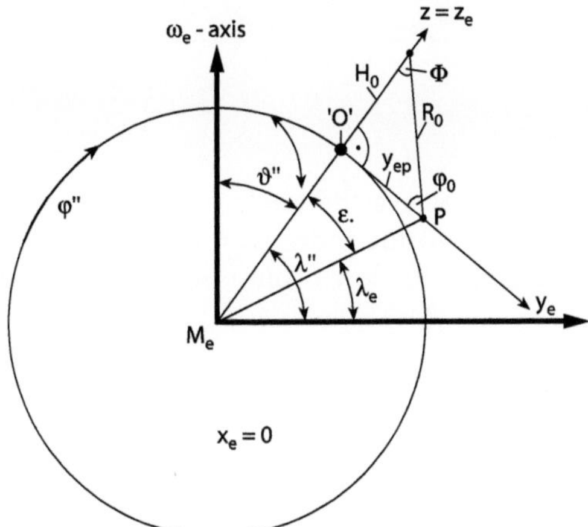

Image 2.31: Projection of point P

We are now in the position to transform point P. As the x_e-component for point P has no influence on the calculation of the latitude, we will only describe the y_e-component in Image 2.31 *Projection of point P*.

$$y_{ep} = x_p\sin(\alpha) + y_p\cos(\alpha) \tag{2.179}$$

Presupposing an even earth's surface $\varepsilon.$ becomes:

$$\sin(\varepsilon.) = \frac{y_{ep}}{R_e} \tag{2.180}$$

In fact there is no plain geometry and we enhance the approximation in equation (2.180). For that purpose we have a look at Image 2.32 *Reflection to determine the latitude.*

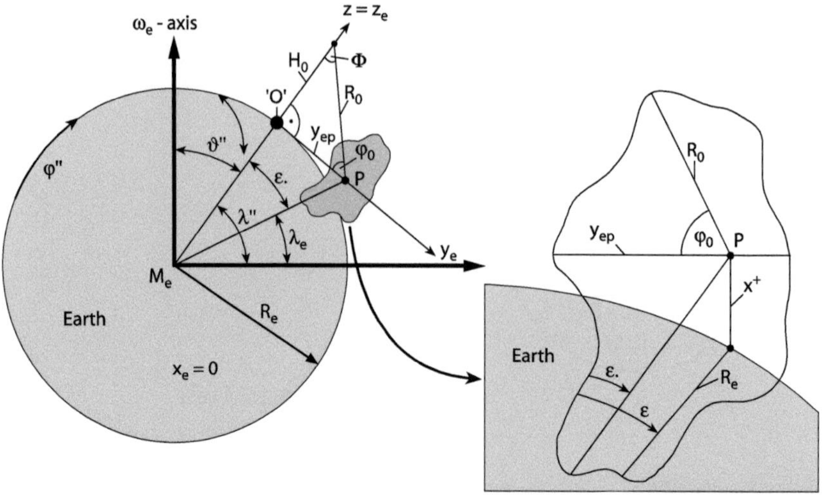

Image 2.32: Description to determine the latitude

Corresponding to Image 2.32, we do not use the angle 'ε.' to determine the respective latitude (point projection) but the more correct angle 'ε'. We calculate for the assumption of parallelity (for low orbits) of R_e and x^+ via circle equation:

$$R_e^2 + y_{ep}^2 = (R_e + x^+)^2 \tag{2.181}$$

with

$$y = y_{ep} = H_0 \cdot \tan(\Phi) \tag{2.182}$$

results for the assisting parameter $R_e + x^+$ in

$$R_e + x^+ = \sqrt{R_e^2 + y_{ep}^2} \tag{2.183}$$

Therefore we get for ε:

$$\sin(\varepsilon) = \frac{y_{ep}}{R_e + x^+} = \frac{y_{ep}}{\sqrt{R_e^2 + y_{ep}^2}} = \frac{H_0 \cdot \tan(\Phi)}{\sqrt{R_e^2 + (H_0 \cdot \tan(\Phi))^2}} \tag{2.184}$$

The latitude λ_e of point P is calculated by subtraction of angle ε from the latitude of the position of the SAR carrier λ'':

$$\lambda_e = \lambda'' - \varepsilon = \frac{\pi}{2} - \vartheta'' - \varepsilon$$

with

$$\lambda'' = \frac{\pi}{2} - \vartheta''$$

$$\varepsilon = \arcsin\left(\frac{y_{ep}}{\sqrt{R_e^2 + y_{ep}^2}}\right) \tag{2.185}$$

$$\lambda_e = \frac{\pi}{2} - \vartheta'' - \arcsin\left(\frac{y_{ep}}{\sqrt{R_e^2 + y_{ep}^2}}\right) \tag{2.186}$$

with y_{ep} (eq. (2.184)) is with the help of equation (2.179) , equation (2.172) and equation (2.174):

$$y_{ep} = x_p \sin(\alpha) + y_p \cos(\alpha)$$

$$y_p = \tan(\Phi) \cdot H_0$$

$$x_p = H_0[\tan(\psi)\tan(\Phi) + \tan(\delta)]$$

$$y_{ep} = H_0[\tan(\psi)\tan(\Phi) \cdot \sin(\alpha) + \tan(\delta) \cdot \sin(\alpha) + \tan(\Phi) \cdot \cos(\alpha)] \tag{2.187}$$

Knowing from equation (2.175):

$$v_e = R_e \cdot \omega_e \cdot \cos(\lambda_e)$$

Inserting equation (2.186) in equation (2.175) we get for the velocity of the projected point R_0 on the earth's surface:

$$v_e = R_e \cdot \omega_e \cdot \cos\left\{\frac{\pi}{2} - \vartheta'' - \arcsin\left(\frac{y_{ep}}{\sqrt{R_e^2 + y_{ep}^2}}\right)\right\} \tag{2.188}$$

$$v_e = R_e \cdot \omega_e \cdot \cos\left\{\frac{\pi}{2} - \vartheta'' - \arcsin\left(\frac{H_0 \cdot [\tan(\psi)\tan(\Phi) \cdot \sin(\alpha) + \tan(\delta) \cdot \sin(\alpha) + \tan(\Phi) \cdot \cos(\alpha)]}{\sqrt{R_e^2 - [H_0 \cdot [\tan(\psi)\tan(\Phi) \cdot \sin(\alpha) + \tan(\delta) \cdot \sin(\alpha) + \tan(\Phi) \cdot \cos(\alpha)]]^2}}\right)\right\} \tag{2.189}$$

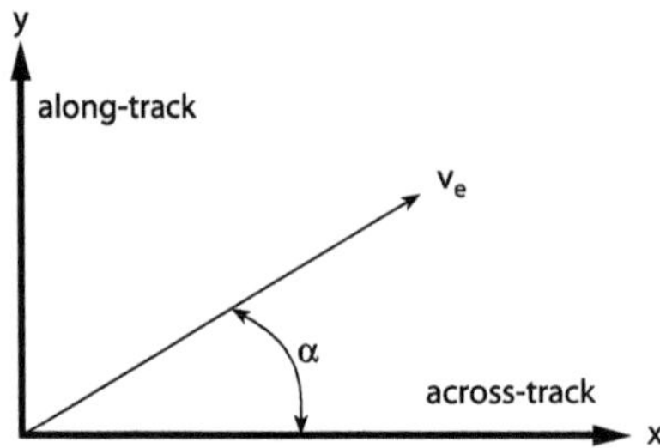

Image 2.33: Velocity components

With the velocity illustrated in equation (2.189) we now can ascertain the respective components. We differentiate between the along-track component

$$v_x = v_e \cos(\alpha) \tag{2.190}$$

and the across-track component (Image 2.33 *Velocity components*)

$$v_y = v_e \sin(\alpha) \tag{2.191}$$

As claimed in the beginning, the respective velocity components of a projected point can be determined in dependency of the local recording geometry parameters and the earth's rotation. Let us discuss now the geometric distortion effects and the resulting imaging failures which are especially used in the interferometric data processing and thus exercised throughout the course of this work.

2.2.1.5 Layover and shadowing

2.2.1.5.1 Basics

In principle, the time a signal needs to travel from a transmitter to an object and back contains the information on the distance from the sensor to the object. If a signal is transmitted and two echoes from different positions are recorded then we can use the time difference between the echoes to determine the distance to the objects to be observed. This way, aircraft or satellite-based SAR is able to measure how far the single objects are away from the satellite or aircraft and also the distance between the objects. These distances along the viewing line to the object are called recordings in "slant range" and measured in direction to the angle of incidence φ_0 (see at Image 2.34 *SAR geometry* depressions angle).

Often, the distance of the object to the sensor is not of interest. It is more interesting to know the distances on the ground, the so-called "ground-range distances". Real ground distances are required to determine for instance the number of agriculturally used areas, the parts of the oceans covered with ice or the topography of a certain area (the corresponding dependencies of altitudes and distances on the ground).

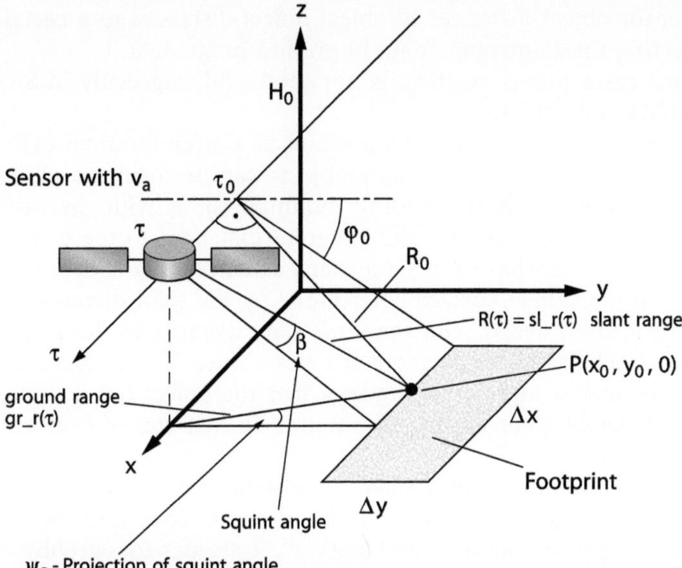

Image 2.34: SAR geometry

In Image 2.34 mean:

τ = Azimuth time; τ_0= point of time of the shortest slant distance

R_0 = nearest slant-range distance

φ_0 = incidence angle

β = squint angle

The SAR sensor receives backscattered radar signals in equidistant time slices that correspond with discrete slant-range distances from the aircraft or satellite. That means, the received raw data is measured in slant range. The transversion of the measured slant-range distances to ground-range means the conversion of slant-range data by regarding different parameters, and afterwards re-sampling, in order to receive equidistantly distributed pixels on the ground. After that, the data can be described in ground range.

The process of the conversion of slant-range distances to ground-range distances can also be regarded as a projection of the slant-range geometry to a respective rectangular geometry and is called ortho-projection (see Image 2.35 *Layover*).

2.2.1.5.2 Foreshortening, layover and shadowing

SAR systems measure very exactly the slant-range distances between the SAR sensor and the corresponding illuminated object on the ground. Sometimes these measurements differ from the actually interesting parameters. We used to work with ground-related distances, that means we know how far we have to move from an object to reach another. We can transform the SAR information from the sensor-object distances to object-object distances to a certain degree by converting the slant-range data in ground-range data.

There are some cases this converting is not successful, especially in SAR scenes with mountainous areas.

When a SAR sensor looks down and sidewards at a steep mountain (Image 2.35 *Layover and shaded areas*), many objects can be localized alongside the ascending respectively descending mountain sides with the same distances to the sensor. Two objects with different distances to the ground and different elevations can have the same slant distances to the sensor as long as they are on the sphere surface determined by the slant distance. As these objects are localised nearly with the same slant distance to the sensor, their backscattered echoes arrive at about the same time. The SAR sensor, relating all information to one point, suggests that the object has a higher backscattering coefficient whereas the information in fact comes from different objects.

According to that, mountain sides tilted against the sensor will be displayed shorter whereas in the reverse case lengthened. Therefore, mountains in SAR scenes appear sort of "overhanging". A steeper topography or a more narrow perspective of the SAR sensor can enforce this shortening effect. This effect, when ascending mountain sides look shorter than descending, is called **foreshortening**. Another extreme, for instance when a mountain peak is positioned in front of the surrounding area (exchanging pixels), is called **layover**.

Layover areas are defined as follows:

Due to the geometry illustrated in Image 2.36 *Layover* results:

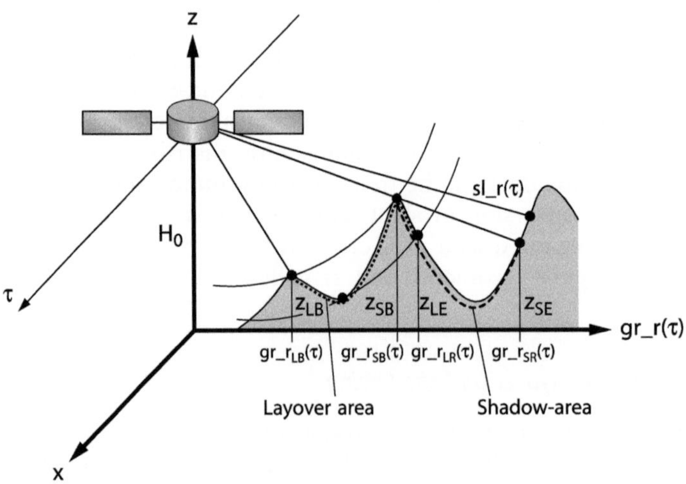

Image 2.35: Layover and shaded areas
gr_r_{LB}(τ): ground range layover begin; gr_r_{LE}(τ): ground range layover end; gr_r_{SB}(τ): ground range shadow begin; gr_r_{LE}(τ): ground range shadow end

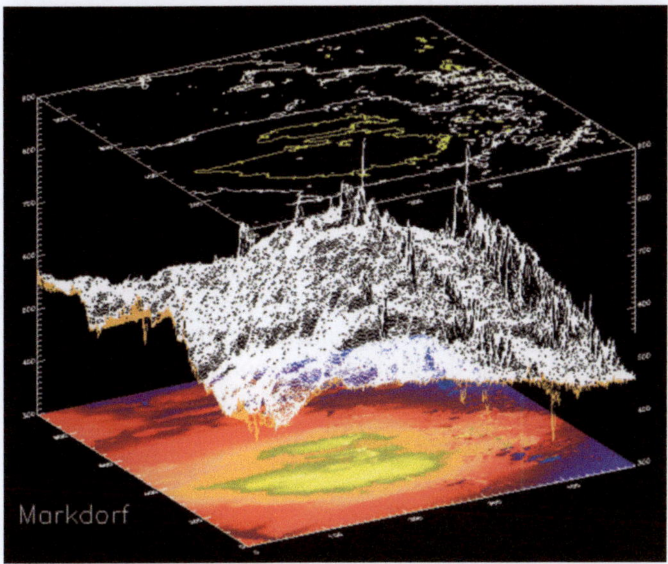

Image 2.35 a: Elevation errors due to shadowing, DO-SAR, Scene: Markdorf

$$r_L = \sqrt{gr\_r_L^2 + (H_0 - z_L)^2} \qquad (2.192)$$

When conversing the slant-range distance R to the ground-range distance gr_r, the layover effect can be defined as pixel-passing effect.

As already discussed, a pixel pass is caused by the layover effect in a form that backscattered targets at a specific elevation can be seen closer to the recording sensor as targets at lower elevations with the same and even shorter ground-range distance. When conversing the corresponding elevation values, the described passing effect has to be detected by watching the corresponding derivation. The derivation of the slant-range distance to the ground-range distance must not become negative. Vice versa, layover areas are characterised by

$$\frac{dr_L}{dgr\_r_L} \leq 0! \qquad (2.193)$$

$$\frac{dr_L}{dgr\_r_L} = \frac{1}{2 \cdot r_L}\left[2 \cdot gr\_r_L - 2 \cdot (H_0 - z_L) \cdot \frac{dz_L}{dgr\_r_L}\right] \qquad (2.194)$$

$$\frac{dr_L}{dgr\_r_L} = \frac{gr\_r_L}{r_L} - \left[\frac{H_0 - z_L}{r_L} \cdot \frac{dz_L}{dgr\_r_L}\right] \qquad (2.195)$$

with the condition of equation (2.193) follows:

$$\frac{\mathrm{gr\_r_L}}{\mathrm{r_L}} - \left[\frac{\mathrm{H_0 - z_L}}{\mathrm{r_L}} \cdot \frac{\mathrm{dz_L}}{\mathrm{dgr\_r_L}}\right] \leq 0! \tag{2.196}$$

$$\boxed{\frac{\mathrm{dz_L}}{\mathrm{dgr\_r_L}} \geq \frac{\mathrm{gr\_r_L}}{\mathrm{H_0 - z_L}}} \tag{2.197}$$

Equation (2.197) describes the layover conditions and is displayed dependent on the ground-distance derivation from the corresponding object's altitude. The derivation of the corresponding altitude to the ground distance can be interpreted as gradient of the altitude. As soon as this gradient gets a critical value which is proportional to the angle of incidence, layover effects are to be expected.

2.2.1.5.3 *Pixel passing*

Pixel passing occurs when altitude values calculated in the slant geometry are converted into the corresponding ground geometry and thus the relative position changes because of their different altitudes. The reason is the above described layover effect. Pixel passing can be illustrated as follows:

Starting-point is Image 2.36. The resulting geometric correlations are used to determine the layover limit per pixel.

We set the corresponding relations according to Pythagoras:

$$\mathrm{gr\_r_1^2} = \mathrm{sl\_r_1^2} - (\mathrm{H} - \mathrm{z_1})^2 \tag{2.198}$$

$$\mathrm{gr\_r_2^2} = \mathrm{sl\_r_2^2} - (\mathrm{H} - \mathrm{z_2})^2 \tag{2.199}$$

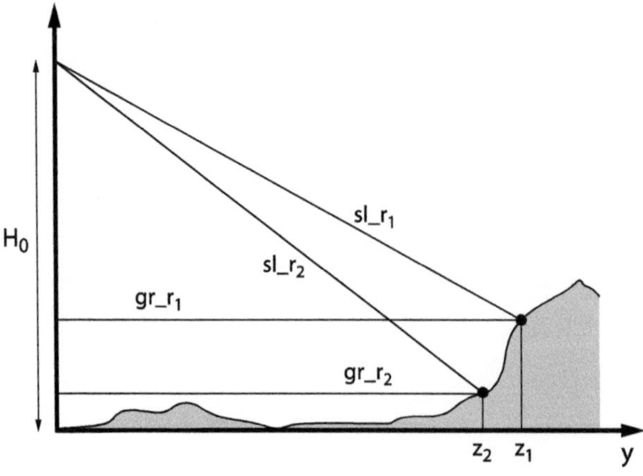

Image 2.36: Layover

The layover limit applies when both ground distances get identical:

$$\boxed{gr\_r_1 \overset{!}{=} gr\_r_2}$$

Using equations (2.198) and (2.199) follows:

$$sl\_r_1^2 - (H - z_1)^2 = sl\_r_2^2 - (H - z_2)^2 \tag{2.200}$$

$$sl\_r_1^2 - (-2 \cdot H \cdot z_1 + z_1^2) = sl\_r_2^2 - (-2 \cdot H \cdot z_2 + z_2^2) \tag{2.201}$$

$$sl\_r_1^2 + 2 \cdot H \cdot z_1 - z_1^2 = sl\_r_2^2 + 2 \cdot H \cdot z_2 - z_2^2 \tag{2.201}$$

$$(H - z_1)^2 - sl\_r_1^2 = (H - z_2)^2 - sl\_r_2^2 \tag{2.202}$$

$$\sqrt{(H - z_1)^2 - sl\_r_1^2 + sl\_r_2^2} = (H - z_2) \tag{2.203}$$

$$\boxed{z_2 = H - \sqrt{(H - z_1)^2 - sl\_r_1^2 + sl\_r_2^2}} \tag{2.204}$$

If a pixel z_2 with an altitude of 10 m in a given aircraft parameter (DO-SAR parameter: $H_0 = 3748$ m, $dsl\_r = .,875$ m, $r_0 = 4475$ m) was determined, a proceeding pixel z_1 with an altitude of only 7.76 m would lead to the pixel-passing effect

$$z_1 = 10 \text{ m} \rightarrow z_2 = 7.7555 \text{ m}$$

To set the absolute altitude of the layover, we formulate:

$$sl\_r_2 = sl\_r_1 + dsl\_r \tag{2.205}$$

$$sl\_r_2^2 = sl\_r_1^2 + 2sl\_r_1 dsl\_r + dsl\_r^2$$

$$z_2 = H - \sqrt{(H - z_1)^2 - sl\_r_1^2 + (sl\_r_1 + dsl\_r)^2} \tag{2.206}$$

$$z_2 = H - \sqrt{(H - z_1)^2 - sl\_r_1^2 + sl\_r_1^2 + 2 \cdot sl\_r_1 \cdot dsl\_r + dsl\_r^2} \tag{2.207}$$

and get as a result with the generally known parameters slant distance and slant resolution as layover for a given "pixel" altitude:

$$z_2 = H - \sqrt{(H - z_1)^2 + 2 \cdot sl\_r_1 \cdot dsl\_r + dsl\_r^2} \tag{2.208}$$

To determine the limit of the relative layover per pixel, we can calculate $\Delta z_{L\,max}$:

$$\Delta z_{L\,max} = (H_0 - z) - \sqrt{(H_0 - z)^2 + dsl\_r^2 + 2 \cdot R(z) \cdot dsl\_r} \qquad (2.209)$$

i.e. with $H_0 = 3680$ m; $R_0 = 4475$ m; $dsl\_r = 1.5$ m $\Rightarrow \Delta z_{L\,max} \leq 1.8$ m.

2.2.1.5.4 Shadowing

SAR illumination is quite 'similar' to the solar radiation in a way that the backside of a mountain can not to be reached by the radiation. As well as the solar radiation, radar sources generally are radiant sources that cause shadows. Thus, and also through phase errors, certain information is lost that have to be re-gained especially when using SAR interferometry.

SAR-images recorded from aircrafts mostly cause serious foreshortening errors or layover errors and also shaded areas. In contrast to that, orbital SAR systems with their relative steep slants deliver by foreshortening and layovering moderately distorted images that only very rarely contain shade areas.

Generally, the conversion from slant-range into ground range applies for the whole topography so that the occurring interferences are also transfered into the ground-range images. The corresponding ground topography is falsified by the described effects and has to be adjusted using different methods to establish a digital height model.

2.2.1.5.5 Slant-ground correction

To solve the slant-to-ground problem we have to calculate a ground-range axis that represents the new ground-related distances depending on the slant-range height. Any distance on the ground gr_r can be calculated by using the geometry shown in Image 2.37 *Slant Range/ground range*.

For any slant distance we can calculate the corresponding new ground distance with following term:

$$gr\_r = \sqrt{sl\_r^2 - (H_0 - z)^2} \qquad (2.210)$$

These new ground distances result from the corresponding slant heights, so the calculated ground distances are calculated equidistantly. Each ground distance is shifted in contrast to the equidistant slant-range axis by the respective slant height. When the slant height values differ so strongly that two different pixels are shifted against each other, the result is the layover effect.

In this case, the resulting ground distance that is supposed to increase in a linear way (not necessarily equidistantly) shows a negative gradient meaning (as already described) that the top of a building is seen with a shorter distance as the ground of the same building. Thus, at first we have to sort the ground distances in a way to get a constantly increasing ground axis (ground-range axis).

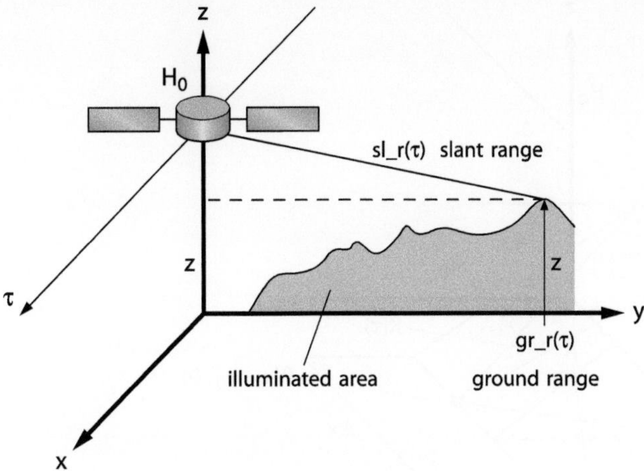

Image 2.37: Slant-range/ground-range conversion

The corresponding shaded terrains have to be treated separately as they would cause the pixel passing which would result in an incorrect allocation of the different backscattering amplitudes to the corresponding ground-range heights. After sorting the non-equidistant ground distances we have to interpolate the values of the ground distances and the corresponding altitudes to achieve an equidistant classification.

Finally, the pixel distances can be changed by re-sampling.

Further reading concerning the geometric effects can be done in the interferometric part of this work (chapter 5).

2.2.2 Doppler effect and frequency shift

2.2.2.1 Vivid derivation of the doppler-frequency

The backscattered transmitting signal received from a punctiform object shows a frequency shift because of the doppler effect that is proportional to the relative velocity v_R between the antenna and object P on the ground and reciprocal to the velocity c the waves spread. When the antenna of the SAR carrier is turned by the so-called squint angle a frequency displacement results due to the doppler effect of the received radar signals. Not changing this squint angle during the flight results in a constant frequency displacement. In this case, the frequency displacement is a linear, additive frequency function named doppler-centroid frequency.

Without an exact knowledge of the doppler-centroid frequency neither a reasonable range-migration correction nor an approximative azimuth compression can be realised. The knowledge of the doppler-centroid frequency is essential for SAR-signal processing. A moving transmitter with the velocity v_a, a squint angle β and an incidence angle φ_0 to the object P re-

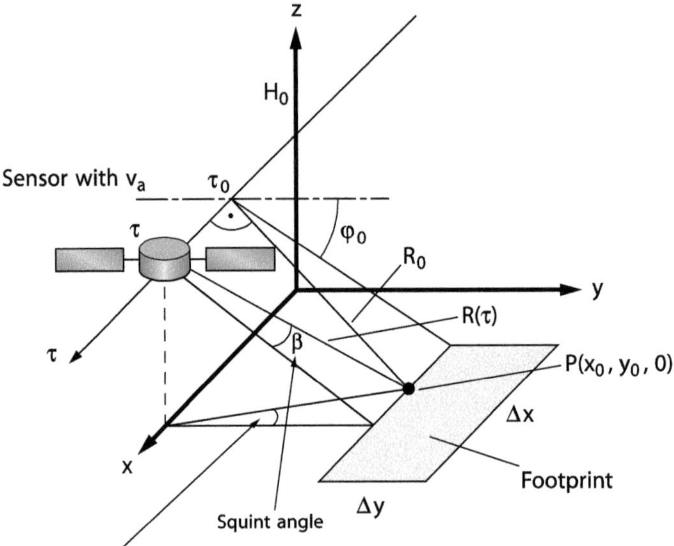

Image 2.38: Geometry for derivation of the doppler

spectively the footprint, radiates an electromagnetic wave via the antenna with a constant transmitting frequency f_0. When the wave hits the object P, the frequency f_p is:

$$f_p = f_0 \cdot \frac{1 + \dfrac{v_R}{c}}{\sqrt{1 - \left(\dfrac{v_R}{c}\right)^2}} \tag{2.211}$$

Looking at Image 2.38 *Geometry for derivation of the doppler,* for the relative velocity of the sensor on the surface dependent on the squint angle follows:

$$v_R = v_a \cdot \sin(\beta) \tag{2.212}$$

When observing the earth's rotation we have to accelerate the sensor's velocity by the rate caused by the earth's rotation (see Image 2.39) and calculate an effective velocity v_{res}, the difference of the corresponding azimuth components:

$$v_{res} = v_a - v_x$$

Please re-consider Image 2.26 *Yaw angle:*

$$\tan(\beta) = \frac{x_P}{R_0} \tag{2.213}$$

$$\sin(\varphi_0) = \frac{H_0}{R_0} \tag{2.214}$$

$$\tan(\psi_0) = \frac{x_p}{y_p} \tag{2.215}$$

The squint angle can be expressed as follows:

$$\tan(\beta) = \frac{x_p}{R_0} = \frac{x_p \cdot \sin(\varphi_0)}{H_0} = \frac{\tan(\psi_p) \cdot \sin(\varphi_0) \cdot y_p}{H_0} = \frac{\tan(\psi_p) \cdot \sin(\varphi_0)}{\tan(\varphi_0)} \tag{2.216}$$

$$\tan(\beta) = \tan(\psi_p) \cdot \cos(\varphi_0) \tag{2.217}$$

$$\sin(\beta) = \sin(\psi_p) \cdot \cos(\varphi_0) \tag{2.218}$$

Inserting in equation (2.212) results in:

$$v_R = v_{res} \cdot \sin(\psi_p) \cdot \cos(\varphi_0) \tag{2.219}$$

the received frequency from the illuminated point P is calculated as follows:

$$f_p = f_0 \cdot \frac{1 + \frac{v_{res}}{c} \cdot \sin(\psi_p) \cdot \cos(\varphi_0)}{\sqrt{1 - \left(\frac{v_R}{c}\right)^2}} \tag{2.220}$$

The wave arriving at object P is reflected in direction of the transmitter. The frequency of the arriving wave is characteristically doppler-displaced. For the receiving frequency f_{PE} applies:

$$f_{PE} = f_0 \cdot \frac{\left(1 + \frac{v_{res}}{c} \cdot \sin(\psi_p) \cdot \cos(\varphi_0)\right)^2}{\left(1 - \left(\frac{v_R}{c}\right)^2\right)} \tag{2.221}$$

In this work we define a non-relative approach, as $v_{res}^2 \ll c^2$. We approximate the fraction in equation (2.221):

$$f_{PE} = f_0 \cdot \frac{1 + \dfrac{v_{res}}{c} \cdot \sin(\psi_p) \cdot \cos(\varphi_0)}{1 - \dfrac{v_{res}}{c} \cdot \sin(\psi_p) \cdot \cos(\varphi_0)}$$

$$= f_0 \cdot \left(1 + 2 \cdot \frac{v_{res}}{c} \cdot \sin(\psi_p) \cdot \cos(\varphi_0)\right)$$

The doppler frequency f_D is calculated as the difference between the receiving and the transmitting frequency:

$$f_D = f_{PE} - f_0 = 2 \cdot f_0 \cdot \frac{v_{res}}{c} \cdot \sin(\psi_p) \cdot \cos(\varphi_0) \tag{2.223}$$

With the wave length $\lambda = \dfrac{c}{f_0}$ and the relative velocity v_R follows:

$$\boxed{f_D = 2 \cdot \frac{v_R}{\lambda} = 2 \cdot \frac{v_{res}}{\lambda} \cdot \sin(\psi_p) \cdot \cos(\varphi_0) = 2 \cdot \frac{v_{res}}{\lambda} \cdot \sin(\beta)} \tag{2.224}$$

The function for the doppler frequency reveals that it does not depend on the flying altitude but is only a term of the flight velocity, the angle of inclination and squint, and the wave length of the radar. As soon as the flight velocity, the angles and the wave length can exactly be defined, the doppler frequency is exactly determinable. As the position of the SAR carrier can not be regulated to a fixed position, the SAR carrier additionally drifts around its three-body axis. The doppler frequency also depends on the drifting angles. The temporal lapse of the drifting angle can neither be forecasted nor calculated. Therefore, f_D can be interpreted as a stochastic process where the actual values have to be generated respectively estimated by means of the measurement data. Besides the mentioned influencing parameters we have to consider the corresponding relative velocity of the earth's rotation. This rotation and also the drifting of the carrier are implemented as a simple geometric description as shown in Image 2.38 *Geometry for derivation of the doppler*. The squint angle illustrated in Image 2.38 does not represent static but a kind of dynamic angle resulting from the corresponding doppler frequency.

In fact, the corresponding doppler frequency is estimated whilst doing the SAR-signal processing. Due to the raw data, a squint angle is estimated via the doppler frequency. Thus, the squint angle is not only a geometric measurement but above all an equivalent description of the doppler effect.

2.2.2.2 Precise derivation via the distance derivations

In chapter 2.2.2.1 *Descriptive derivation of the doppler frequency* we realised that the corresponding doppler frequency depends on the relative velocity v_R between the transmitter and the target. To determine the exact relative velocity we take the way via the corresponding distance derivations.

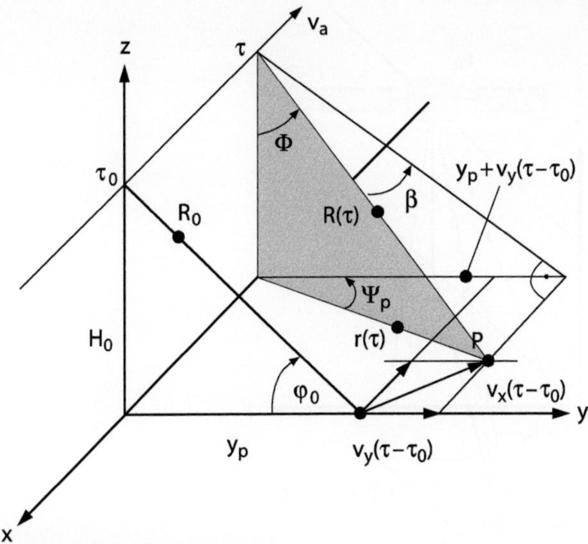

Image 2.39: Sensoric geometry

The velocity 'v_R' is determined by the first derivation of the slant-range trend $R(\tau)$ as shown in Image 2.39 *Sensoric geometry.*

$$v_R(\tau) = \frac{\partial R(\tau)}{\partial \tau} \tag{2.225}$$

$R(\tau)$ applies to a universal point P and considering the earth's rotation:

$$R(\tau) = \sqrt{H_0^2 + R_0^2 + x(\tau)^2} \tag{2.226}$$

The shortest distance R_0 is substituted with a universal distance:

$$R_p = \sqrt{y_p^2 + (\tau - \tau_c)^2 v_y^2}$$

For the general distance trend [BAM89] we get:

$$\boxed{R(\tau) = \sqrt{H_0^2 + (y_p + (\tau - \tau_c) \cdot v_y)^2 + (x_p + (v_x - v_a) \cdot (\tau - \tau_c))^2}} \tag{2.227}$$

Later the along-track velocity (velocity component along the sensor track will be set: $v_{res} = v_a - v_x$. The satellite's velocity is reagarded as constant like the along-track velocity of the earth's rotation. The universal distance $R_p(\tau)$ is given for the center of the footprint τ_c, as in opposite to the doppler frequency, for this point the given doppler-centroid frequency applies (see Image 2.40).

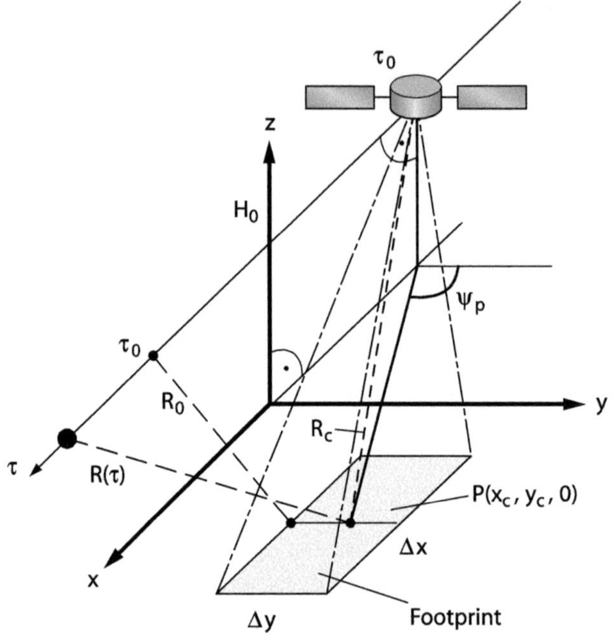

Image 2.40: Doppler-centroid-Position

$$f_{DC} = f_D(\tau = \tau_c) \tag{2.228}$$

The velocity v_R, the target displacement, the phase displacement and the doppler frequency are related as follows:

$$R(\tau) = \sqrt{H_0^2 + R_0^2 + x(\tau)^2} \tag{2.229}$$

$$t_v(\tau) = \frac{2R(\tau)}{c} \tag{2.230}$$

Via phase displacement:

$$\rho(\tau) = -2\pi f_0 t_v(\tau) = -4\pi \frac{R(\tau)}{\lambda} \tag{2.231 a}$$

$$v_R = \frac{\partial R(\tau)}{\partial \tau} \tag{2.231 b}$$

$$f(\tau) = \frac{\dot{\rho}(\tau)}{2\pi} \tag{2.231 c}$$

The development of the phase displacement in a Taylor series around the centroid point τ_c results in:

$$\rho(\tau) = -\frac{4\pi}{\lambda}\left[R(\tau_c) + \dot{R}(\tau_c)(\tau - \tau_c) + \frac{1}{2}\ddot{R}(\tau_c)(\tau - \tau_c)^2\right] \tag{2.232}$$

$$f(\tau) = \dot{\rho}(\tau) = -\frac{2}{\lambda}[\dot{R}(\tau_c) + \ddot{R}(\tau_c)(\tau - \tau_c)]$$

$$f(\tau) = f_{DC} + k_{az}(\tau - \tau_c)$$

$$\uparrow \qquad \uparrow$$
$$\text{doppler} \quad \text{sweep rate}$$
$$\text{centroid}$$

The doppler-centroid frequency and the sweep rate can be calculated according to equation (2.233) via the corresponding derivation of the slant-range distance. Initially, we calculate the distance derivation needed several times in the following.

2.2.2.2.1 First derivative

Due to equation (2.227) the first distance derivation is calculated as

$$R(\tau) = \sqrt{H_0^2 + (y_p + (\tau - \tau_c)v_y)^2 + (x_p + (v_x - v_a)(\tau - \tau_c))^2}$$

with

$$R(\tau) = \sqrt{H_0^2 + (y_p + (\tau - \tau_c)v_y)^2 + (x_p - v_{res}(\tau - \tau_c))^2} \tag{2.234}$$

Please note: All along-track ('x') components are constant. Thus

$$\frac{\partial R(\tau)}{\partial \tau} = \frac{1}{2\sqrt{\cdots}}$$
$$\cdot \left[2 \cdot (y_p + (\tau - \tau_c)v_y) \cdot \left(v_y + (\tau - \tau_c)\frac{\partial v_y}{\partial \tau}\right)\right.$$
$$\left. - 2(x_p - v_{res}(\tau - \tau_c)) \cdot v_{res}\right] \tag{2.235}$$

$$\frac{\partial R(\tau)}{\partial \tau} = \frac{1}{\sqrt{H_0^2 + (y_p + (\tau - \tau_c) \cdot v_y)^2 + (x_p - v_{res}(\tau - \tau_c))^2}}$$
$$\cdot \left[v_y y_p + (\tau - \tau_c v_y^2 + y_p(\tau - \tau_c)\frac{\partial v_y}{\partial \tau} + (\tau - \tau_c)^2 v_y \frac{\partial v_y}{\partial \tau} - x_p v_{res}\right.$$
$$\left. + (\tau - \tau_c)v_{res}^2\right] \tag{2.236}$$

in $\tau = \tau_c$ becomes

$$\left.\frac{\partial R(\tau)}{\partial \tau}\right|_{\tau=\tau_c} = \frac{1}{R(\tau_c)}(v_y y_p - v_{res} x_p) \tag{2.237}$$

and with the results of chapter 2.2.1.4 *Local sensor geometry*, equations (2.166) and (2.173a):

$$\left.\frac{\partial R(\tau)}{\partial \tau}\right|_{\tau=\tau_c} = \frac{1}{R(\tau_c)}(-v_{res} \cdot r(\tau) \cdot (\sin(\psi)\sin(\Phi) + \tan(\delta)\cos(\Phi) \\ - v_y\cos(\psi)\sin(\Phi)))) \tag{2.238}$$

Please note:

$$R(\tau_c) = R(\tau_c) \cdot \sin(\Phi)$$

$$\dot{R}(\tau_c) = \sin(\Phi)(-v_{res}\sin(\psi)\sin(\Phi) + \tan(\delta)\cos(\Phi) + v_y\cos(\psi)\sin(\Phi)) \tag{2.239}$$

$$f_{DC} = -2\frac{v_R(\tau = \tau_c)}{\lambda} = -\frac{2\partial R(\tau = \tau_c)}{\lambda \partial \tau} = -\frac{2}{\lambda}\frac{1}{R(\tau)}(v_y y_p - v_{res} x_p) \tag{2.240}$$

$$f_{DC} = \frac{2}{\lambda}\sin(\Phi)(v_{res}\sin(\psi)\sin(\Phi) + \tan(\delta)\cos(\Phi) - v_y\cos(\psi)\sin(\Phi)) \tag{2.241}$$

$$\boxed{f_{DC} = \frac{2}{\lambda}v_{res}\sin(\Phi)\left[\sin(\psi)\sin(\Phi) + \tan(\delta)\cos(\Phi) - \frac{v_y}{v_{res}}\cos(\psi)\cos(\varphi_0)\right]} \tag{2.242}$$

$$\boxed{f_{DC} = \frac{2}{\lambda}v_{res}\sin(\Phi)\left[\sin(\psi)\sin(\Phi) + \tan(\delta)\cos(\Phi) - \frac{v_y}{v_{res}}\sin(\Phi)\cos(\psi)\right]} \tag{2.243}$$

The obtained doppler-centroid frequency (eq. (2.242)) is defined within the local sensoric geometry. To derivate a respective reflection for an analogue squint geometry, we have to regard the following approximations:

$$\cos(\varphi_0) = \sin(\Phi) \cong 1 \tag{2.244}$$

$$\frac{v_y}{v_{res}} = \tan(\psi_p) \tag{2.245}$$

$$\tan(\psi_p) \cong \tan(\psi) \cong \sin(\psi) \cong \tan(\beta) \tag{2.246}$$

We formulate equation (2.243) with the help of the correlations from chapter 2.2.1.3 *Local sensor geometry*:

$$f_{DC} = \frac{2}{\lambda} \cdot v_{res} \cdot \left[\frac{x_p \cdot y_p}{y_p \cdot R_0} + \frac{x_p \cdot H_0}{H_0 \cdot R_0} - \tan(\psi_p) \right] \tag{2.247}$$

$$f_{DC} = \frac{2}{\lambda} \cdot v_{res}[\tan(\beta) + \tan(\beta) - \tan(\beta)] = \frac{2}{\lambda} \cdot v_{res} \cdot \tan(\beta) \tag{2.248}$$

For a small squint angle again applies:

$$\boxed{f_{DC} \cong \frac{2}{\lambda} \cdot v_{res} \cdot \sin(\beta) \cong \frac{2}{\lambda} \cdot v_{res} \cdot \sin(\psi_p) \cdot \cos(\varphi_0)} \tag{2.249}$$

2.2.2.2.2 Second derivation

To determine the sweep rate in chapter 2.2.4 now the second derivation of the slant-range distance is determined as the calculation appears reasonable in the direct sequence after the first derivation:

$$\frac{\partial R(\tau)}{\partial \tau} =$$

$$= \frac{v_y y_p + (\tau - \tau_c)v_y^2 + y_p(\tau - \tau_c)\frac{\partial v_y}{\partial \tau} + (\tau - \tau_c)^2 v_y \frac{\partial v_y}{\partial \tau} - x_p v_{res} + (\tau - \tau_c)v_{res}^2}{\sqrt{H_0^2 + (y_p + (\tau - \tau_c)v_y)^2 + (x_p - v_{res}(\tau - \tau_c))^2}}$$

$$\tag{2.250}$$

$$\frac{\partial R(\tau)}{\partial \tau} = \frac{u}{v} = \frac{u}{R(\tau)}$$

$$\frac{\partial^2 R(\tau)}{\partial \tau^2} = \frac{\frac{\partial u}{\partial \tau} \cdot v - \frac{\partial R(\tau)}{\partial \tau} u}{R^2(\tau)} \tag{2.251}$$

$$u = v_y y_p + (\tau - \tau_c)v_y^2 + y_p(\tau - \tau_c)\frac{\partial v_y}{\partial \tau} + (\tau - \tau_c)^2 v_y \frac{\partial v_y}{\partial \tau} - x_p v_{res} + (\tau - \tau_c)v_{res}^2$$

$$u = v_y y_p - x_p v_{res} + (\tau - \tau_c)\left[v_y^2 + y_p \frac{\partial v_y}{\partial \tau} + v_{res}^2 \right] + (\tau - \tau_c)^2 v_y \frac{\partial v_y}{\partial \tau}$$

$$\frac{\partial u}{\partial \tau} = y_p \frac{\partial v_y}{\partial \tau} - x_p \frac{\partial v_{res}}{\partial \tau} + v_y^2 + y_p \frac{\partial v_y}{\partial \tau} + v_{res}^2 + (\tau - \tau_c)$$

$$\cdot \left[\frac{\partial v_y^2}{\partial \tau} + y_p \frac{\partial^2 v_y}{\partial \tau^2} + \frac{\partial v_{res}^2}{\partial \tau} \right] + 2(\tau - \tau_c) v_y \frac{\partial v_y}{\partial \tau} + (\tau - \tau_c)^2$$

$$\cdot \left[\frac{\partial v_y}{\partial \tau} \cdot \frac{\partial v_y}{\partial \tau} + v_y \frac{\partial^2 v_y}{\partial \tau^2} \right]$$

with

$$\frac{\partial v_{res}}{\partial \tau} \sim 0$$

$$\frac{\partial u}{\partial \tau} = v_y^2 + v_{res}^2 + 2y_p \frac{\partial v_y}{\partial \tau} + (\tau - \tau_c) \left[\frac{\partial v_y^2}{\partial \tau} + y_p \frac{\partial^2 v_y}{\partial \tau^2} + 2v_y \frac{\partial v_y}{\partial \tau} \right]$$

$$+ (\tau - \tau_c)^2 \left[\left(\frac{\partial v_y}{\partial \tau} \right)^2 + v_y \frac{\partial^2 v_y}{\partial \tau^2} \right]$$

Inserting

$$\frac{\partial^2 R(\tau)}{\partial \tau^2} = \frac{1}{R(\tau)} \frac{\partial u}{\partial \tau} - \frac{\frac{\partial R(\tau)}{\partial \tau} u}{R^2(\tau)} \tag{2.252}$$

$$\frac{\partial^2 R(\tau)}{\partial \tau^2} = \frac{v_y^2 + v_{res}^2 + 2y_p \frac{\partial v_y}{\partial \tau} + (\tau - \tau_c) \left[\frac{\partial v_y^2}{\partial \tau} + y_p \frac{\partial^2 v_y}{\partial \tau^2} + 2v_y \frac{\partial v_y}{\partial \tau} \right]}{\sqrt{H_0^2 + (y_p + (\tau - \tau_c)v_y)^2 + (x_p - v_{res}(\tau - \tau_c))^2}} \ldots$$

$$\ldots \frac{+(\tau - \tau_c)^2 \left[\left(\frac{\partial v_y}{\partial \tau} \right)^2 + v_y \frac{\partial^2 v_y}{\partial \tau^2} \right]}{}$$

$$- \frac{\left[v_y y_p + (\tau - \tau_c)v_y^2 + y_p(\tau - \tau_c) \frac{\partial v_y}{\partial \tau} + \right.}{H_0^2 + (y_p + (\tau - \tau_c)v_y)^2 + (x_p - v_{res}(\tau - \tau_c))^2 \cdot} \ldots$$

$$\ldots \frac{\left. +(\tau - \tau_c)^2 v_y \frac{\partial v_y}{\partial \tau} - x_p v_{res} + \left(\frac{\partial v_y}{\partial \tau} \right) v_{res}^2 \right]^2}{\cdot \sqrt{H_0^2 + (y_p + (\tau - \tau_c)v_y)^2 + (x_p - v_{res}(\tau - \tau_c))^2}}$$

with $\tau = \tau_c$ follows:

$$\left. \frac{\partial^2 R(\tau)}{\partial \tau^2} \right|_{\tau=\tau_c} = \frac{v_y^2 + v_{res}^2 + 2y_p \dfrac{\partial v_y}{\partial \tau}}{R(\tau_c)} - \frac{[v_y y_p - x_p v_{res}]^2}{R^3(\tau_c)} \tag{2.253}$$

Neglecting

$$\frac{[v_y y_p - x_p v_{res}]^2}{R^3(\tau_c)} \cong 0$$

$$v_y^2 < v_{res}^2$$

$$2y_p \frac{\partial v_y}{\partial \tau} < v_y^2$$

we get the following approximation for the second distance derivation (compare with chapter 3.1 *Common dynamic system description*)

$$\ddot{R}(\tau_c) = \frac{v_{res}^2}{R(\tau_c)} \cong \frac{v_a^2}{R(\tau_c)} \tag{2.254}$$

This result is necessary for the determination of the sweep rate in chapter 2.2.3.

2.2.2.3 Alternative derivations

The following description uses derived correlations on the position of the rectangular azimuth-illuminating envelope (see chapter 3.4 *Point-target spectrum*). The doppler displacement can also result in the definition of the envelope of the transmitting signal. The corresponding parameters to determine the doppler-centroid frequency (see "calculation of the 2-dimensional point-target spectrum: Determination of the rectangular position") are defined as follows (we approximatively insert the velocity as v_{res}):

Centroid frequency:

$$f_{DC} = \frac{1}{2}(f_0 + f_u) \cong \frac{2v_{res}^2}{c} \cdot \frac{1}{R(\tau_c)} \cdot (\tau_0 - \tau_c) \cdot (f + f_0) \tag{2.255}$$

Bandwidth:

$$B_{az} = f_0 - f_u \cong \frac{2v_{res}^2}{c} \cdot \frac{1}{R(\tau_c)} \cdot (f + f_0) \cdot \Delta\tau \tag{2.256}$$

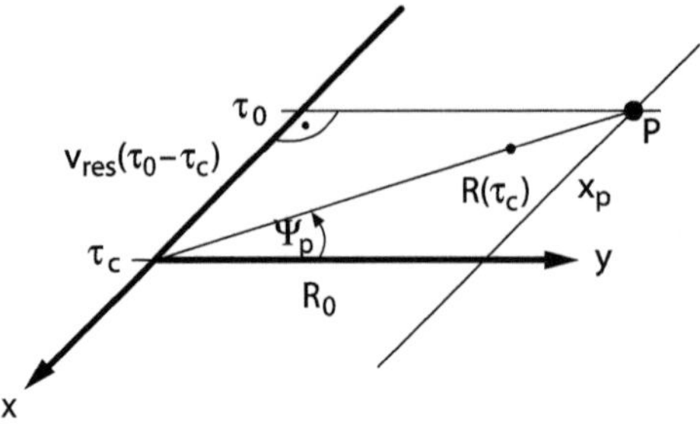

Image 2.41: Simplified SAR geometry

Regard the centroid frequency:

$$f_{DC} \cong \frac{2v_{res}^2}{c} \cdot \frac{1}{R(\tau_c)} \cdot \tau_0 - \tau_c) \cdot (f + f_0)$$

with $\lambda = \dfrac{c}{f_0}$ follows:

$$f_{DC} \cong \frac{2v_{res}^2}{\lambda} \cdot \frac{1}{R(\tau_c)} \cdot (\tau_0 - \tau_c) \cdot \left(1 + \frac{f}{f_0}\right) \tag{2.257}$$

Please have a look at the chapter *SAR geometry* including the motion angle and Image 2.38 *Geometry for derivation of the doppler.*

ψ_p: projection of the squint angle
β: squint angle

The projection of the squint angle is defined as

$$\frac{v_{res} \cdot (\tau_0 - \tau_c)}{R(\tau_c)} = \sin(\psi_p) \cdot \cos(\varphi_0) \tag{2.258}$$

As a result of the doppler displacement by neglection of the additive frequency term follows:

$$\boxed{\begin{aligned} f_{DC} &\cong \frac{2v_{res}}{\lambda} \cdot \left(1 + \frac{f}{f_0}\right) \cdot \sin(\psi_p) \cdot \cos(\varphi_0) \cong \frac{2v_{res}}{\lambda} \cdot \sin(\psi_p)\cos(\varphi_0) \\ &\cong \frac{2v_{res}}{\lambda} \cdot \sin(\beta) \end{aligned}} \tag{2.259}$$

Simplified description on the distance derivations calculated in the part "side- looking geometry"

In chapter 2.2.1 *Recording geometry*, a momentary frequency is derivated from the carrier's movements:

$$f(\tau) = \frac{\dot{\rho}}{2\pi} = -\frac{2}{\lambda} \cdot [\dot{R}(\tau_c) + \ddot{R}(\tau_c) \cdot (\tau - \tau_c)]$$

The momentary frequency can be rephrased according to equation (2.233) to

$$f(\tau) = f_{DC} + k_{az} \cdot (\tau - \tau_c)$$

with

$$f_{DC} = -\frac{2 \cdot \dot{R}(\tau_c)}{\lambda} \tag{2.260}$$

and

$$k_{az} = -\frac{2 \cdot \ddot{R}(\tau_c)}{\lambda} \tag{2.260\,a}$$

whereas 'f_{DC}' is the doppler-centroid frequency and 'k_{az}' the doppler rate (sweep rate) (see Image 2.42 *SAR-geometry*). The distance derivations are defined from:

$$R(\tau) = [(\tau - \tau_0)^2 \cdot v_{res}^2 + y_0^2 + H_0^2]^{\frac{1}{2}} \tag{2.261}$$

$$\dot{R}(\tau) = (\tau - \tau_0) \cdot v_{res}^2 \cdot [(\tau - \tau_0)^2 \cdot v_{res}^2 + y_0^2 + H_0^2]^{-\frac{1}{2}} \tag{2.262}$$

$$\dot{R}(\tau) = (\tau - \tau_0) \cdot \frac{v_{res}^2}{R(\tau_c)} \tag{2.263}$$

$$\ddot{R}(\tau) = \frac{v_{res}^2}{R(\tau)} - \frac{(\tau - \tau_0)^2 \cdot v_{res}^4}{R^3(\tau)} \tag{2.264}$$

$$\ddot{R}(\tau_c) \approx \frac{v_{res}^2}{R(\tau_c)} \simeq \frac{v_a^2}{R(\tau_c)} \tag{2.265}$$

Inserting equation (2.265) in equation (2.260) results in

$$k_{az} = -\frac{2 \cdot v_{res}^2}{\lambda \cdot R(\tau_c)} \tag{2.266}$$

Due to Image 2.42 *SAR geometry* for small squint angles (have a look at the circular arc with the radius y_0) approximatively applies:

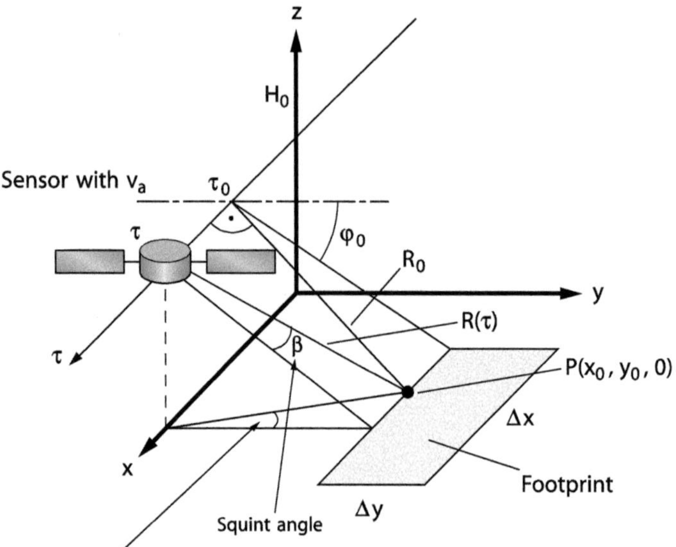

Image 2.42: SAR geometry

$$\tan(\varphi_0) = \frac{H_0}{y_0} \qquad (2.267)$$

$$\sin(\psi_p) \cong \frac{x_0}{y_0} \qquad (2.268)$$

$$\sin(\varphi_0) = \frac{H_0}{R_0} \qquad (2.269)$$

$$y_0 = H_0 \cdot \frac{\cos(\psi_p)}{\tan(\varphi_0)} \qquad (2.270)$$

and

$$x_0 = H_0 \cdot \frac{\sin(\psi_p)}{\tan(\varphi_0)} \qquad (2.271)$$

Without restriction of the generality, the center point of the aperture τ_c can be supposed as zero. If you put $\tau_0 = x_0/v_{res}$, $\tau_c = 0$, 'x₀' and 'y₀' and as well equation (2.263) in equation (2.260), we get the correlation illustrated in equations (2.224) and (2.259).

$$f_{DC} = -\frac{2 \cdot v_{res}^2}{\lambda} \cdot \frac{(\tau_c - \tau_0)}{R_c} = 2 \cdot v_{res} \cdot x_0 \cdot \frac{1}{\lambda \cdot R_0} \qquad (2.272)$$

$$\boxed{f_{DC} = 2 \cdot v_{res} \cdot \frac{\sin(\psi_p) \cdot H_0}{\tan(\varphi_0) \cdot \lambda \cdot R_0} = 2 \cdot v_{res} \cdot \cos(\varphi_0) \cdot \frac{\sin(\psi_p)}{\lambda} = \frac{2 \cdot v_{res}}{\lambda} \cdot \sin(\beta)}$$

$$(2.273)$$

All presented ways of reflections lead to identical results and this way confirm the correctness of the different starting points as well as the result itself.

The next step will describe further important parameters necessary for SAR-image processing like the so-called sweep rate, the doppler bandwidth and the maximum doppler-centroid rate.

2.2.3 The sweep rate

In the field of SAR-signal processing the so-called sweep rate plays an important role along with the doppler frequency. This signal parameter describes the change in modulation per time or, in other words, a kind of modulating velocity of the corresponding signal structure. We differentiate between the range-sweep rate k_r and the azimuth-sweep rate k_{az}.

2.2.3.1 Range-sweep rate k_r

As explained in chapter 2.1.2 *Transmitted and received pulse*, k_r describes the phase-modulating velocity of the transmitting chirp signal. Along with the transmitting period T, the range bandwidth $B_r = k_r \cdot T$ and thus the resolution in range direction is defined via the range-sweep rate. See chapter 3.2 *Geometrical resolution*.

The defined characteristics of the phase difference by k_r are already described in chapter 2.1.3.

2.2.3.2 Azimuth-sweep rate k_{az}

Unlike the range-sweep rate, the azimuth-sweep rate is defined via the corresponding recording geometry. Like k_r, k_{az} describes the phase characteristics of an observed point on the earth within the antenna's lobe. Regarding the distance trend $r(\tau)$ of a point P within the antenna's lobe results from Image 2.42 a *Distance progression*, that during the overflight of the SAR carrier the distance $r(\tau)$ decreases and after reaching the shortest distance k_0 increases again.

This distance progress follows a quadratic regularity:

$$R(\tau) = \sqrt{v_{res}^2(\tau - \tau_0) + R_0^2}$$

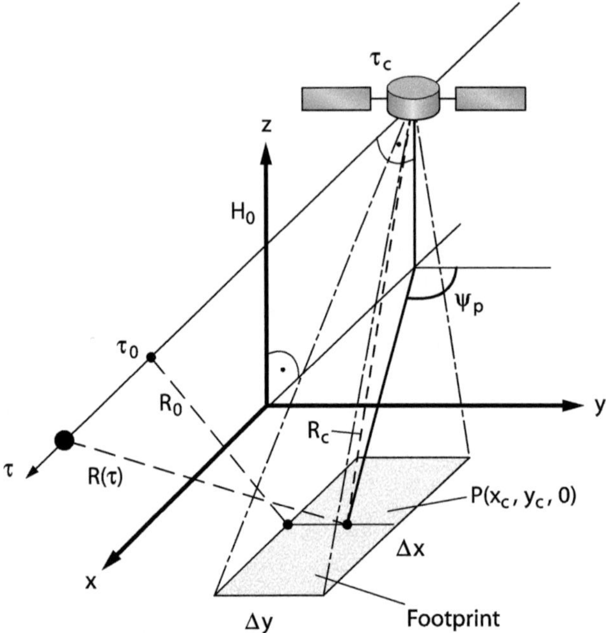

Image 2.42 a: Distance progress

The corresponding currency of signals is calculated as

$$t_v(\tau) = \frac{2R(\tau)}{c}$$

The currency of signals projects to the received signal as phase history:

$$\rho(\tau) = -2\pi f_0 t_v(\tau) = -4\pi \frac{R(\tau)}{\lambda}$$

and thus results, like the transmitted chirp (in range direction), in a signal with quadratic phase – a chirp.

Neglecting the backscattering coefficient, the azimuth chirp can be described as follows:

$$r_T(t, \tau) = S_T(t - t_v(\tau)) \cdot e^{-j2\pi f_0 t_v(\tau)} \tag{2.274}$$

Using the correlations shown in chapter 2.2.2.2 *Exact derivation via the distance derivatives*:

$$\rho(\tau) = -4\pi\frac{R(\tau)}{\lambda} = -\frac{4\pi}{\lambda}\left[R(\tau_c) + \dot{R}(\tau_c)(\tau - \tau_c) + \frac{1}{2}\ddot{R}(\tau_c)(\tau - \tau_c)\right]$$

$$(2.275)$$

and regarding analogous with the range-sweep rate (eq. (2.275)) in the frequency range

$$f(\tau) = \dot{\rho}(\tau) = -\frac{2}{\lambda}\left[\dot{R}(\tau_c) + \frac{1}{2}\ddot{R}(\tau_c)(\tau - \tau_c)\right] \tag{2.276}$$

we get

$$f(\tau) = f_{DC} + k_{az}(\tau - \tau_c) \tag{2.277}$$

as against in range direction:

$$f(t) = f_0 + k_r t \tag{2.278}$$

Comparing equation (2.276) and equation (2.277), a clear analogy is obvious. On the one hand, the doppler-centroid frequency corresponds to the carrier frequency, on the other hand k_{az} is equal with k_r thus is equivalent to the modulating velocity. The special kind of the signal, which also is chosen in range direction because of the better signal recognition, also appears in azimuth direction by the quadratic distance dependency on the target. The resulting processing-advantages apply both in range direction and in azimuth direction. As the sweep rate k_{az} is defined by geometric conditions besides the signal-theoretical description also a short geometrical description shall follow. The comparison between the coefficients in equation (2.276) and (2.277) reveals:

$$k_{az} = -\frac{2}{\lambda}\ddot{R}(\tau_c) \tag{2.279}$$

The second derivation of the distance was already calculated in chapter 2.2.2.2.2 *Second derivation:*

$$\left.\frac{\partial^2 R(\tau)}{\partial\tau^2}\right|_{\tau=\tau_c} = \frac{v_y^2 + v_{res} + 2y_p\dfrac{\partial v_y}{\partial\tau}}{R(\tau_c)} - \frac{[v_y y_p - x_p v_{res}]^2}{R^3(\tau_c)}$$

Neglecting

$$\frac{[v_y y_p - x_p v_{res}]^2}{R^3(\tau_c)} \cong 0$$

$$v_y^2 < v_{res}$$

$$2y_p \frac{\partial v_y}{\partial \tau} < v_y^2$$

for the second distance derivation results

$$\ddot{R}(\tau_c) = \frac{v_{res}^2}{R(\tau_c)} \cong \frac{v_a^2}{R(\tau_c)}$$

and for k_{az}

$$k_{az} = -\frac{2v_{res}^2}{\lambda R(\tau_c)} \cong -\frac{2v_a^2}{\lambda R(\tau_c)} \, . \tag{2.280}$$

2.2.4 Additional parameters

2.2.4.1 Azimuth bandwidth

From the signal-theoretical consideration of the rectangular illumination envelope in azimuth we get from equation (2.272):

$$f_{DC} = -\frac{2}{\lambda} \frac{v_{res}^2(\tau_c - \tau_0)}{R_c} \tag{2.272}$$

$$B_{az} = f_0 - f_u \cong \frac{2v_{res}^2}{c \cdot R(\tau_c)}(f + f_0)\Delta\tau \tag{2.281}$$

$\Delta\tau$ represents the pulse length of the azimuth signal and can also be defined geometrically via the antenna's lobe (chapter 3 Image 3.9 *Side looking geometry*):

$$\sin(\Phi_{HB}) = \frac{\Delta x}{2R_0} \tag{2.282}$$

$$\Delta\tau = \frac{\Delta x}{v_{res}} \tag{2.283}$$

$$B_{az} = \frac{2v_{res}^2}{c} \frac{1}{R(\tau_c)}\left((f + f_0)\frac{\Delta x}{v_{res}}\right) \tag{2.284}$$

Using from chapter 3.2.2 *Geometric resolution in azimuth-direction*

$$\Delta x = \frac{R(\tau_c)\lambda}{D_x} \quad \text{and} \quad R(\tau_c) \approx R_0$$

we get

$$B_{az} \approx \frac{2v_a}{D_x} = \frac{2 \cdot 7548 \text{ m/s}}{12 \text{ m}} \approx 1.3 \text{ kHz} \quad (\text{ERS1}) \tag{2.285}$$

2.2.4.2 Maximum doppler-centroid rate

During the given duration of an azimuth chirp influenced by the illuminating geometry, the doppler-centroid frequency f_{DC} changes by

$$\Delta f_{DC} = \frac{\partial f_{DC}}{\partial \tau} \cdot \Delta \tau \tag{2.286}$$

We insert the temporal derivation of the doppler-frequency for the azimuth drift:

$$\frac{\partial f_{DC}}{\partial \tau} = \frac{2v_{res}\sin(\Phi)}{\lambda}\left\{ \sin(\Phi)\frac{\partial \sin(\psi)}{\partial \tau} + \cos(\Phi)\frac{\partial \tan(\delta)}{\partial \tau} - \frac{\cos(\psi)\sin(\Phi)}{v_{res}}\frac{\partial v_y}{\partial \tau} \right\} \tag{2.286 a}$$

in equation (2.286). We regard $v_y = v_e \cdot \sin(\alpha)$ where to applies for the derivation of the velocity v_y to the time τ:

$$\frac{\partial v_y}{\partial \tau} = -v_{y\,equator} \cdot \cos(\alpha) \cdot \frac{2\pi}{T_{Uml.}} \sin\left(2\pi \frac{\tau}{T_{Uml.}} \right) \tag{2.287}$$

$$\frac{\partial f_{DC}}{\partial \tau} = \frac{2v_{res}\sin(\Phi)}{\lambda}\left\{ \frac{\sin(\Phi)}{v_{res}} v_{y\,equator} \cdot \cos(\alpha) \cdot \frac{2\pi}{T_{Uml.}} \sin\left(2\pi \frac{\tau}{T_{Uml.}} \right) \right\} \tag{2.288}$$

$$\Delta \tau = \frac{\Delta x}{v_{res}} = \frac{R(\tau_c)\lambda}{D_x v_{res}}$$

$$T_{Uml.} = 5370 \text{ s}; \quad \alpha = 57°$$

$$\Delta f_{DC} = \frac{2R}{D_x} \cdot \frac{\sin[\Phi]^2 \cdot \cos(\alpha) \cdot 2\pi}{T_{Uml.}} \approx 60 \text{ Hz} \quad (\text{ERS1}) \tag{2.289}$$

A comparison with equation (2.285) reveals that the maximum doppler-centroid frequency in the ERS-1 SAR is within the so-called doppler bandwidth and thus no ambiguities are to be expected.

2.2.4.3 Criteria for the pulse-repetition frequency

According to Image 2.43, the radar transmits a series of identical pulses of the length T_{pulse}. Each pulse can be monochromatic or modulated. The number of pulses per second is called pulse-repetition frequency

$$\mathrm{PRF} = \frac{1}{T_{repeat}} \tag{2.290}$$

with ERS parameters

T_{pulse} $= 37\ \mu s$
T_{repeat} $= 595\ \mu s$
$N \cdot T_{repeat} = 28000 \cdot 595\ \mu s = 16.6\ s$

Regarding monochromatic pulses (in Image 2.43 *Transmitting pulse series* modulated pulses are displayed) and not modulated pulses results in the spectrum shown in Image 2.44 *Spectrum of the transmitting-pulse series.*

The reflected transmitting signal received from a punctiform object features a frequency displacement according to the doppler effect, which is proportional to the relative velocity v_R between the antenna and the object P on the ground and reciprocal to the wave propagation's velocity. Turning the antenna of the SAR carrier by a correcting angle, the so-called squint angle, results in a frequency displacement due to the doppler effect of the received radar signals. Fixing this squint angle during the flight, the frequency displacement remains constant. Now it is a linear, additive frequency term called doppler-centroid frequency.

The spectrum consists of many narrow-band spectral areas resulting from $1/(N \cdot T_{repeat})$ and is windowed by the envelope of the Fourier transformation of the transmitting pulse. As reflecting spectrum we get due to the doppler a spectral result characteristically doppler-displaced in the frequency range.

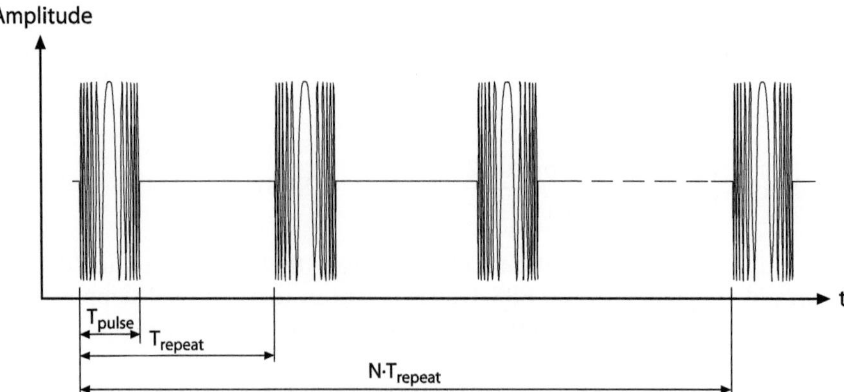

Image 2.43: Transmitting pulse series

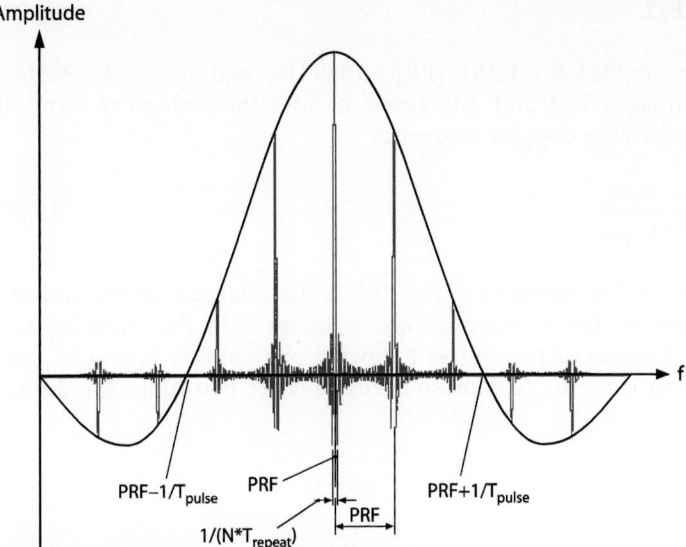

Image 2.44: Spectrum of receiving pulses

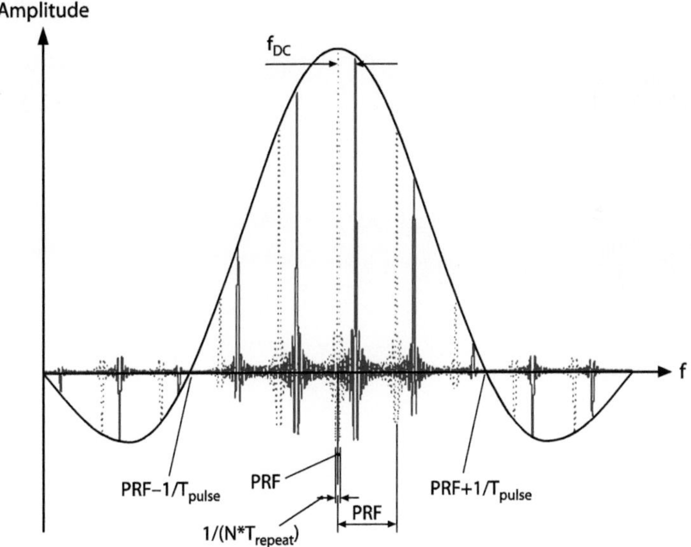

Image 2.45: Spectrum of receiving pulses

If the frequency displacement caused by the doppler effect is just as wide as the PRF, the transmitting and receiving spectrum can not be separated from each other anymore. Ambiguities occur that have to be corrected. To avoid the ambiguities, the following term must apply:

$$f_{DC_{max}} < PRF/2 \tag{2.291}$$

This condition applies for ERS1 (PRF = 1680 Hz) and $f_{DC_{max}} = f_{DC} + \Delta f_{DC}/2$ (eq. (2.289)). Images 2.44 and 2.45 reveal besides the ambiguity term also the smallest resolvable doppler frequency

$$\Delta f_{DC_{min}} = \frac{1}{N \cdot T_{repeat}} \tag{2.292}$$

A second criterion for selection of the PRF is the clearness of the arrival of the received pulses. Let us have a look at Image 2.46 *Puls-running time.* The temporal distance of two pulses following one another has to be larger than the running time the first pulse needs to cover two times the distance ΔR.

$$T_{repeat} > \frac{2 \cdot \Delta R}{c} \tag{2.293}$$

Regard as an analogy to the geometric resolution in azimuth direction:

$$\Phi_{HBe} \cong \frac{\lambda}{D_y} \tag{2.294}$$

$$\frac{h}{2} = R_0 \cdot \tan\left(\frac{\Phi_{HBe}}{2}\right) \tag{2.295}$$

for small angles applies

$$\tan\left(\frac{\Phi_{HBe}}{2}\right) \cong \frac{\Phi_{HBe}}{2}$$

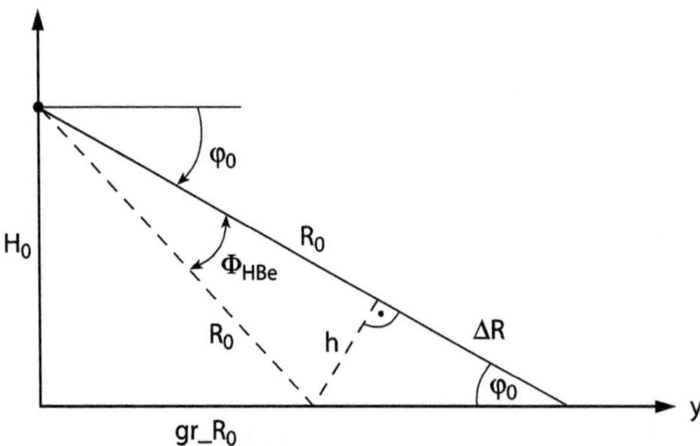

Image 2.46: Puls-running time

becoming:

$$h = R_0 \cdot \Phi_{HBe} \tag{2.296}$$

$$\Delta R = h \cdot \tan(\varphi_0) = R_0 \cdot \Phi_{HBe} \cdot \tan(\varphi_0) \tag{2.297}$$

$$PRF < \frac{c}{2 \cdot \Delta R} = \frac{c}{2 \cdot h \cdot \tan(\varphi_0)}$$

$$PRF < \frac{c \cdot D_y}{2 \cdot R_0 \cdot \lambda \cdot \tan(\varphi_0)} \approx 1620 \text{ Hz} \tag{2.298}$$

The derivated condition (here for ERS-1) is fulfilled approximatly (approximated concerning the antenna's parameters) as a PRF of 1680 Hz is used. Increasing the PRF would lead to ambiguities when receiving if the explicit correlation between transmitted and received pulse got lost.

Besides the general description of basic, geometrical, and signal-theoretical observations, we want to regard, after the discussion of the development of the SAR-raw data matrix, the SAR system from the point of view of signal-processing.

The system-theoretical point of view via transfer function of the SAR-sensor will be derivated, that means after vivid reflection of the different resolution limits and the corresponding processing-steps in range direction and azimuth direction, the way of a complete description of the transfer functions in the 2-dimensional frequency domain.

The point-target spectrum will serve as a general starting-point when discussing different SAR-processing algorithms.

First of all let us have a look at the origins of raw data.

2.3 Development of SAR-raw data

At a certain time, a radar impulse is transmitted, weighted with the backscattering coefficient of the corresponding area, reflected and finally received again. The transmitted pulse is reflected and recorded by all points located within the antenna taper (footprint). Considering any of these points as a point target, every measured value is the result of an interaction of all point-target echoes located in the footprint. In the radar receiving system generally a range-gating comes into operation. From a certain time on, a gate opens and after a certain time determined by the far-field characteristics of the antenna depending on the expansion of the footprints in range direction, closes again. The range gate basically is chosen in a way that the reflected targets outside the 3-dB range of the footprint are suppressed by not recording them.

The signals recorded during a range gate Δy represent the backscattering coefficient of the just illuminated range line Δy. As at every azimuth

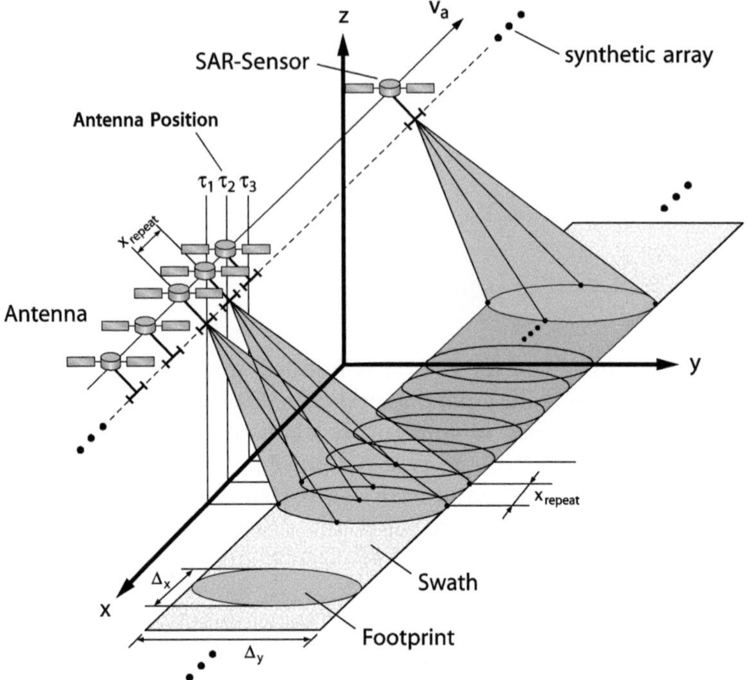

Image 2.47: Development of a two-dimensional data field

point of time $\tau = i/PRF$ ($i = 1, 2, \ldots$) the corresponding range line is sampled, discretised and saved, the resulting raw-data matrix results in a two-dimensional field containing the necessary backscattering information.

To describe the problem vividly and clearly, the coordinates of the antennas according to Image 2.48 *Recording geometry* are used along with a rectangular footprint. The 3-dB width of the footprint in azimuth direction corresponds approximately (have a close view at chapter 3.2 *Geometric resolution/range compression*):

$$\Delta x = \frac{R_c \cdot \lambda}{D_x} \tag{2.299}$$

R_c represents the distance to the center of the antenna's footprint. The expansion of the footprint in ground-range direction amounts to approximately

$$\Delta y = \frac{R_c \cdot \lambda}{D_y} \tag{2.300}$$

Please note that the expansions both in ground-range direction and in azimuth direction represent an approximated value. As described in chapter

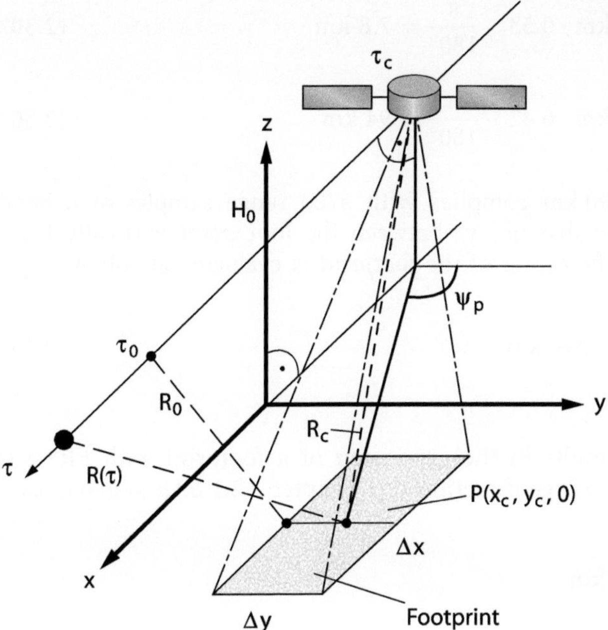

Image 2.48: Recording geometry

3.2 *Geometric resolution*, a rectangular antenna coating (an even field distribution in the aperture of the antenna) results in an antenna's function which complies with a two-dimensional Fourier-transformated signal of the corresponding form of the rectangular antenna. Thereby, in addition to the scattering coefficient with the belonging value of the Fourier transformation, echoes are created, by means of a rectangular antenna weighted with a si-function. The described local expansions (eq. (2.299)) Δx and (eq. (2.300)) Δy comply with a 3-dB width in half height of the directivity pattern characteristic of the antenna.

In fact, the area illuminated by the antenna (footprint) is infinitely extended due to a vigorous drop-off of the far-field characteristics (concerning the si-function, the first side-lobe maximum is 13.6 dB below the main-lobe-maximum) the side-lobe-maxima provide only small contributions. Echoes outside the *3-dB width* of the antenna's far field can be ignored because of their small values. Considering the ERS-SAR, the 3-dB widths in range direction and azimuth direction are 6.42° resp. 0.53°. For the mean distance R_c and the expansion of the footprints with $H_0 = 780$ km, $\beta = 0°$ and a tilt angle of $\varphi_0 = 67°$ the following is effective:

$$R_c = \frac{H_0}{\sin(\varphi_0)} \cong 847 \text{ km} \tag{2.301}$$

$$\Delta x = \frac{R_c \cdot \lambda}{D_x} = 875 \text{ km} \cdot 0.53° \cdot \frac{\pi}{180°} = 7.8 \text{ km} \tag{2.302}$$

$$\Delta y = \frac{R_c \cdot \lambda}{D_y} = 875 \text{ km} \cdot 6.42° \cdot \frac{\pi}{180°} = 94 \text{ km} \tag{2.303}$$

The distance Δy of 90 km complies with 5700 range samples at a bandwidth of 19 MHz. The distance y_0 between the foot point vertically below the SAR carrier and the center of the *footprint* is evaluated as follows:

$$y_0 = \frac{H_0 \cdot \cos(\psi_p)}{\tan(\varphi_0)} \approx 390 \text{ km} \tag{2.304}$$

Thus the following results in the expansion of a footprint with ERS-SAR parameters in ground-range direction (cp. chapter 2.2.1 *SAR system/recording geometry*):

$$310 \text{ km} \leq y \leq 470 \text{ km}$$

and in azimuth direction:

$$-4 \text{ km} \leq x \leq 4 \text{ km}$$

The expansion of the antenna's footprint can be estimated closely by considering the above mentioned instructions.

In the receiving channel the backscattering signal is already treated in a signal-theoretical way to simplify the further steps.

As the carrier frequency of the SAR is very high, e.g. 9.6 GHz at the X-SAR, the received signal usually is mixed down. After that, the real signal is transformed in a complex depiction, the so-called equivalent low-pass signal. The transformation can be realised both in the time domain and

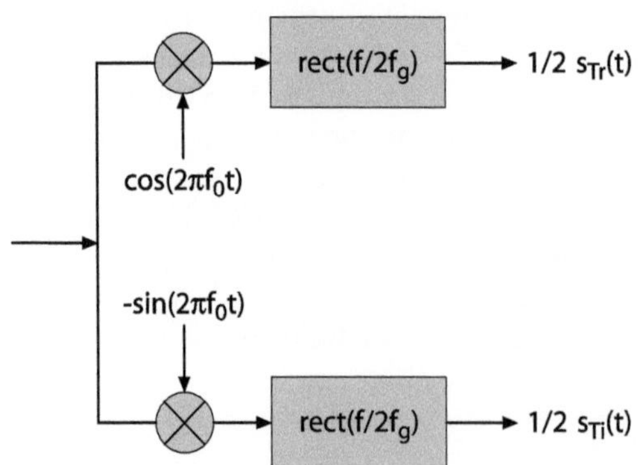

Image 2.49: Evaluation of the equivalent lowpass signal

the frequency domain and is schematically described in Image 2.49 *Evaluation of the equivalent low-pass signal*. The quadrature components $s_{Tr}(t)$ and $s_{Ti}(t)$ for creating the equivalent low-pass signal can be obtained by watching the real band pass signal $s(t) = s_0(t) \cdot \cos[2\pi f_0 t + \varphi(t)]$ outlined in Image 2.49. The data then are available for the SAR signal processing.

SAR-signal processing

3.1 General dynamic system description

If only one point target $P(x_0,y_0,0)$ on the earth is to be looked at, Image 3.1 *SAR geometry* represents a simple SAR geometry, which considers neither the earth's curvature nor the curved flight path of the satellite. If we follow the flight path of the satellite in Image 3.1 in τ-direction and watch the behaviour of the slant range $R(\tau)$, we will realize that the slant range of the satellite's motion towards the object $P(x_0,y_0,0)$ is decreasing. Putting some distance between the SAR carrier and the illuminated object increases the distance $R(\tau)$ (see Image 3.11). So the slant range is a function of the azimuth time and reaches its minimum at point τ_0. With the aid of the length of the antenna array respectively the synthetic aperture which is a result of $v_{res} \cdot \tau$, the slant range $R(\tau)$ between the antenna and a point target on the ground during the fly-by can be calculated by means of a geometric addition, demonstrated as follows:

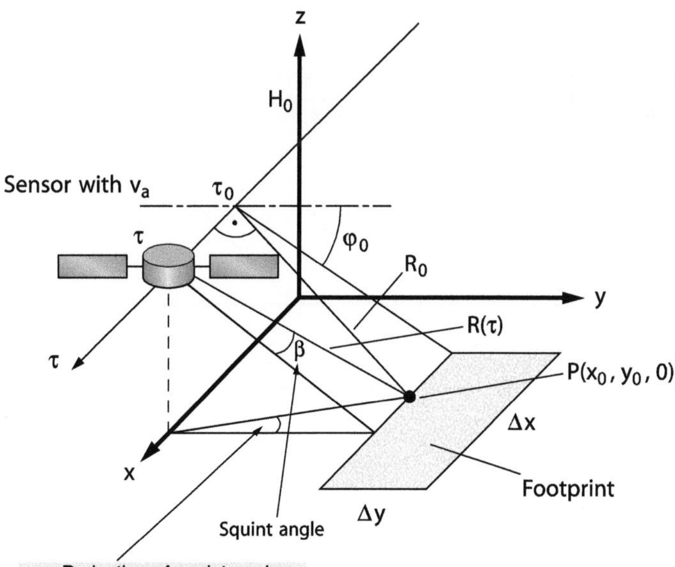

Image 3.1: SAR geometry

$$R(\tau) = \sqrt{\left(v_{res} \cdot (\tau - \tau_0)\right)^2 + R_0^2} \tag{3.1}$$

The slant range changes almost parabolically according to the azimuth position of the SAR carrier which is a main problem in the field of SAR-signal processing, for the parabolic "deformation", also called range migration, is a distance variant function. In other words, according to the range distance, the range migration shows in addition to the actual "distorted Image" other different parabolic slopes for every single range distance.

On the other hand, the parabolic range dependency is absolutely necessary to achieve a correspondingly high resolution in azimuth direction by corresponding signal processing. The "distance" parabola reproduces itself, as already demonstrated, on the phase response of the received signal, thus is similar to a chirp and suitable for correlation reception. After a time lag $t_v(\tau)$ depending on the distance $R(\tau)$, the signal is received again by the radar. Here, a start-stop approximation comes into operation presupposing that the satellite's position whilst sending and receiving does not change substantially. The distortion shown in Image 3.2 makes evident that start-stop approximation can be neglected.

Now we are looking for a description of the phase response. We are beginning with re-formulating the equation (3.1). Concerning the time lag, there is the following correlation:

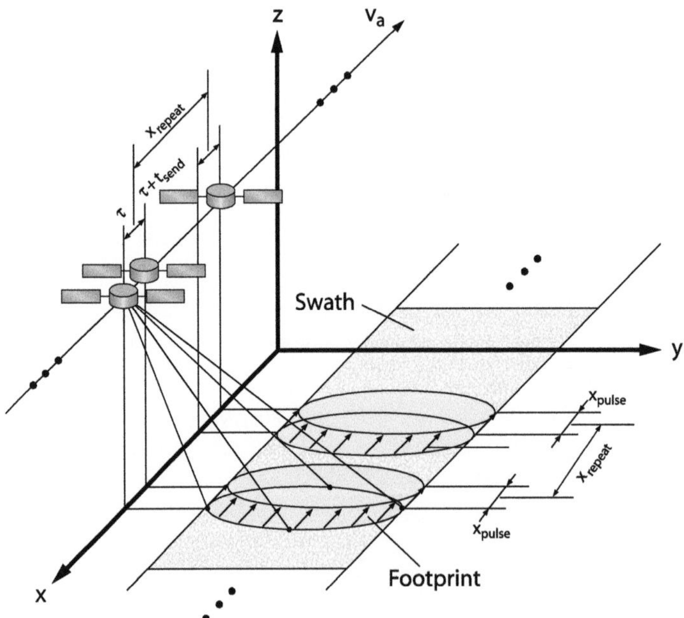

Image 3.2: Start-stop approximation

$$t_v(\tau) = \frac{2 \cdot R(\tau)}{c} = \frac{2}{c} \cdot \sqrt{(x - x_0)^2 + y_0^2 + H_0^2} \tag{3.2}$$

$$t_v(\tau) = \frac{2}{c} \cdot \sqrt{(\tau - \tau_0)^2 \cdot v_{res}^2 + y_0^2 + H_0^2} \tag{3.3}$$

The received signal shows a phase shift in comparison with the transmitted signal:

$$\rho = -2\pi \cdot f_0 \cdot t_v(\tau) = -4\pi \cdot \frac{R(\tau)}{\lambda} \tag{3.4}$$

At any time τ, the phase shift results in the effective antenna's occupancy of the radar with synthetic aperture. If the Taylor Row of the in equation (3.4) described phase shift is developed to the second classification by the azimuth time in which the target is in the middle of the footprint, the result is:

$$\rho = -\frac{4\pi}{\lambda} \left[R(\tau_c) + \dot{R}(\tau_c) \cdot (\tau - \tau_c) + \frac{1}{2} \cdot \ddot{R}(\tau_c) \cdot (\tau - \tau_c)^2 \right] \tag{3.5}$$

The time-dependent frequency by neglecting of $\ddot{R}(\tau_c)$ results in a temporal derivation of the phase shift:

$$f(\tau) = \frac{\dot{\rho}}{2\pi} = -\frac{2}{\lambda} \cdot \left[\dot{R}(\tau_c) + \ddot{R}(\tau_c) \cdot (\tau - \tau_c) \right] \tag{3.6}$$

As the frequency is a linear function of the azimuth time τ, the bandwidth in azimuth direction can be expressed as follows:

$$B_{az} = f(\tau)\big|_{\tau=\tau_c - \frac{\Delta x}{2v_{res}}} - f(\tau)\big|_{\tau=\tau_c + \frac{\Delta x}{2v_{res}}}$$

$$= -\frac{2}{\lambda} \cdot \left[\dot{R}(\tau_c) - \ddot{R}(\tau_c) \cdot \frac{\Delta x}{2v_a} \right] + \frac{2}{\lambda} \cdot \left[\dot{R}(\tau_c) + \ddot{R}(\tau_c) \cdot \frac{\Delta x}{2v_a} \right] \tag{3.7}$$

$$B_{az} = \frac{2 \cdot \ddot{R}(\tau_c) \cdot \Delta x}{\lambda \cdot v_{res}} \tag{3.8}$$

If we calculate the first derivation of the in equation (3.1) described slant range, the result is:

$$\dot{R}(\tau) = (\tau - \tau_0) \cdot v_a^2 \cdot \left[(\tau - \tau_0)^2 \cdot v_{res}^2 + y_0^2 + H_0^2 \right]^{-\frac{1}{2}} \tag{3.9}$$

After another derivation follows:

$$\ddot{R}(\tau) = \frac{v_{res}^2}{R(\tau)} - \frac{(\tau - \tau_0)^2 \cdot v_{res}^4}{R^3(\tau)} \tag{3.10}$$

As the second term of the equation (3.10) of remote sensing applications is neglectfully small, $\ddot{R}(\tau_c)$ can be approached to:

$$\ddot{R}(\tau_c) \approx \frac{v_{res}^2}{R(\tau_c)} \approx \frac{v_a^2}{R(\tau_c)} \tag{3.11}$$

By calculating the second derivation the phase response according to equation (3.5) is successfully described. Now let us have a look at the range resolution of an SAR system.

3.2 Range resolution

Any kind of sensors able to map terrain is supposed to achieve the best possible range resolution.

This is what we call the ability of a sensor to detect two neighbouring objects independently from one another. For the radar with synthetic aperture, which is associated with the imaging radar systems, the range resolution is a very important quality criteria. A two-dimensional SAR image is differentiated between the resolution in range direction and azimuth direction.

SAR systems usually are used for imaging of natural, non-metallic targets, which, because of their lower electrical conductivity, have got a significant lower backscattering profile as metal surfaces. Furthermore, the area illuminated by the antenna is much larger compared with conventional radar. This would result in a more inaccurate image of the target. That means, an SAR system would have to work with a significantly higher output rating to achieve an acceptable signal-to-noise ratio (> 10 dB). Conventional electronics would soon face its own economic and technical limits. Increasing the output rating of the radar could be done by sending longer pulses or by increasing the pulse-repeating frequency (PRF). However, both methods can hardly be realized concerning the SAR system as longer pulses would result in a worse resolution in range direction. Increasing the PRF could lead to an ambiguity of the receiving signal as the explicit allocation of the transmitted and the received pulse could be lost (chapter 2.2.4).

That is why the method of pulse compression comes into operation. With the aid of the so-called matched-filter operation, this method maximises the sensitivity of the receiver. According to [LÜK92], the achievable signal-to-noise ratio only depends on the signal energy. That is the reason why radar signals with high energy are to be aimed at. The level of the pulse performance is not important as optimal filtering (signal processing) changes a received signal (in consideration of certain energy) into an output signal. The proportion between the momentary signal power and the momentary interfering power is maximised. Now, provided that the energy level of the output signal is maintained, this is equivalent to a temporary compression of the input signal. The energy content of the input signal is reflected in the maximum amplitude of the compression.

The derivation of the range compression should be done with the aid of the autocorrelation function.

3.2.1 Range resolution in range direction/range compression

Starting point of the following reflections is a radar with real aperture working with conventional pulse process. Using this method, monochromatic pulses with a rectangular envelope of duration t_{send} and the pulse repetition frequency PRF are transmitted and received. The delay of the electromagnetic waves spreading with the speed of light from the radar antenna to a reflector and back delivers a value to determine the distance to that target with. However, different targets in range direction can only be differentiated if their distance to one another is large enough to avoid an overlapping of the received echoes. The resolution in range ΔR therefore is limited by the duration T of the transmitted pulse and is calculated as follows:

$$\Delta R = \frac{c \cdot T}{2} \tag{3.12}$$

Thus, the resolution in range direction can be decreased by shortening the pulse duration T respectively by increasing the bandwidth B_r of the pulse.

3.2.1.1 Range resolution without using chirp-signals

The meaning of a range compression is revealed by comparing the limiting resolution in range direction. The radar transmits a chirp with a pulse length T and the bandwidth B_r. Considering Image 3.3 *SAR-range resolution in range-height level,* after the running times t_{01} and t_{02} the two receiving signals $r_T(t_{01})$ and $r_T(t_{02})$ appear (Image 3.4).

The limiting resolution of an SAR image in range directions is reached at the moment the two received signals begin to overlap each other. Let us have a look at Image 3.4 *Two non-compressed receiving signals backscattered by two targets with different distances.*

The borderline case of the resolution is reached when $t_{02} - t_{01} = T$.

With the running times:

$$t_{01} = \frac{2R_1}{c} \quad \text{and} \quad t_{02} = \frac{2R_2}{c} \tag{3.13}$$

result in a range resolution for non-compressed receiving signals:

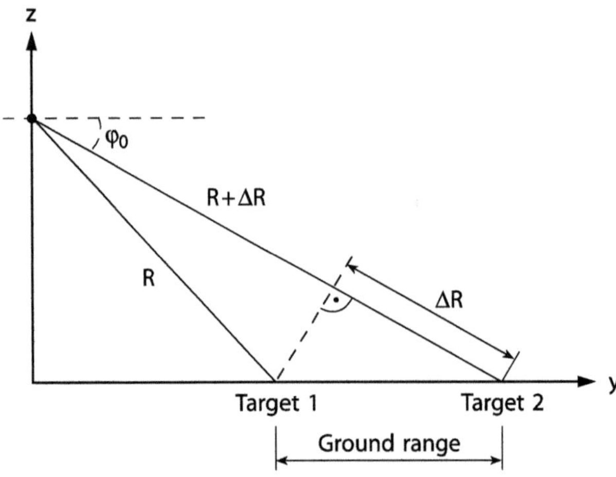

Image 3.3: SAR geometry in range height

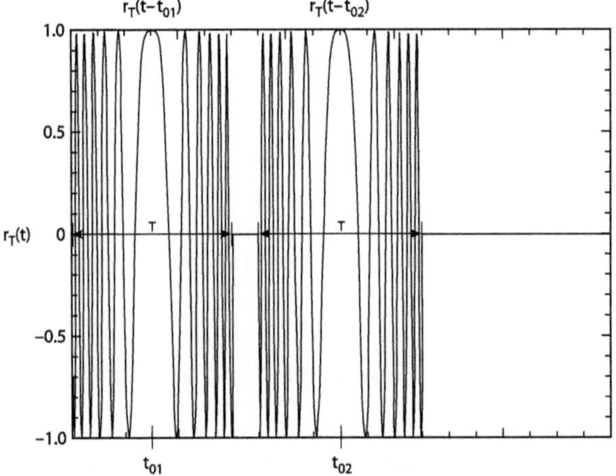

Image 3.4: Two non-compressed receiving signals backscattered by two targets with different distances.

$$\Delta R_u = \frac{c \cdot T}{2} \qquad (3.14)$$

If ERS-1 parameters are considered, then the range resolution without range compression is calculated with approximately 5.5 km. This unacceptable high value reveals that an SAR system without range compression can not be utilized in a reasonable way.

Now let us compare the improvement of the resolution with the aid of the so-called range compression.

3.2.1.2 Range-compression resolution using chirp signals

To explain the necessity of compression let us have a look at the benefit of a better resolution with the aid of the range compression described below.

As a comparison to the just calculated resolution without pulse compression, the minimum theoretical border distance two neighbouring targets have got to one another in range direction after the range compression in order to detect them separately will be calculated. Let us presume as follows:

The two compressed receiving signals $g_T(t_{01})$ and $g_T(t_{02})$ are almost si-functions close to their maximums.

The first zero-point detected corresponds to the reciprocal value of the bandwidth B_r.

Two si-functions in the beginning can be separated from each other when the maximum of the second si-function begins to overlap the zero-point of the first si-function.

The range-compressed signals shown in image 3.5 *Two range-compressed receiving signals*, can be shifted that way, the maximum and the zero point of the particular other function will overlap. Out of previous presumptions the following terms can be derivated if the received signals are range-compressed:

For Image 3.5 is valid:

$$t_{02} = t_{01} + \frac{1}{B_r} \tag{3.15}$$

Thus

$$t_{02} - t_{01} = \frac{1}{B_r} = \frac{1}{k_r \cdot T} \tag{3.16}$$

and the resolvable distance after the compression becomes:

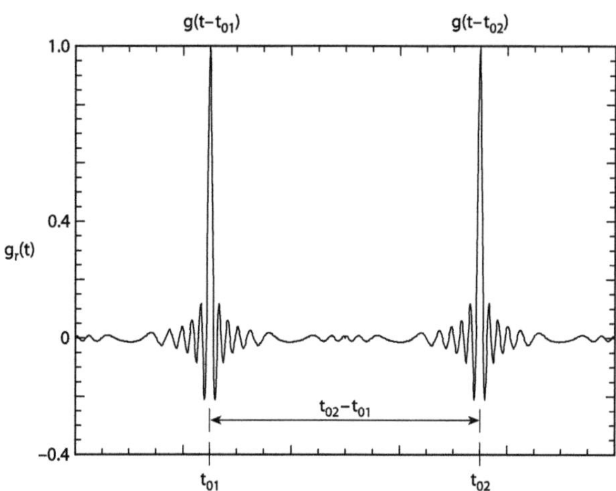

Image 3.5: Two range-compressed receiving signals

$$\Delta R_k = \frac{c}{2 \cdot B_r} = \frac{c}{2 \cdot k_r \cdot T} \tag{3.17}$$

As value for the ERS-1 parameters results a distance of about 9 m! By pulse compression, the still resolvable distance of two point targets is improved by more than the factor 600.

The result of equation (3.17) reveals that the range resolution increases both with rising chirp length and increasing frequency ratio respectively bandwith. In equation (3.17) we see also *the range resolution does not depend on the distance of the SAR-carrier to the target!*

The previous considerations towards the resolution were only made for the slant distance $R(\tau)$. To get the real resolution on the earth's surface let us use once again Image 3.2. It is easy to see that only equation (3.17) has to be divided by the cosine of the antenna's angle of inclination to achieve a real ground-range resolution value.

$$\Delta R_y = \frac{\Delta R_k}{\cos(\varphi_0)} = \frac{c}{2 \cdot B_r \cdot \cos(\varphi_0)} \tag{3.18}$$

The effects of the inclination to the ground-range resolution can be read in equation (3.18). The larger the inclination the worse, that means greater, is the resolvable distance of two neighbouring targets. In the field of range compression the correlation of chirp signals displayed in chapter 2 SAR-basics applies.

The transmitting signal is regarded:

$$s(t) = \mathrm{Re}\{s_T(t) \cdot e^{j2\pi f_0 t}\} = \mathrm{rect}\left(\frac{t}{T}\right) \cdot \cos(2\pi f_0 t + \pi k_r t^2) \tag{3.19}$$

whereas

$$s_T(t) = \mathrm{rect}\left(\frac{t}{T}\right) \cdot e^{j\pi k_r t^2} \tag{3.20}$$

The received signal is temporally delayed by '$t_v(\tau)$', on the other hand weighed with a scattering coefficient $\sigma(R_0, \tau_0)$:

$$r(t) = \sigma(R_0, \tau_0) \cdot s(t - t_v) = \sigma(R_0, \tau_0) \cdot \mathrm{Re}\{s_T(t - t_v) \cdot e^{j2\pi f_0(t - t_v)}\} \tag{3.21}$$

$$r_T(t) = \sigma(R_0, \tau_0) \cdot s_T(t - t_v) \cdot e^{-j2\pi f_0 t_v} \tag{3.22}$$

The correlation of transmitting and receiving signal leads to:

$$g_T(t) = 1/2 \cdot s_T(-t)^* \cdot r_T(t) = 1/2 \cdot s_T(-t)^* \cdot s_T(t - t_v)$$
$$\cdot \sigma(R_0, \tau_0) \cdot e^{-j2\pi f_0 t_v}$$

$$g_T(t) = 1/2 \cdot \{[s_T(-t)^* \cdot s_T(t)] \cdot \delta(t - t_v)\} \cdot \sigma(R_0, \tau_0) \cdot e^{-j2\pi f_0 t_v} \tag{3.24}$$

and considering the results of the SAR basics results in:

$$g_T(t) = \Phi_{ss_T}(t - t_v) \cdot \sigma(R_0, \tau_0) \cdot e^{-j2\pi f_0 t_v} \tag{3.25}$$

If the temporally delay t_0 is described as a term of the range distance, then:

$$t_v(\tau) = \frac{2 \cdot R(\tau)}{c} \tag{3.26}$$

then equation (3.25) is changing to:

$$\boxed{g_T(t) = \Phi_{ss_T}(t - t_v) \cdot \sigma(R_0, \tau_0) \cdot e^{-j4\pi R(\tau)/\lambda}} \tag{3.27}$$

In the frequency domain it has to be transformed in:

$$G_T(f) = 1/2 \cdot S_T(f)^* \cdot R_T(f) \tag{3.28}$$

With equation (3.27) the range compression is described completely. The result reveals the range-time delayed autocorrelation function, which is multiplied with the scattering coefficient of the point-target and in addition with the range-dependent phase term.

After the range compression, in other words the signal processing in visual direction to the antenna, it is necessary to do signal processing in azimuth or flight direction in order to achieve the resolution needed.

But first let us have a look at the specific geometric characteristics of an antenna to regard the resolution limits of a radar with synthetic aperture in order to reveal further steps to describe the resolution capabilities of the radar with synthetic aperture.

Therefore geometric considerations as well as signal-theoretical considerations are going to be used.

3.2.2 Range resolution in azimuth direction

In contrast to the range resolution, which depends on the bandwidth of the transmitted pulse, the resolution in azimuth is exclusively determined by the form of the antenna's diagram of the radar antenna. To detect via the antenna more point targets with the same radial distance separately, an antenna with distinctive directional radio patterns, that means with a preferably narrow main-lobe, is required.

3.2.2.1 Radar with real aperture

The directional radio patterns of an antenna can roughly be described as the Fourier transformation of the electric loading of the antenna's aperture. This is known as the Fraunhofer-approximation and is valid for the far field of an antenna. If the electric loading function of a one-dimensional antenna is assumed as constant, thus a rectangular aperture, so the result is a si-function as Fraunhofer approximation, which helps to calculate the resolution.

The resolution can be defined as the 3dB-full width at half maximum of the antenna's main lobe. This is equivalent to an angle range, within itself the emittance does not decline to the half of the maximum value. To determine the azimuth resolution, the directional radio pattern of the antenna is regarded in azimuth, in other words the field strength is calculated as a term of the angle ϕ between the aperture's normal and the object. The field strength $E(\phi)$ is calculated for a line emitter with a length D_x, consisting of a certain number of dipoles and the coordinate x along the real aperture. All dipoles transmit respectively receive simultaneously. Every single element receives the backscattered energy which is emitted by all dipoles. The backscattered signal of a punctual object with the angle ϕ has for $R_c \gg D_x$ (Fraunhofer approximation) the phase shift:

$$\phi(x) = \frac{2 \cdot \pi}{\lambda} \cdot \Delta R = \frac{2 \cdot \pi}{\lambda} \cdot \sin(\phi) \cdot x \tag{3.29}$$

The field strength is calculated via the antenna's function:

$$E(\phi) = \int_{-\frac{D_x}{2}}^{+\frac{D_x}{2}} A(x) \cdot e^{j \cdot \left(\frac{2\pi}{\lambda} \cdot x \cdot \sin(\phi)\right)} \cdot dx \tag{3.30}$$

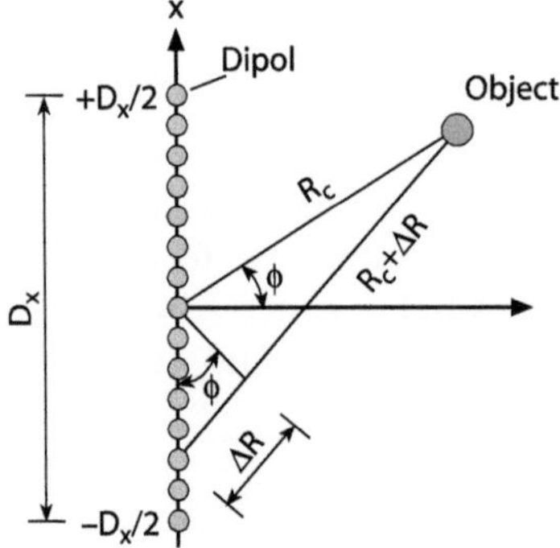

Image 3.6: Angle resolution of the antenna

As the simplest case, the antenna's function shall be assumed as equally distributed, that means $A(x) = A_0$ for $|x| \leq D_x/2$ and else 0. It is following for $D_x \gg \lambda$:

$$E(\phi) = A_0 \cdot D_x \cdot \text{si}\left(\frac{\pi}{\lambda} \cdot D_x \cdot \sin(\phi)\right) \tag{3.31}$$

Standardizing the energy $E(\phi)$ at its maximum $E(0)$ results in the directional characteristic $C(\phi)$:

$$C(\phi) = \frac{E(\phi)}{E_0} \tag{3.32}$$

The relative power trend results as follows:

$$|C(\phi)|^2 = \text{si}^2\left(\frac{\pi}{\lambda} \cdot D_x \cdot \sin(\phi)\right) \tag{3.33}$$

The 3dB-bandwidth is indicated with the half power angle:

$$\text{si}\left(\frac{\pi}{\lambda} \cdot D_x \cdot \sin(\phi_{HW})\right) = \frac{1}{2} \tag{3.34}$$

$$\sin(\phi_{HW}) \approx 0.44 \cdot \frac{\lambda}{D_x} \tag{3.35}$$

For small angles ϕ_{HW} is valid $\sin(\phi_{HW}) \approx \phi_{HW}$ and therefore the 3dB-full width at half maximum ϕ_{HB} can be stated as:

$$\phi_{HB} = 2 \cdot \sin(\phi_{HW}) \approx 0.88 \cdot \frac{\lambda}{D_x} \tag{3.36}$$

The azimuth resolution corresponding to the full width at half maximum has to be calculated at the distance R_c:

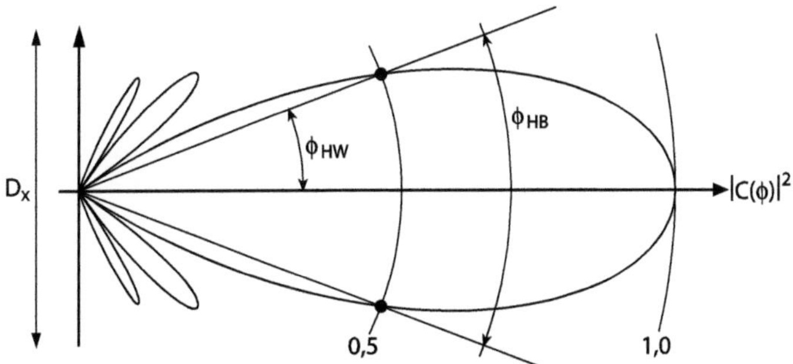

Image 3.7: Antenna characteristics in polar coordinates

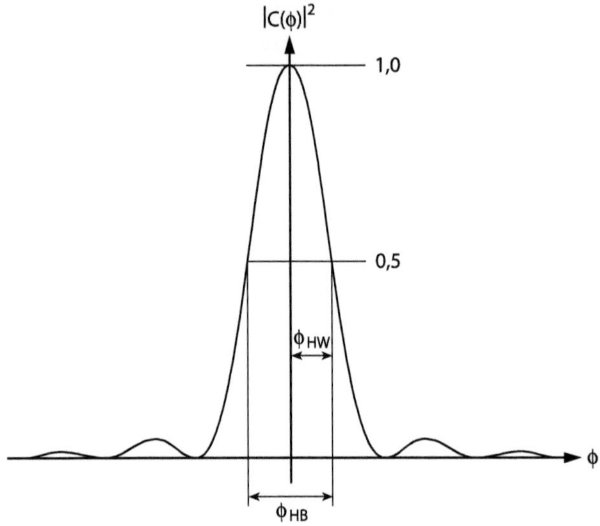

Image 3.8: Antenna's directional characteristics in cartesian coordinates

$$\Delta x_e \approx \phi_{HB} \cdot R_c = 0.88 \cdot \frac{\lambda}{D_x} \cdot R_c \tag{3.37}$$

If the same antenna is used for transmitting and receiving, the signal is passing twice the antenna's diagram and results in a double 3dB-aperture:

$$\Delta x_z \approx 0.64 \cdot R_c \cdot \frac{\lambda}{D_x} \tag{3.38}$$

Here, R_c is the distance to the target, D_x the antenna's length and ϕ_{HB} the 3-dB-aperture angle of the antenna's emitting diagram in azimuth direction. For ERS parameters ($R_c = 830$ km; $D_x = 12$ m) the best resolution results in:

$$\Delta x_z \approx 3 \text{ km}$$

As well as with the range resolution for real apertures, this value is unacceptably high and is significantly improved by using the concept of synthetic aperture.

3.2.2.2 Radar with synthetic aperture

To achieve the aim of improving the resolution in azimuth direction, either the wave length of the transmitting signal can be reduced or the length of the aperture of the radar antenna can be increased. The first solution is not the best one as the information content of a radar echo backscattered from an illuminated target depends on the transmitted wavelength. The second solution has got constructive limits.

Concerning airborne and satellite-based SAR systems, the required long aperture is generated on a calculative way. Thus it is possible to generate synthetic apertures with a length of several kilometers. The synthetic aperture is created when a punctual object inside the lobe of a real antenna (which moves alongside the flight path) is illuminated – and all received echoes recorded at their absolute value and phase. The flight path covered during the illumination is called synthetic aperture.

Because of the changing distance between antenna and object, the received echo signal shows a doppler frequency shift in opposition to the transmitting signal. Each illuminated object is characterised by its typical doppler frequency characteristics and can principally be distinguished from neighbouring objects.

Concerning the SAR processing, a cross correlation receiving between the receiving signal and a reference function is evaluated. The reference function is derivated from the lighting geometry of a complete scene on the ground during the fly-by within a predefined distance interval.

Images 3.9 *Lighting geometry* and 3.10 *Geometry for the synthetic aperture* show the lighting geometries for derivating the resolution in azimuth.

A SAR carrier moves with constant speed v_a along a linear flight path and illuminates both objects on the positions x_1 and x_2 from a lateral distance R_c. Both objects reflect the energy emitted by the radar, dependent on their radar backscattering cross section. To simplify things we are assuming that both objects have got the same backscattering coefficient.

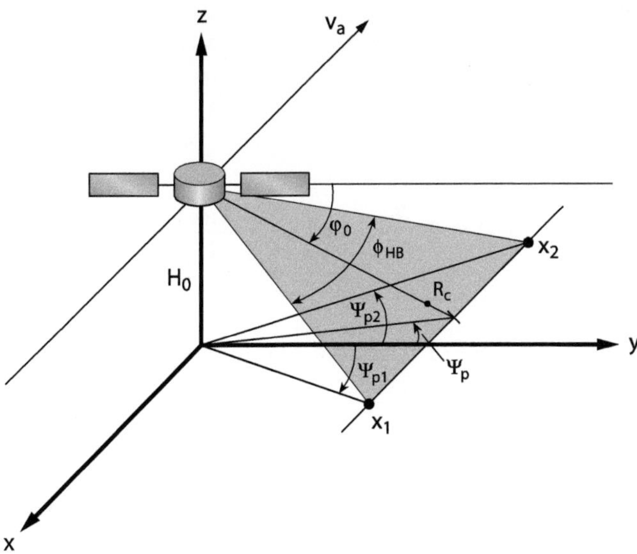

Image 3.9: Lighting geometry

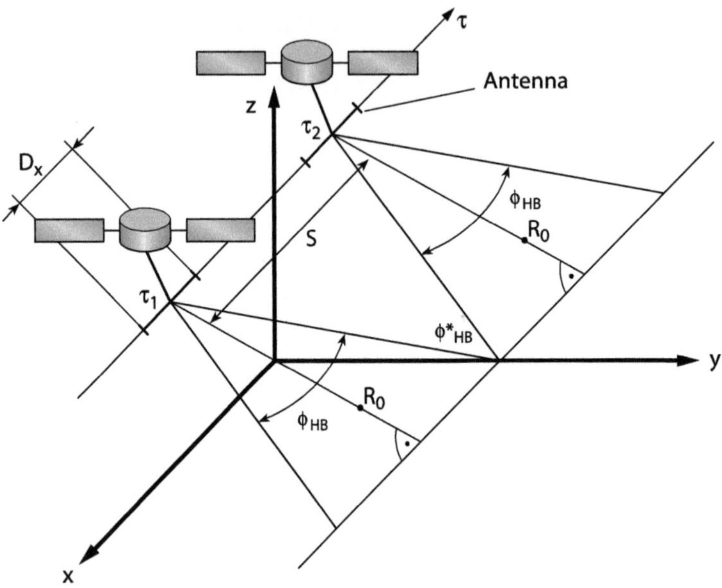

Image 3.10: Geometry of synthetic aperture

Both objects have got the antenna's apex angle ϕ_{HW} within the average distance R_c. Thus, they have got within the shortest distance R_0 the distance Δx to one another which is equivalent to the dopplershift Δf_D.

According to Image 3.9 *Lighting geometry*, both objects are equivalent to following doppler shifts:

$$f_{D_{x1}} = 2 \cdot \frac{v_a}{\lambda} \cdot \sin(\psi_{p1}) \cdot \cos(\varphi_0) = 2 \cdot \frac{v_{res}}{\lambda} \cdot \sin\left(\psi_p + \frac{\Phi_{HB}}{2}\right) \cdot \cos(\varphi_0)$$

$$(3.39)$$

$$f_{D_{x2}} = 2 \cdot \frac{v_a}{\lambda} \cdot \sin(\psi_{p2}) \cdot \cos(\varphi_0) = 2 \cdot \frac{v_{res}}{\lambda} \cdot \sin\left(\psi_p - \frac{\Phi_{HB}}{2}\right) \cdot \cos(\varphi_0)$$

$$(3.40)$$

By means of the addition theorem of trigonometric functions:

$$\sin\left(\psi_p \pm \frac{\Phi_{HB}}{2}\right) = \sin(\psi_p) \cdot \cos\left(\frac{\Phi_{HB}}{2}\right) \pm \cos(\psi_p) \cdot \sin\left(\frac{\Phi_{HB}}{2}\right) \qquad (3.41)$$

and the approximations for small angles $\cos\left(\frac{\Phi_{HB}}{2}\right) \approx 1$ and $\sin\left(\frac{\Phi_{HB}}{2}\right) \approx \frac{\Phi_{HB}}{2}$ the doppler frequency resolution follows as:

$$|\Delta f_D| = |f_{D_{x2}} - f_{D_{x1}}| = \frac{2 \cdot v_{res}}{\lambda} \cdot \cos(\psi_p) \cdot \Phi_{HB} \tag{3.42}$$

The best line of sight of an antenna for SAR applications, corresponding to the optimal phase shift during the fly-by, is given at the Yaw angle $\psi = 0°$: within this mode of operation we get:

$$|\Delta f_D| = \frac{2 \cdot v_a}{\lambda} \cdot \Phi_{HB} = \frac{2 \cdot v_{res}}{\lambda} \cdot \frac{\Delta x}{R_0} \tag{3.43}$$

and with

$$\Phi_{HB} \approx \frac{\Delta x}{R_0} = 2 \cdot \sin(\Phi_{HB}) \tag{3.44}$$

$$\Delta x = \frac{|\Delta f_D| \cdot R_0 \cdot \lambda}{2 \cdot v_{res}} \tag{3.45}$$

The frequency resolution of a signal is equivalent to the reciprocal value of the measurement period of the transmitted radar signal, respectively to the reciprocal value of the time signal within the half-power bandwidth.

$$|\Delta f_D| = \frac{1}{T_{transmit}} \tag{3.46}$$

The signal duration $T_{separate}$ is determined by the half-power bandwidth ϕ_{HB} and can be determined from the actual transmitting-signal sequence period $T_{sende} = T$ by approximate, proportional regarding of equation (3.35 /3.36) via a "pseudo 3dB limit":

$$T_{sep} = 0.88 \cdot T_{transmit} \tag{3.47}$$

The consideration leads to a doppler-frequency resolution of

$$|\Delta f_D| = \frac{0.88}{T_{sep}} \tag{3.48}$$

In addition to the doppler-frequency resolution, which is given by the half-power bandwidth of the antenna's lobe, the signal's resolution due to the overlapping of two transmitting pulses must be taken into account. Please notice the angle Φ_{HB}^* in Image 3.10 *Geometry of synthetic aperture*. This angle is equivalent to the half-power bandwidth of the antenna's lobe (it is here called differently just because of reasons of plausibility). The angle Φ_{HB}^* is the determining angle of the azimuth resolution as it can be regarded as critical angle of the overlapping of two successively transmitted antenna lobes.

If the transmitting sequence period is defined via the distance S between the transmitting pulses:

$$T_{sep} = \frac{S}{v_a}$$

(3.49)

so is the result:

$$S = \Phi_{HB} \cdot R_0 = 0.88 \cdot \frac{\lambda}{D_x} \cdot R_0$$

(3.50)

$$|\Delta f_D| = \frac{v_{res} \cdot D_x}{\lambda \cdot R_0}$$

(3.51)

and deployed in equation (3.45) we get as an amazing result of the azimuth resolution a formula, **which is neither depending on the wave length nor the distance to the target object:**

$$\boxed{\Delta x = \frac{D_x}{2}}$$

(3.52)

The theoretically achievable resolution of a radar with synthetic aperture working as a side-locking radar, amounts to the half of the aperture's length of the used real antenna and no longer depends on the wave length and the distance to the target. If we compare these results for ERS-parameter (theoretically achievable resolution of $\Delta x \cong 6$ m) with the results of chapter 3.2.2.1 Radar with real aperture, an improvement of approximately factor 500 can be determined.

However, the resolution can not be increased by simply downscaling the real aperture, as with decreasing dimensions of the antenna the sensitivity of the radar diminishes because of a too low directivity.

The smaller the antenna's length D_x is, the bigger the half-power bandwidth Φ_{HB} gets, the length S of the synthetic aperture increases and the better the azimuth resolution Δx becomes.

Presupposition for the results according to equation (3.52) is the assumption that during transmitting and receiving the same antenna is used. In addition the illuminated object has to be in the far field of the antenna.

3.2.2.3 Signal-theoretical view

Please have a look at chapter 2.2.1 Side looking geometry:

To remember and because of the thematic affiliation, let us have a look again at some relations:

$$R(\tau) = \sqrt{(v_{res} \cdot (\tau - \tau_0))^2 + R_0^2}$$

(3.53)

After transformation the time-delay results as follows:

$$t_v(\tau) = \frac{2 \cdot R(\tau)}{c} = \frac{2}{c} \cdot \sqrt{(x - x_0)^2 + y_0^2 + H_0^2} \tag{3.54}$$

$$t_v(\tau) = \frac{2}{c} \cdot \sqrt{(\tau - \tau_0)^2 \cdot v_{res}^2 + y_0^2 + H_0^2} \tag{3.55}$$

The received signal shows a phase shift in opposite to the transmitted signal.

$$\rho = -2\pi \cdot f_0 \cdot t_v(\tau) = -4\pi \cdot \frac{R(\tau)}{\lambda} \tag{3.56}$$

If the Taylor approximation is developed to the second order, we get:

$$\rho = -\frac{4\pi}{\lambda} \left[R(\tau_c) + \dot{R}(\tau_c) \cdot (\tau - \tau_c) + \frac{1}{2} \cdot \ddot{R}(\tau_c) \cdot (\tau - \tau_c)^2 \right] \tag{3.57}$$

The frequency depending on the time results in a temporal derivation of the phase shift:

$$f(\tau) = \frac{\dot{\rho}}{2\pi} = -\frac{2}{\lambda} \cdot [\dot{R}(\tau_c) + \ddot{R}(\tau_c) \cdot (\tau - \tau_c)] \tag{3.58}$$

As the frequency is a linear term of the azimuth time τ, the bandwidth in the azimuth direction can be described as follows:

$$B_{az} = \frac{2 \cdot \ddot{R}(\tau_c) \cdot \Delta x}{\lambda \cdot v_{res}} \tag{3.59}$$

If we calculate the first distance derivation we get:

$$\dot{R}(\tau) = (\tau - \tau_0) \cdot v_{res}^2 \cdot [(\tau - \tau_0)^2 \cdot v_{res}^2 + y_0^2 + H_0^2]^{-\frac{1}{2}} \tag{3.60}$$

After another derivation follows:

$$\ddot{R}(\tau) = \frac{v_{res}^2}{R(\tau)} - \frac{(\tau - \tau_0)^2 \cdot v_{res}^4}{R^3(\tau)} \tag{3.61}$$

As the second term of equation (3.61) is negligible concerning remote-sensing applications, $\ddot{R}(\tau_c)$ can be approximated as:

$$\ddot{R}(\tau_c) \approx \frac{v_{res}^2}{R(\tau_c)} \tag{3.62}$$

If we deploy $\Delta x = R(\tau_c) \cdot \dfrac{\lambda}{D_x}$ (see also equation (3.37)) and the result from equation (3.62) in equation (3.59) we get:

$$B_{az} = \frac{2 \cdot v_{res}}{D_x} \tag{3.63}$$

If we regard the resolution as speed/pulse-repetition frequency, follows:

$$\boxed{\Delta x = \frac{v_{res}}{B_{az}} = \frac{D_x}{2}} \tag{3.64}$$

This way the resolution of a radar with synthetic aperture in azimuth direction can be derivated.

After having derivated the limiting resolution of a radar with synthetic aperture, we will now have a closer look at the steps concerning the compression which are necessary to achieve the relative resolutions.

3.3 Azimuth compression

3.3.1 Range parabola

As already mentioned, the so called range migration results in the azimuth time τ which depends on the time delay of the autocorrelation function (range parabola). The phase-term ρ (chapter 3.1 General dynamic system description equation (3.4)) which depends on the azimuth time τ creates the "doppler history". Eliminating the range migration as well as taking advantage of the doppler history to the signal compression in flight direction are the substantial tasks of the azimuth compression.

3.3.2 Range migration

For every range line arises a range-compressed signal. The maximum of the different auto-correlation terms is not constantly at its minimum distance $t_0 = 2R(\tau_0)/c$ (the middle of the antenna's aperture). Caused by the range migration, the time delay $t_v(\tau) = 2R(\tau)/c$ of a point target at each azimuth position τ inside an antenna's aperture moves via the so-called range parabola. This range migration must be compensated before executing the actual azimuth compression.

Image 3.11 shows the autocorrelation term as well as its displacement along the range parabola in dependence of the azimuth time τ.

Starting point of this consideration is the azimuth-time depending delay (see chapter 3.1 General dynamic system description equation (3.2)). If we

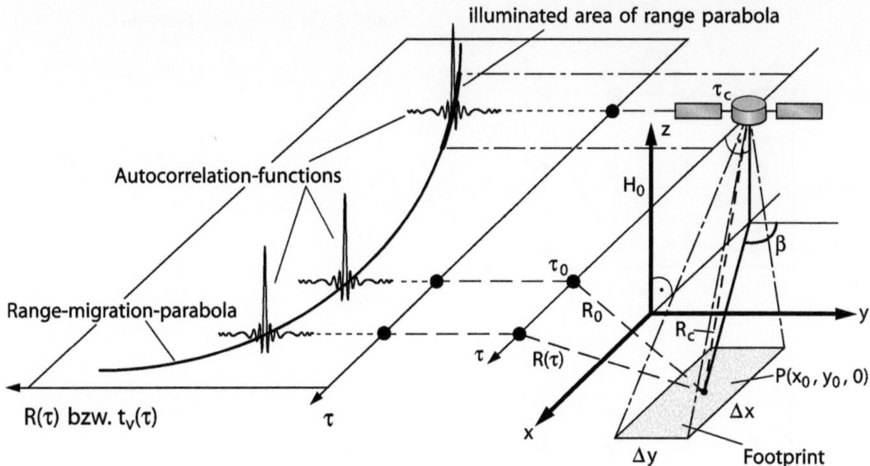

Image 3.11: Illustration of the range migration

develop the Taylor approximation for the time delay and consider the terms higher than second order as neglegible, we get:

$$t_v(\tau) = \frac{2 \cdot R_0}{c} + \frac{2 \cdot \dot{R}(\tau_c)}{c} \cdot (\tau - \tau_c) + \frac{\ddot{R}(\tau_c)}{c} \cdot (\tau - \tau_c)^2 \tag{3.65}$$

If the Taylor approximation is supposed to be developed up to the 3rd and 4th order, the terms of this higher order must be estimated. The range migration in equation (3.65) is equivalent to a parabola. At any time τ the main maximum of the autocorrelation function (ACF) is at another place. Images 3.12 and 3.13 illustrate qualitatively the effect of the range migration on a target-point. The influence of the squint angle is not taken into consideration here.

Images 3.12 and 3.13 illustrate the maximum "track" of the ACF of a point target. Depending on the SAR parameters, particularly also from the slant range of the target object to the SAR carrier, the parabola can extend over the range-sample intervals $1/B_r$ (within the aperture where the point target can be seen) (Image 3.12). Another opportunity is shown in Image 3.12 – the range parabola variation is less than one range-sample interval and thus not visible. The smaller the distance $R(\tau)$ the "steeper" is the range parabola. This short distance inside the SAR image is called near range (Image 3.12), the farthest point is called far range.

The necessary data to execute the azimuth compression unfortunately are to be found along the range parabola and are not directly accessible as the different parabolas can not be registrated automatically by the respective "sample matrix" (Images 3.12, 3.13).

The sample matrix is marked with dots in Image 3.12 and 3.13, which are given by the different sampling rates and the pulse repetition frequency (PRF). The case illustrated in Image 3.13 describes that the range parabola

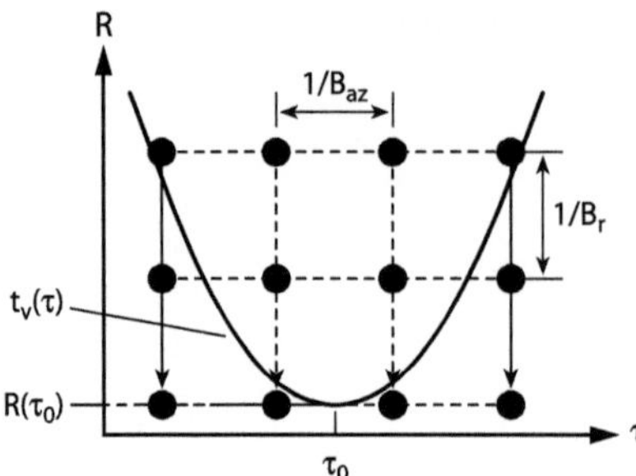

Image 3.12: Near-range parabola

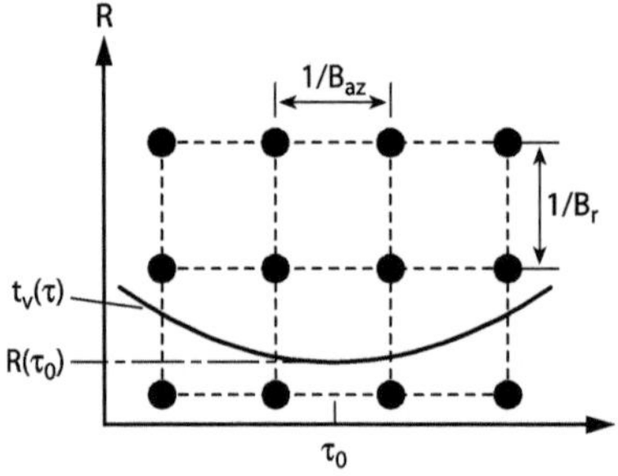

Image 3.13: Far-range parabola

is not registrated by the sampling matrix. That is why the range-compressed data have to be interpolated.

After increasing the number of scanning values, it is possible to shift the range parabola pixel by pixel to get a single-range line. In Image 3.12 this operation, which normally is used in conventional SAR sensors, is illustrated schematical with arrows. After the shifting, the range line results in a horizontal line parallelly to the azimuth axis τ.

Therefore the range migration is revised by shifting the migration parabola to a line parallel to the azimuth axis.

3.3.3 Azimuth compression

After eliminating the range migration (chapter 3.3.2 Range migration), the necessary data for the azimuth compression are along a line in the data-matrix column $t_v(\tau_0) = t_0$ and are described as follows:

$$g_{ra_T}(t, \tau, R_0, \tau_0) = \varphi_{SS_T}(t - t_0) \cdot \sigma(R_0, \tau_0) \cdot e^{-j4\pi R(\tau)/\lambda} \tag{3.66}$$

$$g_{ra_T}(t, \tau, R_0, \tau_0) = \varphi_{SS_T}\left(t - \frac{2 \cdot R_0}{c}\right) \cdot \sigma(R_0, \tau_0) \cdot e^{-j4\pi R(\tau)\lambda} \tag{3.67}$$

$$g_{ra_T}(t, \tau, R_0, \tau_0) = \varphi_{SS_T}\left(t - \frac{2 \cdot R_0}{c}\right) \cdot \sigma(R_0, \tau_0) \cdot e^{j \cdot p(\tau)} \tag{3.68}$$

The phase term $\rho(\tau)$, which represents the so-called doppler history, contains the second information (coordinate) of a point target and is used for the azimuth compression. The phase function is approximated by Taylor up to the second order as shown in chapter 3.1 General dynamic system description.

At the same time, the doppler-centroid frequency and the doppler rate mentioned in chapter 2.2.2.2 Precise derivation via the distance derivative is used. The result of the approximation is as follows:

$$\rho(\tau) \approx \rho_0 + 2\pi \cdot f_{dc} \cdot (\tau - \tau_c) + \pi \cdot k_{az} \cdot (\tau - \tau_c)^2 \tag{3.69}$$

This phase expression contains besides the constant and linear phase term also a quadratic one. Thus, this signal is one with a quadratic phase – again, a chirp.

Assuming that the distance function is generating a chirp in azimuth direction (in other words generating a quadratic term in azimuth direction) the azimuth compression takes place in a manner analogous to the range compression and results (taking the discussed relations in chapter 3.2.1.2 Range compression/resolution using chirps into account) in:

$$\begin{aligned}
g_{ra_T}(t, \tau, R_0, \tau_0) = {}& \frac{\sigma(R_0, \tau_0) \cdot e^{j\rho_0} \cdot T}{2} \cdot e^{j2\pi \cdot f_{dc} \cdot (\tau - \tau_c)} \\
& \cdot \Lambda\left(\frac{t - t_0}{T}\right) \cdot \mathrm{si}\left[\pi B_r(t - t_0) \cdot \Lambda\left(\frac{t - t_0}{T}\right)\right] \\
& \cdot \Lambda\left(\frac{\tau - \tau_c}{\Delta\tau}\right) \cdot \mathrm{si}\left[\pi B_{az}(\tau - \tau_c) \cdot \Lambda\left(\frac{\tau - \tau_c}{\Delta\tau}\right)\right]
\end{aligned} \tag{3.70}$$

Now the azimuth bandwidth is:

$$B_{az} = k_{az} \cdot \Delta\tau \tag{3.71}$$

with $\Delta\tau$ as length of aperture.

The result corresponds to a 2-dimensional autocorrelation function whose maximum determines the target position of the object. The result illustrated in equation (3.70), however, still has to be shifted in azimuth direction to the position $\tau = \tau_0$.

The final result both range-compressed and azimuth-compressed, the envelope of equation (3.70), is illustrated in Image 3.14.

The final result the range- and azimuth-compressed point-target is shown in Image 3.14. It is the envelope of equation (3.70). In Image 3.14 the range bandwith as well as the azimuth bandwidth are identical. The resulting function has not been windowed and represents only a part of the complete compression function.

After having illustrated the corresponding limiting resolutions and the necessary work steps, let us now derivate the 2-dimensional point-target response in the 2-dimensional frequency domain.

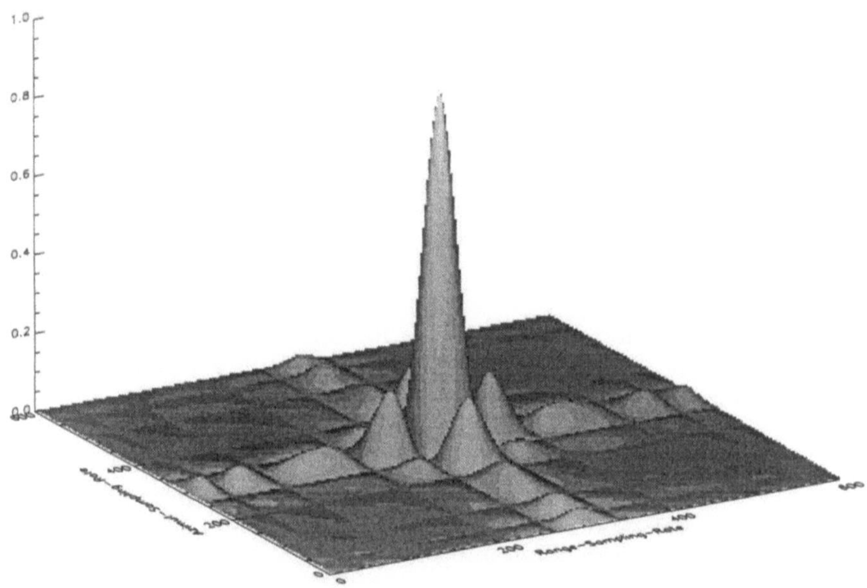

Image 3.14: Absolute value cutout of a range-compressed and azimuth-compressed point target

3.4 Point-target spectrum in the 2D-frequency domain

3.4.1 Derivation of the 2-dimensional point-target response

3.4.1.1 Introduction

SAR processors are different particularly in the way of their compression. There exist processing algorithms in the time domain as well as in the frequency domain. To derivate the process of the SAR raw-data compression in the frequency-domain, the spectrum of an entire SAR scenario has to be calculated. Therefore the 2-dimensional point-target spectrum in the 2-dimensional frequency domain is necessary.

At the same time the development of the 2-dimensional point-target response in the 2D-frequency domain serves as a common starting point for different derivations of varying processing algorithms and this way characterize the last step of the throughout system-theoretical description of SAR-signal processing.

We begin with a short explanation of the necessary basic correlations and parameters to perform afterwards the required mathematical calculations. The wanted result, the 2-dimensional point-target spectrum, is illustrated in equation 3.227.

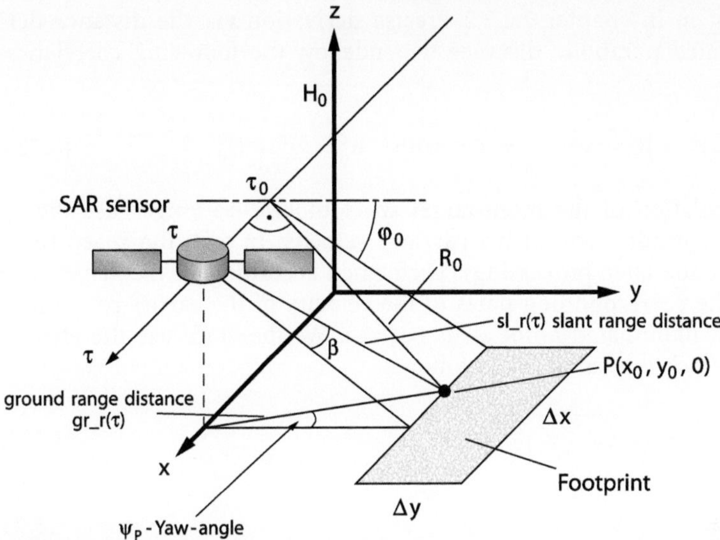

Image 3.15: SAR-geometry

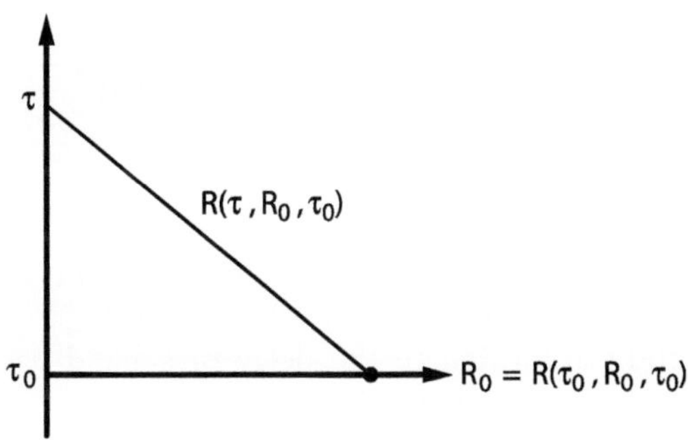

Image 3.16 a: Simplified geometry

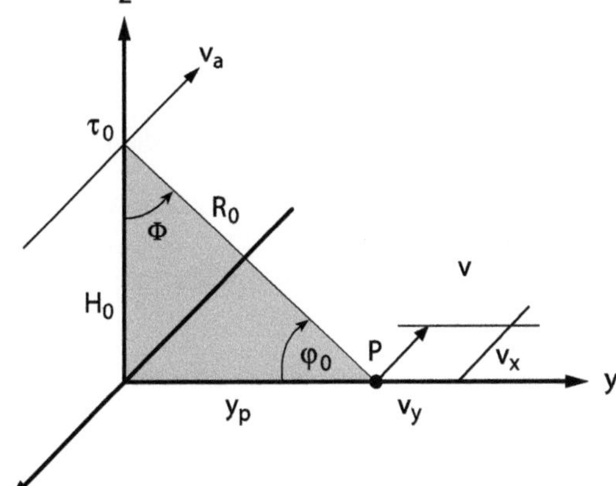

Image 3.16 b: Effective velocity

According to in chapter 2.2.2.2 Precise derivation via the distance derivative derivated parabolic distance dependency, the following correlations result:

$$R(\tau, R_0, \tau_0)^2 = R_0^2 + v^2(\tau - \tau_0)^2 \quad \text{with} \quad R_0^2 = y_0^2 + H_0^2 \qquad (3.72\,\text{a})$$

For the calculation of the point-target spectrum we no longer use the approximative consideration of the carrier's velocity. In fact, the speed (especially during the later processing), performed according to the sensor's velocity and the corresponding parts of the velocity of the earth's rotation velocity for an illuminated point at its surface. We therefore use the effective velocity v, which is assembled as follows:

$$v = \sqrt{(v_x - v_a)^2 + v_y^2} \qquad (3.72\,\text{b})$$

$$R(\tau, R_0, \tau_0) = \sqrt{R_0^2 + v^2(\tau - \tau_0)^2} \qquad (3.73)$$

The signal received by the SAR carrier can, as described in chapter 2 SAR basics, be performed in the equivalant low-pass domain as follows:

$$r_T(t, \tau, \tau_0, R_0) = \sigma(\tau_0, R_0) \cdot s_T(t - t_v(\tau, R_0, \tau_0)) \cdot e^{-j2\pi f_0 \cdot t_v(\tau, R_0, \tau_0)} \cdot \text{rect}\left(\frac{\tau - \tau_c}{\Delta\tau}\right)$$

This means:

$\sigma(\tau_0, R_0)$	Backscattering coefficient
$s_T(t - t_v(\tau, R_0, \tau_0))$	Time-delayed transmitting signal depending on the distance
$e^{-j2\pi f_0 \cdot t_v(\tau, R_0, \tau_0)}$	Phase function according to the delay of a band-pass signal in the equivalent low-pass domain
$\text{rect}\left(\dfrac{\tau - \tau_c}{\Delta\tau}\right)$	Rectangular function characterizing the part of a transmitting pulse in azimuth direction.

3.4.1.2 Evaluation of the point-target spectrum

3.4.1.2.1 Simplifying of the integral using the principle of the stationary phase

Obtaining the point-target spectrum works via the 2-dimensional Fourier transformation of the back-scattered signal received by the SAR carrier. Now the already mentioned start-stop approximation comes into operation (transmitting time and receiving time τ is identical):

$$R_T(f, f_\tau, \tau_0, R_0) = \mathscr{F}\mathscr{F}_{t,T}\{r_T(t, \tau, \tau_0, R_0)\} \qquad (3.74)$$

$$R_T(f, f_\tau, \tau_0, R_0) = \int\limits_{-\infty}^{\infty} \int\limits_{-\infty}^{\infty} \sigma(\tau_0, R_0) \cdot s_T\big(t - t_v(\tau, R_0, \tau_0)\big) \cdot e^{-j2\pi f_0 \cdot t_v(\tau, R_0, \tau_0)}$$
$$\cdot \operatorname{rect}\left(\frac{\tau - \tau_c}{\Delta\tau}\right) \cdot e^{-j2\pi f \cdot t} \cdot e^{-j2\pi f_\tau \cdot \tau} dt d\tau$$

$$(3.75)$$

Substitution:

$$t - t_v(\tau, R_0, \tau_0) = u \qquad dt = du \qquad t = u + t_v(\tau, R_0, \tau_0)$$

$$R_T(f, f_\tau, \tau_0, R_0) = \sigma(\tau_0, R_0) \cdot \int\limits_{-\infty}^{\infty} \int\limits_{-\infty}^{\infty} s_T(u) \cdot e^{-j2\pi f_0 \cdot t_v(\tau, R_0, \tau_0)} \cdot \operatorname{rect}\left(\frac{\tau - \tau_c}{\Delta\tau}\right)$$
$$\cdot e^{-j2\pi f \cdot (u + t_v(\tau, R_0, \tau_0))} \cdot e^{-j2\pi f_\tau \cdot \tau} du\, d\tau \qquad (3.76)$$

$$R_T(f, f_\tau, \tau_0, R_0) = \sigma(\tau_0, R_0) \cdot \int\limits_{-\infty}^{\infty} s_T(u) \cdot e^{-j2\pi f \cdot u} du$$
$$\cdot \int\limits_{-\infty}^{\infty} \operatorname{rect}\left(\frac{\tau - \tau_c}{\Delta\tau}\right) \cdot e^{-j2\pi \cdot (f + f_0) \cdot t_v(\tau, R_0, \tau_0)} \cdot e^{-j2\pi f_\tau \cdot \tau} d\tau \qquad (3.77)$$

With $S_T(f) = \int\limits_{-\infty}^{+\infty}$ as Fourier transformation of the transmitting signal in the equivalent low-pass domain follows:

$$R_T(f, f_\tau, \tau_0, R_0) = \sigma(\tau_0, R_0) \cdot S_T(f) \cdot \int\limits_{-\infty}^{\infty} \operatorname{rect}\frac{(\tau - \tau_c)}{\Delta\tau}$$
$$\cdot e^{-j2\pi \cdot (f + f_0) \cdot t_v(\tau, R_0, \tau_0)} \cdot e^{-j2\pi f_\tau \cdot \tau} d\tau \qquad (3.78)$$

Let us now have a look at the integral:

$$I(f, f_\tau, R_0, \tau_0) = \int\limits_{-\infty}^{\infty} \operatorname{rect}\left(\frac{\tau - \tau_c}{\Delta\tau}\right) \cdot e^{-j2\pi \cdot [(f + f_0) \cdot t_v(\tau, R_0, \tau_0)]} \cdot e^{-j2\pi f_\tau \cdot \tau} d\tau$$

$$I(f, f_\tau, R_0, \tau_0) = \int\limits_{-\infty}^{\infty} \text{rect}\left(\frac{\tau - \tau_c}{\Delta\tau}\right) \cdot e^{-j2\pi \cdot [(f+f_0) \cdot t_v(\tau, R_0, \tau_0) + f_\tau \cdot \tau]} d\tau$$

$$= \int\limits_{-\infty}^{\infty} \text{rect}\left(\frac{\tau - \tau_c}{\Delta\tau}\right) \cdot e^{-j\Phi(\tau, R_0, \tau_0)} d\tau \tag{3.79}$$

whereas $\Phi(\tau, R_0, \tau_0) = 2\pi \cdot [(f + f_0) \cdot t_v(\tau, R_0, \tau_0) + f_\tau \cdot \tau]$ is a quick varying phase term and is simplified to:

$$\Phi(\tau, R_0, \tau_0) = \Phi(\tau) = 2\pi \cdot \left[(f + f_0) \cdot \frac{2}{c} \cdot R(\tau, R_0, \tau_0) + f_\tau \cdot \tau\right] \tag{3.80}$$

The integral including the quick varying phase function can be solved approximatively according to chapter 2 using the method of the stationary phase [PAP68].

The integral over the range a function is oscillating quickly, is approximatively zero. We can compare the integration over complete durations of periods. Positive and negative parts of the function cancel each other. The main parts in the Fourier spectrum occur when the changing rate of the oscillation is at its minimum. This condition is fulfilled at the point of the stationary phase. The integral makes a contribution of the function where the derivation is $(\tau^*) = 0$.

Therefore we develop $\Phi(\tau)$ around the stationary point τ^* with $\dot{\Phi}(\tau^*) = 0$ where the first derivation disappears (see also "Calculation of the chirp spectrum")

$$\Phi(\tau) \cong \Phi(\tau^*) + \frac{1}{2}\ddot{\Phi}(\tau^*) \cdot (\tau - \tau^*)^2 \tag{3.81}$$

To keep things clear we insert $I(f, f_\tau, R_0, \tau_0) = I$:

$$I = \int\limits_{-\infty}^{\infty} \text{rect}\left(\frac{\tau - \tau_c}{\Delta\tau}\right) \cdot e^{-j\Phi(\tau)} d\tau \tag{3.82}$$

We insert τ^*:

$$I \cong \text{rect}\left(\frac{\tau^* - \tau_c}{\Delta\tau}\right) \cdot \int\limits_{-\infty}^{\infty} e^{-j\Phi(\tau^*)} \cdot e^{-j\frac{1}{2}\ddot{\Phi}(\tau^*) \cdot (\tau - \tau^*)^2} d\tau$$

$$= \text{rect}\left(\frac{\tau^* - \tau_c}{\Delta\tau}\right) \cdot e^{-j\Phi(\tau^*)} \cdot \int\limits_{-\infty}^{\infty} e^{-j\frac{1}{2}\ddot{\Phi}(\tau^*) \cdot (\tau - \tau^*)^2} d\tau \tag{3.83}$$

Substitution:

$$\frac{1}{2}\ddot{\Phi}(\tau^*)\cdot(\tau-\tau^*)^2 = u^2 \qquad u = \frac{1}{\sqrt{2}}\cdot\sqrt{\ddot{\Phi}(\tau^*)}\cdot(\tau-\tau^*)$$

$$du = \frac{1}{\sqrt{2}}\cdot\sqrt{\ddot{\Phi}(\tau^*)}\cdot d\tau \qquad d\tau = \frac{\sqrt{2}\cdot du}{\sqrt{\ddot{\Phi}(\tau^*)}}$$

$$I = \mathrm{rect}\left(\frac{\tau^*-\tau_c}{\Delta\tau}\right)\cdot e^{-j\Phi(\tau^*)}\cdot\frac{\sqrt{2}}{\sqrt{\ddot{\Phi}(\tau^*)}}\cdot\int_{-\infty}^{\infty}e^{-ju^2}du \tag{3.84}$$

Please notice:

$$\int_{-\infty}^{\infty}e^{-ju^2}du = \int_{-\infty}^{\infty}\cos(u^2)du - j\cdot\int_{-\infty}^{\infty}\sin(u^2)du$$

$$= 2\cdot\int_{0}^{\infty}\cos(u^2)du - j\cdot 2\cdot\int_{0}^{\infty}\sin(u^2)du \tag{3.85}$$

with

$$u^2 = v \qquad \frac{dv}{du} = 2u = 2\cdot\sqrt{v} \qquad du = \frac{dv}{2\cdot\sqrt{v}}$$

$$\int_{-\infty}^{\infty}e^{-ju^2}du = \int_{0}^{\infty}\frac{\cos(v)}{\sqrt{v}}dv - j\cdot\int_{0}^{\infty}\frac{\sin(v)}{\sqrt{v}}dv = \sqrt{\frac{\pi}{2}} - j\cdot\sqrt{\frac{\pi}{2}} = \sqrt{\pi}\cdot e^{-j\frac{\pi}{4}}$$

$$\tag{3.86}$$

[BRO91], p. 68, no. 16: The integral I is performed as:

$$\boxed{I \cong \mathrm{rect}\left(\frac{\tau^*-\tau_c}{\Delta\tau}\right)\cdot e^{-j\Phi(\tau^*)}\cdot\frac{\sqrt{2}}{\sqrt{\ddot{\Phi}(\tau^*)}}\cdot\sqrt{\pi}\cdot e^{-j\frac{\pi}{4}}} \tag{3.87}$$

In the next step the necessary parameters τ^*, $\Phi(\tau^*)$ and $\ddot{\Phi}(\tau^*)$ for solving the integral are calculated in order to determine the stationary point afterwards.

3.4.1.2.2 Calculation of the phase derivatives

$$\Phi(\tau) = 2\pi \cdot \left[(f + f_0) \cdot \frac{2}{c} \cdot R(\tau, \tau_0, R_0) + f_\tau \cdot \tau \right] \tag{3.88}$$

3.4.1.2.2.1 First phase derivative

$$\dot{\Phi}(\tau) = 2\pi \cdot \left[(f + f_0) \cdot \frac{2}{c} \cdot \dot{R}(\tau, \tau_0, R_0) + f_\tau \right] \tag{3.89}$$

with

$$\dot{R}(\tau, \tau_0, R_0) = \frac{d}{d\tau} R(\tau, \tau_0, R_0)$$

the condition for the stationary phase:

$$\boxed{\dot{R}(\tau^*) = 2\pi \cdot \left[(f + f_0) \cdot \frac{2}{c} \cdot \dot{R}(\tau^*, \tau_0, R_0) + f_\tau \right] \overset{!}{=} 0} \tag{3.90}$$

$$\Rightarrow (f + f_0) \cdot \frac{2}{c} \cdot \dot{R}(\tau^*, \tau_0, R_0) + f_\tau \overset{!}{=} 0 \tag{3.91}$$

with

$$R(\tau) = \sqrt{(\tau - \tau_c)^2 \cdot v^2 + R_0^2} \; .$$

3.4.1.2.2.2 Second phase derivative

$$\ddot{\Phi}(\tau) = 2\pi \cdot (f + f_0) \cdot \frac{2}{c} \cdot \ddot{R}(\tau, \tau_0, R_0) \tag{3.92}$$

$$\ddot{\Phi}(\tau^*) = 2\pi \cdot (f + f_0) \cdot \frac{2}{c} \cdot \ddot{R}(\tau^*, \tau_0, R_0) \ . \tag{3.93}$$

3.4.1.2.2.3 Third phase derivative

$$\dddot{\Phi}(\tau) = 2\pi \cdot (f + f_0) \cdot \frac{2}{c} \cdot R^{(3)}(\tau, \tau_0, R_0) \tag{3.94}$$

$$\dddot{\Phi}(\tau^*) = 2\pi \cdot (f + f_0) \cdot \frac{2}{c} \cdot R^{(3)}(\tau^*, \tau_0, R_0) \tag{3.95}$$

After calculating the phase derivatives the distance derivatives have to be determined:

$$\dot{R}(\tau), \ddot{R}(\tau), \dddot{R}(\tau) \ .$$

3.4.1.3 Calculation of the distance derivatives

In the beginning we introduce the following abbreviated form:

$$R^2(\tau) = R^2(\tau, \tau_0, R_0) = (\tau - \tau_0)^2 \cdot v^2 + R_0^2 \ . \tag{3.96}$$

3.4.1.3.1 First distance derivative

$$\frac{d}{d\tau} R^2(\tau) = 2 \cdot R(\tau) \cdot \dot{R}(\tau) = 2 \cdot (\tau - \tau_0) \cdot v^2 \tag{3.97}$$

$$\dot{R}(\tau) = \frac{(\tau - \tau_0) \cdot v^2}{R(\tau)} \tag{3.98}$$

1. Derivative at the stationary point

$$\dot{R}(\tau^*) = \frac{(\tau^* - \tau_0) \cdot v^2}{R(\tau^*)} \ . \tag{3.99}$$

3.4.1.3.2 Second distance derivative

$$\frac{d^2}{d\tau^2}R^2(\tau) = 2 \cdot \frac{d}{d\tau}(R(\tau) \cdot \dot{R}(\tau)) = 2 \cdot \frac{d}{d\tau}(\tau - \tau_0) \cdot v^2 \tag{3.100}$$

$$\dot{R}^2(\tau) + R(\tau) \cdot \ddot{R}(\tau) = v^2 \tag{3.101}$$

$$\boxed{\ddot{R}(\tau) = \frac{v^2 - \dot{R}^2(\tau)}{R(\tau)}} \tag{3.102}$$

2. Derivative at the stationary point

$$\ddot{R}(\tau^*) = \frac{v^2 - \dot{R}^2(\tau^*)}{R(\tau^*)} \ . \tag{3.103}$$

3.4.1.3.3 Third distance derivative

$$\frac{d^3}{d\tau^3}R^2(\tau) = \frac{d^3}{d\tau^3}((\tau - \tau_0)^2 \cdot v^2 + R_0^2) \tag{3.104}$$

$$\frac{d^3}{d\tau^3}R^2(\tau) = \frac{d^2}{d\tau^2}(2 \cdot R(\tau) \cdot \dot{R}(\tau)) \quad \text{Notice: } \frac{dv^2}{d\tau} = 0!$$

$$= \frac{d}{d\tau}(2 \cdot \dot{R}^2(\tau) + 2 \cdot R(\tau) \cdot \ddot{R}(\tau))$$

$$= 4 \cdot \dot{R}(\tau) \cdot \ddot{R}(\tau) + 2 \cdot \dot{R}(\tau) \cdot \ddot{R}(\tau) + 2 \cdot R(\tau) \cdot R^{(3)}(\tau)$$

$$R^{(3)}(\tau) = -\frac{6 \cdot \dot{R}(\tau) \cdot \ddot{R}(\tau)}{2 \cdot R(\tau)} = -3 \cdot \frac{\dot{R}(\tau) \cdot \ddot{R}(\tau)}{R(\tau)} \tag{3.105}$$

$$R^{(3)}(\tau) = -3 \cdot \frac{\dot{R}(\tau) \cdot (v^2 - \dot{R}^2(\tau))}{R^2(\tau)} = -3 \cdot \left(\frac{v^2 \cdot \dot{R}(\tau)}{R^2(\tau)} + \frac{\dot{R}^3(\tau)}{R^2(\tau)}\right) \tag{3.106}$$

3. Derivative at the stationary point

$$\boxed{R^{(3)}(\tau^*) = 3 \cdot \dot{R}(\tau^*) \cdot \frac{\dot{R}^2(\tau^*) - v^2}{R^2(\tau^*)}} \tag{3.107}$$

With the distance derivatives, the phase derivatives can be calculated and the stationary point can be determined.

3.4.1.4 Evaluation of the stationary point

Valid was:

$$(f + f_0) \cdot \frac{2}{c} \cdot \dot{R}(\tau^*, \tau_0, R_0) + f_\tau \overset{!}{=} 0 \tag{3.109}$$

$$(f + f_0) \cdot \frac{2}{c} \cdot \frac{(\tau^* - \tau_0) \cdot v^2}{R(\tau^*)} + f_\tau = 0 \tag{3.110}$$

$$(\tau^* - \tau_0) \cdot 2v^2 = -f_\tau \cdot \frac{R(\tau^*) \cdot c}{(f + f_0)} \tag{3.111}$$

$$(\tau^* - \tau_0) = -f_\tau \cdot \frac{R(\tau^*) \cdot c}{2v^2 \cdot (f + f_0)} \tag{3.112}$$

The stationary point τ^* is determined by the implicit equation. By squaring we get:

$$(\tau^* - \tau_0)^2 = f_\tau^2 \cdot \frac{R^2(\tau^*) \cdot c^2}{4v^4 \cdot (f + f_0)^2} \tag{3.113}$$

$$(\tau^* - \tau_0)^2 \cdot 4v^4 \cdot (f + f_0)^2 = f_\tau^2 \cdot R^2(\tau^*) \cdot c^2 \tag{3.114}$$

with

$$R^2(\tau^*) = R_0^2 + v^2 \cdot (\tau^* - \tau_0)^2$$

follows:

$$v^2 \cdot (\tau^* - \tau_0)^2 = R^2(\tau^*) - R_0^2 \tag{3.115}$$

Inserting equation (3.115) in equation (3.114) results in:

$$4 \cdot [R^2(\tau^*) - R_0^2] \cdot v^2 \cdot (f + f_0)^2 = f_\tau^2 \cdot R^2(\tau^*) \cdot c^2 \tag{3.116}$$

$$R^2(\tau^*) \cdot [4v^2 \cdot (f + f_0)^2 - f_\tau^2 \cdot c^2] = 4 \cdot R_0^2 \cdot v^2 \cdot (f + f_0)^2 \tag{3.117}$$

$$R^2(\tau^*) = \frac{4 \cdot R_0^2 \cdot v^2 \cdot (f + f_0)^2}{4v^2 \cdot (f + f_0)^2 - f_\tau^2 \cdot c^2} \tag{3.118}$$

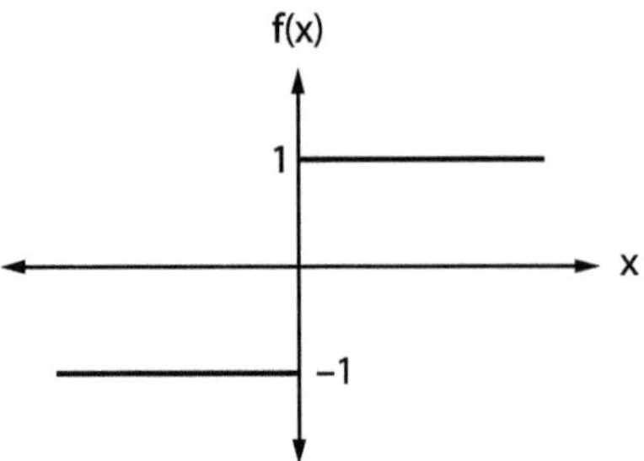

Image 3.17: Signum function

It follows the distance in the stationary point ($R \geq 0$):

$$R(\tau^*) = \frac{R_0 \cdot |f + f_0|}{\left[(f + f_0)^2 - \dfrac{f_\tau^2 \cdot c^2}{4v^2}\right]^{\frac{1}{2}}} \tag{3.119}$$

With equation (3.119) in equation (3.112) follows:

$$\tau^* - \tau_0 = -f_\tau \cdot \frac{R_0 \cdot |f + f_0| \cdot c}{\left[(f + f_0)^2 - \dfrac{f_\tau^2 \cdot c^2}{4v^2}\right]^{\frac{1}{2}} \cdot 2v^2 \cdot (f + f_0)} \tag{3.120}$$

We implement the signum function with:

$$\mathrm{sgn}(x) = \mathrm{sgn}^{-1}(x) = \frac{|x|}{x}$$

After conversion and considering the signum function we get the point of stationary phase, the stationary point:

$$\tau^* = \tau_0 - f_\tau \cdot \frac{R_0 \cdot \mathrm{sgn}(f + f_0) \cdot \dfrac{c}{2v^2}}{\left[(f + f_0)^2 - \dfrac{f_\tau^2 \cdot c^2}{4v^2}\right]^{\frac{1}{2}}} \tag{3.121}$$

3.4.1.5 Distance derivatives at the stationary point

3.4.1.5.1 First distance derivative at the stationary point

Therefore, we insert equation (3.121) and equation (3.119) in equation (3.99).

$$\dot{R}(\tau^*) = \frac{(\tau^* - \tau_0) \cdot v^2}{R(\tau^*)} = -f_\tau \cdot \frac{R_0 \cdot \mathrm{sgn}(f + f_0) \cdot \dfrac{c}{2}}{\left[(f + f_0)^2 - \dfrac{f_\tau^2 \cdot c^2}{4v^2}\right]^{\frac{1}{2}} \cdot \dfrac{R_0 \cdot |f + f_0|}{\left[(f + f_0)^2 - \dfrac{f_\tau^2 \cdot c^2}{4v^2}\right]^{\frac{1}{2}}}} \tag{3.122}$$

$$\boxed{\dot{R}(\tau^*) = -\frac{f_\tau}{(f + f_0)} \cdot \frac{c}{2}} \tag{3.123}$$

3.4.1.5.2 Second distance derivative at the stationary point

Therefore we insert equation (3.123) and equation (3.119) in equation (3.103) and get:

$$
\begin{aligned}
\ddot{R}(\tau^*) &= \frac{v^2 - \dot{R}^2(\tau^*)}{R(\tau^*)} = \frac{v^2 - \dfrac{f_\tau^2}{(f + f_0)^2} \cdot \dfrac{c^2}{4}}{R_0 \cdot |f + f_0|} \cdot \left[(f + f_0)^2 - \frac{f_\tau^2 \cdot c^2}{4v^2}\right]^{\frac{1}{2}} \\[2ex]
&= \frac{4v^2 \cdot (f + f_0)^2 - f_\tau^2 \cdot c^2}{4 \cdot (f + f_0)^2 \cdot |f + f_0| \cdot R_0} \cdot \left[(f + f_0)^2 - \frac{f_\tau^2 \cdot c^2}{4v^2}\right]^{\frac{1}{2}} \\[2ex]
&= \frac{v^2 \cdot \left[(f + f_0)^2 - \dfrac{f_\tau^2 \cdot c^2}{4 \cdot v^2}\right]}{(f + f_0)^2 \cdot |f + f_0| \cdot R_0} \cdot \left[(f + f_0)^2 - \frac{f_\tau^2 \cdot c^2}{4v^2}\right]^{\frac{1}{2}}
\end{aligned}
\tag{3.124}
$$

We get:

$$\boxed{\ddot{R}(\tau^*) = \frac{v^2}{R.(f + f_0)^2 \cdot |f + f_0|} \left[(f + f_0)^2 - \frac{f_\tau^2 \cdot c^2}{4v^2}\right]^{\frac{3}{2}}} \tag{3.125}$$

3.4.1.6 Evaluation of the phase terms at the stationary point

3.4.1.6.1 Calculation of the first phase constant

Therefore, we insert equation (3.121) and equation (3.119) in equation (3.88):

$$\Phi(\tau^*) = 2\pi \cdot \left[(f + f_0) \cdot \frac{2}{c} \cdot R(\tau^*, \tau_0, R_0) + f_\tau \cdot \tau^*\right] \tag{3.126}$$

$$\Phi(\tau^*) = 2\pi \cdot \left\{ (f + f_0) \cdot \frac{2}{c} \cdot \frac{R_0 \cdot |f + f_0|}{\left[(f + f_0)^2 - \frac{f_\tau^2 \cdot c^2}{4v^2} \right]^{\frac{1}{2}}} + f_\tau \right.$$
$$\left. \cdot \left[\tau_0 - \frac{f_\tau \cdot R_0 \cdot \mathrm{sgn}(f + f_0) \cdot \frac{c}{2v^2}}{\left[(f + f_0)^2 - \frac{f_\tau^2 \cdot c^2}{4v^2} \right]^{\frac{1}{2}}} \right] \right\} \qquad ((3.127)$$

Please note

$$\mathrm{sgn}(x) \cdot x = |x|$$

$$\Phi(\tau^*) = 2\pi \cdot \left[f_\tau \cdot \tau_0 + \frac{R_0 \cdot \mathrm{sgn}(f + f_0) \cdot \left[(f + f_0)^2 \cdot \frac{2}{c} - f_\tau^2 \cdot \frac{c}{2v^2} \right]}{\left[(f + f_0)^2 - \frac{f_\tau^2 \cdot c^2}{4v^2} \right]^{\frac{1}{2}}} \right]$$

$$(3.128)$$

$$\Phi(\tau^*) = 2\pi \cdot \left[f_\tau \cdot \tau_0 + \frac{\frac{2}{c} \cdot R_0 \cdot \mathrm{sgn}(f + f_0) \cdot \left[(f + f_0)^2 - f_\tau^2 \cdot \frac{c^2}{4v^2} \right]}{\left[(f + f_0)^2 - \frac{f_\tau^2 \cdot c^2}{4v^2} \right]^{\frac{1}{2}}} \right]$$
$$= 2\pi \cdot f_\tau \cdot \tau_0 + 4\pi \cdot \frac{R_0}{c} \cdot \mathrm{sgn}(f + f_0) \cdot \left[(f + f_0)^2 - f_\tau^2 \cdot \frac{c^2}{4v^2} \right]^{\frac{1}{2}} \qquad (3.129)$$

The first-phase constant results in:

$$\boxed{\Phi(\tau^*) = 2\pi \cdot f_\tau \cdot \tau_0 + 4\pi \cdot \frac{R_0}{c} \cdot \mathrm{sgn}(f + f_0) \cdot \left[(f + f_0)^2 - f_\tau^2 \cdot \frac{c^2}{4v^2} \right]^{\frac{1}{2}}} \qquad (3.130)$$

3.4.1.6.2 Calculation of the second-phase constant

Valid was:

$$\ddot{\Phi}(\tau^*) = 2\pi \cdot (f + f_0) \cdot \frac{2}{c} \cdot \ddot{R}(\tau^*, \tau_0, R_0) \qquad (3.131)$$

Inserting of equation (3.125) results in:

$$\ddot{\Phi}(\tau^*) = 2\pi \cdot (f + f_0) \cdot \frac{2}{c} \cdot \frac{v^2}{R_0 \cdot (f + f_0)^2 \cdot |f + f_0|} \left[(f + f_0)^2 - \frac{f_\tau^2 \cdot c^2}{4v^2} \right]^{\frac{3}{2}}$$

(3.132)

with $\mathrm{sgn}(x) = \mathrm{sgn}^{-1}(x)$ the second-phase constant is resulting as:

$$\ddot{\Phi}(\tau^*) = 4\pi \cdot \frac{v^2}{c \cdot R_0} \cdot \mathrm{sgn}(f + f_0) \cdot \frac{\left[(f + f_0)^2 - \frac{f_\tau^2 \cdot c^2}{4v^2} \right]^{\frac{3}{2}}}{(f + f_0)^2}$$

(3.133)

With this, the integral can almost completely be described.

3.4.1.7 Solution of the simplified integral

According to equation (3.79) the integral to be solved was defined as follows:

$$I = \int_{-\infty}^{\infty} \mathrm{rect}\left(\frac{\tau - \tau_c}{\Delta\tau}\right) \cdot e^{-j2\pi \cdot [(f+f_0) \cdot t_v(\tau, R_0, \tau_0) + f_\tau \cdot \tau]} d\tau$$

Inserting of the calculated interparameters leads to:

$$I \cong \mathrm{rect}\left(\frac{\tau^* - \tau_c}{\Delta\tau}\right) \cdot e^{-j\Phi(\tau^*)} \cdot \frac{\sqrt{2}}{\sqrt{\ddot{\Phi}(\tau^*)}} \cdot \sqrt{\pi} \cdot e^{-j\frac{\pi}{4}}$$

$$I = \mathrm{rect}\left(\frac{\tau^* - \tau_c}{\Delta\tau}\right) \cdot \frac{\sqrt{2} \cdot \sqrt{\pi} \cdot e^{-j\frac{\pi}{4}} \cdot \sqrt{R_0 \cdot c} \cdot |f + f_0|}{\sqrt{4\pi} \cdot v \cdot \left[(f + f_0)^2 - f_\tau^2 \cdot \frac{c^2}{4v^2} \right]^{\frac{3}{4}}}$$

$$\cdot \frac{1}{\sqrt{\mathrm{sgn}(f + f_0)}} \cdot e^{-j2\pi \cdot f_\tau \cdot \tau_0} \cdot e^{-j4\pi \cdot \frac{R_0}{c} \cdot \mathrm{sgn}(f+f_0) \cdot \left[(f+f_0)^2 - f_\tau^2 \cdot \frac{c^2}{4v^2} \right]^{\frac{1}{2}}}$$

$$= \mathrm{rect}\left(\frac{\tau^* - \tau_c}{\Delta\tau}\right) \cdot \sqrt{\frac{c \cdot R_0}{2}} \cdot \frac{1}{v} \cdot \frac{|f + f_0|}{\left[(f + f_0)^2 - f_\tau^2 \cdot \frac{c^2}{4v^2} \right]^{\frac{3}{4}}}$$

$$\cdot \frac{1}{\sqrt{\mathrm{sgn}(f + f_0)}} \cdot e^{-j\frac{\pi}{4}} \cdot e^{-j2\pi \cdot f_\tau \cdot \tau_0} \cdot e^{-j4\pi \cdot \frac{R_0}{c} \cdot \mathrm{sgn}(f+f_0) \cdot \left[(f+f_0)^2 - f_\tau^2 \cdot \frac{c^2}{4v^2} \right]^{\frac{1}{2}}}$$

(3.134)

Please note:

$$\sqrt{\mathrm{sgn}(f+f_0)} = \sqrt{\pm 1} = \begin{cases} 1 & \text{for} \quad \mathrm{sgn}(f+f_0) = 1 \\ \pm j & \text{for} \quad \mathrm{sgn}(f+f_0) = -1 \end{cases}$$

$$= \begin{cases} e^{j2\pi} \\ e^{\pm j\frac{\pi}{2}} \end{cases}$$

$$\sqrt{\mathrm{sgn}(f+f_0)}^{-1} = \begin{cases} e^{j2\pi} & \text{for} \quad \mathrm{sgn}(f+f_0) = 1 \\ e^{\pm j\frac{\pi}{2}} & \text{for} \quad \mathrm{sgn}(f+f_0) = -1 \end{cases}$$

$$= e^{\pm j\frac{\pi}{4}\cdot[1-\mathrm{sgn}(f+f_0)]}$$

$$= e^{\pm j\frac{\pi}{4}} \cdot e^{j\frac{\pi}{4}\mathrm{sgn}(f+f_0)}$$

For solving the simplified integral follows:

$$\boxed{\begin{aligned} I \cong \; & \mathrm{rect}\left(\frac{\tau^* - \tau_c}{\Delta\tau}\right) \cdot \sqrt{\frac{c \cdot R_0}{2}} \cdot \frac{1}{v} \cdot \frac{|f+f_0|}{\left[(f+f_0)^2 - f_\tau^2 \cdot \dfrac{c^2}{4v^2}\right]^{\frac{3}{4}}} \cdot e^{\pm j\frac{\pi}{4}\cdot[2-\mathrm{sgn}(f+f_0)]} \\ & \cdot e^{-j2\pi\cdot f_\tau\cdot\tau_0} \cdot e^{+j4\pi\cdot\frac{R_0}{c}\cdot\mathrm{sgn}(f+f_0)\cdot\left[(f+f_0)^2-f_\tau^2\cdot\frac{c^2}{4v^2}\right]^{\frac{1}{2}}} \end{aligned}}$$

$$(3.135)$$

3.4.1.8 The point-target spectrum

In the beginning we stated equation (3.78):

$$R_T(f, f_\tau, \tau_0, R_0) = S_T(f) \cdot \sigma(\tau_0, R_0) \cdot I(f, f_\tau, R_0, \tau_0) \tag{3.136}$$

Because of its limited band the Fourier transformation of the transmitted signal has to be described as follows:

$$S_T(f) = S_T(f) \cdot \mathrm{rect}\left(\frac{f}{B_r}\right) \tag{3.137}$$

with the range bandwidth

$$B_r = k_r \cdot T$$

As the bandwidth B_r of the transmitting signal typically is at $B_r = 20$ MHz (ERS-1) and f_0 usually is selected in the GHz-range, we can assume that:

$$\mathrm{sgn}(f+f_0) = \mathrm{sgn}\left(f_0 \pm \frac{B_r}{2}\right) = 1 \quad \text{because} \quad f_0 \approx 5.3 \text{ GHz} \gg B_r \tag{3.138}$$

Therefore the expression for the total spectrum is simplified to:

$$
\begin{aligned}
R_T(f, f_\tau, \tau_0, R_0) &= \sigma(\tau_0, R_0) \cdot S_T(f) \cdot \mathrm{rect}\!\left(\frac{\tau^* - \tau_c}{\Delta\tau}\right) \cdot \sqrt{\frac{c \cdot R_0}{2}} \cdot \frac{1}{v} \\
&\quad \cdot \frac{|f + f_0|}{\left[(f + f_0)^2 - f_\tau^2 \cdot \dfrac{c^2}{4v^2}\right]^{\frac{3}{4}}} \cdot e^{-j\frac{\pi}{4}} \cdot e^{-j2\pi \cdot f_\tau \cdot \tau_0} \\
&\quad \cdot e^{-j4\pi \cdot \frac{R_0}{c} \cdot \left[(f+f_0)^2 - f_\tau^2 \cdot \frac{c^2}{4v^2}\right]^{\frac{1}{2}}}
\end{aligned}
\tag{3.139}
$$

To completely solve the integral in the frequency domain the rectangular function is considered as a term of time in equation (3.139). However, the total spectrum commonly is calculated as a term of the frequency, in particular of the azimuth frequency, too.

The point τ^* is a term of the azimuth frequency according to equation (3.121). With the aid of the stationary point, in other words by inserting in the time-domain rectangular function, we can derivate a rectangular function in the frequency domain. The position of the wanted rectangular function in the frequency domain is described in the appendix. Let us summarize now merely the results of the considerations in the appendix.

Summary of the rectangle position

Centroid frequency:

$$
f_c = \frac{1}{2}(f_0 + f_u) \cong \frac{2v^2}{c} \cdot \frac{1}{R(\tau_c)} \cdot (\tau_0 - \tau_c) \cdot (f + f_0)
\tag{3.223}
$$

Bandwidth

$$
B_{az} = f_0 - f_u \cong \frac{2v^2}{c} \cdot \frac{1}{R(\tau_c)} \cdot (f + f_0) \cdot \Delta\tau
\tag{3.224}
$$

Please note the centroid frequency

$$
f_c \cong \frac{2v^2}{c} \cdot \frac{1}{R(\tau_c)} \cdot (\tau_0 - \tau_c) \cdot (f + f_0)
$$

with $\lambda = \dfrac{c}{f_0}$ follows:

$$
f_c \cong \frac{2v^2}{\lambda} \cdot \frac{1}{R(\tau_c)} \cdot (\tau_0 - \tau_c) \cdot \left(1 + \frac{f}{f_0}\right)
\tag{3.225}
$$

Please have a look at the chapter discussing doppler displacement in SAR geometry including the moving angle and the image SAR geometry.

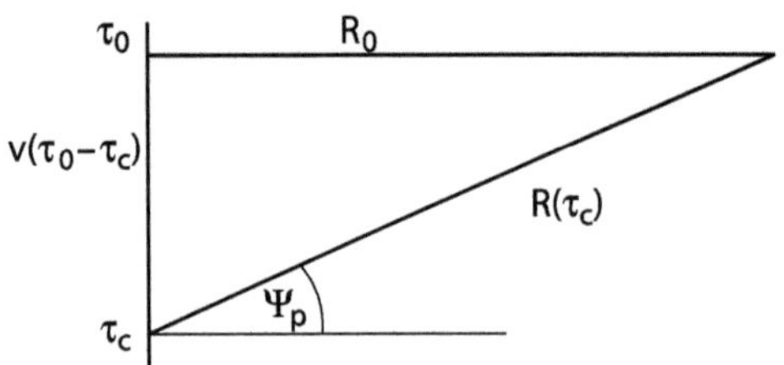

Image 3.18: Simplified SAR geometry

ψ_p: projection of the squint angle
β: squint angle

The projection of the squint angle is defined as:

$$\frac{v \cdot (\tau_0 - \tau_c)}{R(\tau_c)} = \sin(\psi_p)$$

$$f_c \cong \frac{2v}{\lambda} \cdot \left(1 + \frac{f}{f_0}\right) \cdot \sin(\psi_p) \cong \frac{2v}{\lambda} \cdot \sin(\psi_p) \cong \frac{2v}{\lambda} \cdot \sin(\beta) = f_{DC} \qquad (3.226)$$

Considering the 2-dimensional point-target spectrum, in particular the determination of the rectangular function in the frequency domain, follows as already done in earlier calculations to the doppler frequency the dependence of the doppler frequency on the squint angle.

3.4.3 Point-target spectrum in the 2-dimensional frequency domain

As final result we get the following term for the 2-dimensional point-target spectrum:

$$R_T(f, f_\tau, R_0, \tau_0) = S_T(f) \cdot \sigma(R_0, \tau_0) \cdot \mathrm{rect}\left[\frac{f_\tau - f_c}{B_{az}}\right] \cdot \sqrt{\frac{c \cdot R_0}{2}} \cdot \frac{1}{v}$$

$$\cdot \frac{f + f_0}{\left[(f + f_0)^2 - \dfrac{f_\tau^2 \cdot c^2}{4 \cdot v^2}\right]^{\frac{3}{4}}} \cdot e^{-j\cdot\pi/4} \cdot e^{-j\cdot 2\pi f_\tau \tau_0}$$

$$\cdot e^{-j4\pi\frac{R_0}{c}\sqrt{(f+f_0)^2 - \frac{f_\tau^2 c^2}{4v^2}}} \qquad (3.227)$$

Here all necessary influencing parameters of the SAR sensor are contained. Starting from the 2-dimensional point-target spectrum, different sorts of SAR-focussing algorithms can be derivated.

Among others there belongs the range-doppler processor [BEN79], the wave-number-domain processor [ROC89] and also the scaled-inverse-Fourier transformation processor (SIFT processor) [LOF98].

Common starting point of the different signal-processing algorithms and thus the meaning of the obtained derivations is the following equation (3.227).

3.5 Point-target spectrum in the range-doppler domain

3.5.1 Introduction

Different process-relevant influencing parameters can not be adequately displayed in the 2-dimensional frequency domain of the point-target response. Additionally, several comparing derivations require the point-target spectrum as definition in the range time, azimuth frequency domain (range-doppler domain). Therefore we need the ascertainment of the 2-dimensional point-target response of the SAR sensor in the so-called range-doppler domain.

Starting point of the derivation is equation (3.227):

$$
\begin{aligned}
R_T(f, f_\tau, R_0, \tau_0) =\ & s_T(f) \cdot \sigma(R_0, \tau_0) \cdot \mathrm{rect}\left[\frac{f_\tau - f_c}{B_{az}}\right] \cdot \sqrt{\frac{c \cdot R_0}{2}} \cdot \frac{1}{v} \\
& \cdot \frac{f + f_0}{\left[(f + f_0)^2 - \dfrac{f_\tau^2 \cdot c^2}{4 \cdot v^2}\right]^{\frac{3}{4}}} \cdot e^{-j \cdot \pi/4} \cdot e^{-j \cdot 2\pi f_\tau \tau_0} \\
& \cdot e^{-j4\pi \frac{R_0}{c}\sqrt{(f+f_0)^2 - \frac{f_\tau^2 c^2}{4v^2}}}
\end{aligned}
$$

$$\tag{3.227}$$

Wanted is the inverse Fourier transformation of the point-target spectrum in the 2D-frequency domain:

$$
R_T(t, f_\tau, R_0, \tau_0) = \mathscr{F}^{-1}\{R_T(f, f_\tau, R_0, \tau_0)\} \tag{3.228}
$$

$$
R_T(t, f_\tau, R_0, \tau_0) = \int_{-\infty}^{\infty} R_T(f, f_\tau, R_0, \tau_0) \cdot e^{j2\pi ft} \cdot df \tag{3.229}
$$

To remind you:

$$
r_T(t, \tau, \tau_0, R_0) = \sigma(\tau_0, R_0) \cdot s_T(t - t_v(\tau, R_0, \tau_0)) \cdot e^{-j2\pi f_0 \cdot t_v(\tau, R_0, \tau_0)} \cdot \mathrm{rect}\left(\frac{\tau - \tau_c}{\Delta\tau}\right)
$$

$$
s_T(t) = \mathrm{rect}\left(\frac{t}{T}\right) \cdot e^{j\pi k_r \cdot t^2} \circ\!\!-\!\!\bullet\ S_T(f) = \frac{1}{\sqrt{k_r}} \cdot \mathrm{rect}\left(\frac{f}{B_r}\right) \cdot e^{-j\frac{\pi f^2}{k_r}} \cdot e^{j\frac{\pi}{4}}
$$

Inserting the transmitting chirp spectrum in equation (3.227) results in:

$$R_T(f, f_\tau, R_0, \tau_0) = \sigma(R_0, \tau_0) \cdot \frac{1}{\sqrt{k_r}} \cdot \text{rect}\left(\frac{f}{B_r}\right) \cdot e^{j \cdot \pi/4} \cdot e^{-j \frac{\pi f^2}{k_r}} \cdot \text{rect}\left[\frac{f_\tau - f_c}{B_{az}}\right]$$

$$\cdot \sqrt{\frac{c \cdot R_0}{2}} \cdot \frac{1}{v} \cdot \frac{f + f_0}{\left[(f + f_0)^2 - \frac{f_\tau^2 \cdot c^2}{4 \cdot v^2}\right]^{\frac{3}{4}}} \cdot e^{-j \cdot \pi/4}$$

$$\cdot e^{-j \cdot 2\pi f_\tau \tau_0} \cdot e^{-j 4\pi \frac{R_0}{c} \cdot \sqrt{(f+f_0)^2 - \frac{f_\tau^2 c^2}{4v^2}}} \tag{3.230}$$

$$\boxed{\begin{aligned} R_T(f, f_\tau, R_0, \tau_0) &= \sigma(R_0, \tau_0) \cdot \text{rect}\left(\frac{f}{B_r}\right) \cdot \text{rect}\left[\frac{f_\tau - f_c}{B_{az}}\right] \cdot \frac{1}{\sqrt{k_r}} \\ &\cdot \sqrt{\frac{c \cdot R_0}{2}} \cdot \frac{1}{v} \cdot \frac{f + f_0}{\left[(f + f_0)^2 - \frac{f_\tau^2 \cdot c^2}{4 \cdot v^2}\right]^{3/4}} \\ &\cdot e^{-j \cdot 2\pi f_\tau \tau_0} \cdot e^{-j\left(4\pi \frac{R_0}{c} \cdot \sqrt{(f+f_0)^2 - \frac{f_\tau^2 c^2}{4v^2}} + \frac{\pi f^2}{k_r}\right)} \end{aligned}} \tag{3.231}$$

3.5.2 The phase approximation

Considering the exponent in equation (3.231) an inverse Fourier transformation of this non-linear term seems impossible. Thus we are forced to linearise the exponent as follows:

$$E(f, f_\tau) = \pi \cdot \left(\frac{f^2}{k_r} + \frac{4 \cdot R_0}{c} \cdot \sqrt{(f + f_0)^2 - \frac{f_\tau^2 c^2}{4v^2}}\right) \tag{3.232}$$

with

$$c = \lambda \cdot f_0$$

$$E(f, f_\tau) = \pi \cdot \left(\frac{f^2}{k_r} + \frac{4 \cdot R_0}{\lambda} \cdot \sqrt{\left(1 + \frac{f}{f_0}\right)^2 - \frac{f_\tau^2 2\lambda^2}{4v^2}}\right) \tag{3.233}$$

$$E(f, f_\tau) = \pi \cdot \left(\frac{f^2}{k_r} + \frac{4 \cdot R_0}{\lambda} \cdot \sqrt{1 + \frac{2f}{f_0} + \left(\frac{f}{f_0}\right)^2 - \frac{f_\tau^2 \lambda^2}{4v^2}}\right) \tag{3.234}$$

$$E(f, f_\tau) = \pi \cdot \left(\frac{f^2}{k_r} + \frac{4 \cdot R_0}{\lambda} \cdot \sqrt{1 - \frac{f_\tau^2 \lambda^2}{4v^2}} \cdot \sqrt{1 + \frac{\frac{2f}{f_0} + \left(\frac{f}{f_0}\right)^2}{1 - \frac{f_\tau^2 \lambda^2}{4v^2}}} \right) \tag{3.235}$$

where:

$$W(f, f_\tau) = \sqrt{1 + \frac{\frac{2f}{f_0} + \left(\frac{f}{f_0}\right)^2}{1 - \frac{f_\tau^2 \lambda^2}{4v^2}}} = \sqrt{1 + x} \tag{3.236}$$

With the approximation

$$\sqrt{1 + x} \cong 1 + \frac{1}{2} \cdot x - \frac{1}{8} \cdot x^2 \tag{3.237}$$

we obtain

$$\begin{aligned} W(f, f_\tau) = &\sqrt{1 + \frac{\frac{2f}{f_0} + \left(\frac{f}{f_0}\right)^2}{1 - \frac{f_\tau^2 \lambda^2}{4v^2}}} \cong 1 + \frac{1}{2} \cdot \frac{\frac{2f}{f_0} + \left(\frac{f}{f_0}\right)^2}{1 - \frac{f_\tau^2 \lambda^2}{4v^2}} - \frac{1}{8} \\ &\cdot \left[\frac{\frac{2f}{f_0} + \left(\frac{f}{f_0}\right)^2}{1 - \frac{f_\tau^2 \lambda^2}{4v^2}} \right]^2 \end{aligned} \tag{3.238}$$

and the exponent becomes

$$\begin{aligned} E(f, f_\tau) \cong &\pi \cdot \left[\frac{f^2}{k_r} + \frac{4R_0}{\lambda} \cdot \sqrt{1 - \frac{f_\tau^2 \lambda^2}{4v^2}} \cdot \left(1 + \frac{1}{2} \cdot \frac{\frac{2f}{f_0} + \left(\frac{f}{f_0}\right)^2}{1 - \frac{f_\tau^2 \lambda^2}{4v^2}} \right. \right. \\ &\left. \left. - \frac{1}{8} \cdot \frac{\left(\frac{2f}{f_0} + \left(\frac{f}{f_0}\right)^2\right)^2}{\left(1 - \frac{f_\tau^2 \lambda^2}{4v^2}\right)^2} \right) \right] \end{aligned} \tag{3.239}$$

Therefore we write:

$$E(f, f_\tau) \cong \pi \cdot \left\{ \frac{f^2}{k_r} + \frac{4R_0}{\lambda} \cdot \sqrt{1 - \frac{f_\tau^2 \lambda^2}{4v^2}} + \frac{4R_0}{\lambda} \cdot \frac{f}{f_0} \cdot \frac{1}{\sqrt{1 - \frac{f_\tau^2 \lambda^2}{4v^2}}} \right.$$
$$\left. + \frac{4R_0}{\lambda} \cdot \sqrt{1 - \frac{f_\tau^2 \lambda^2}{4v^2}} \cdot \left(\frac{1}{2} \cdot \frac{\left(\frac{f}{f_0}\right)^2}{1 - \frac{f_\tau^2 \lambda^2}{4v^2}} - \frac{1}{8} \cdot \frac{\left(\frac{2f}{f_0} + \left(\frac{f}{f_0}\right)^2\right)^2}{\left(1 - \frac{f_\tau^2 \lambda^2}{4v^2}\right)^2} \right) \right\}$$

$$(3.240)$$

$$E(f, f_\tau) \cong \frac{\pi f^2}{k_r} + \frac{4\pi R_0}{\lambda} \cdot \sqrt{1 - \frac{f_\tau^2 \lambda^2}{4v^2}} + \frac{4R_0}{\lambda} \cdot \frac{f}{f_0} \cdot \frac{1}{\sqrt{1 - \frac{f_\tau^2 \lambda^2}{4v^2}}}$$
$$+ \frac{4\pi R_0}{\lambda} \cdot \sqrt{1 - \frac{f_\tau^2 \lambda^2}{4v^2}} \cdot \left(\frac{\frac{1}{2} \cdot \left(\frac{f}{f_0}\right)^2 \cdot \left(1 - \frac{f_\tau^2 \lambda^2}{4v^2}\right) - \frac{1}{8} \cdot \left(\frac{2f}{f_0} + \left(\frac{f}{f_0}\right)^2\right)^2}{\left(1 - \frac{f_\tau^2 \lambda^2}{4v^2}\right)^2} \right)$$

$$(3.241)$$

$$E(f, f_\tau) \cong \frac{4\pi R_0}{\lambda} \cdot \left(\sqrt{1 - \frac{f_\tau^2 \lambda^2}{4v^2}} + \frac{f}{f_0} \cdot \frac{1}{\sqrt{1 - \frac{f_\tau^2 \lambda^2}{4v^2}}} \right)$$
$$+ \frac{\pi f^2}{k_r} + \frac{4\pi R_0}{\lambda} \cdot \sqrt{1 - \frac{f_\tau^2 \lambda^2}{4v^2}}$$
$$\cdot \left(\frac{\frac{1}{2} \cdot \left(\frac{f}{f_0}\right)^2 - \frac{1}{2} \cdot \frac{f^2 f_\tau^2 \lambda^2}{f_0^2 4v^2} - \frac{1}{2} \cdot \left(\frac{f}{f_0}\right)^2 - \frac{1}{2} \cdot \left(\frac{f}{f_0}\right)^3 - \frac{1}{8} \cdot \left(\frac{f}{f_0}\right)^4}{\left(1 - \frac{f_\tau^2 \lambda^2}{4v^2}\right)^2} \right)$$

$$(3.242)$$

$$E(f, f_\tau) \cong \frac{4\pi R_0}{\lambda} \cdot \left(\sqrt{1 - \frac{f_\tau^2 \lambda^2}{4v^2}} + \frac{f}{f_0} \cdot \frac{1}{\sqrt{1 - \frac{f_\tau^2 \lambda^2}{4v^2}}} \right) + \frac{\pi f^2}{k_r}$$

$$- \frac{4\pi R_0}{\lambda} \cdot \frac{1}{\left(1 - \frac{f_\tau^2 \lambda^2}{4v^2}\right)^{\frac{3}{2}}} \cdot \frac{1}{2} \left(\frac{f}{f_0}\right)^2 \cdot \left(\frac{f_\tau^2 \lambda^2}{4v^2} + \frac{1}{4} \cdot \left(\frac{f}{f_0}\right)^2 + \frac{f}{f_0} \right)$$

$$(3.243)$$

The final result we can go on with is as follows:

$$E(f, f_\tau) \cong \frac{4\pi R_0}{\lambda} \cdot \left(\sqrt{1 - \frac{f_\tau^2 \lambda^2}{4v^2}} + \frac{f}{f_0} \cdot \frac{1}{\sqrt{1 - \frac{f_\tau^2 \lambda^2}{4v^2}}} \right)$$

$$+ \frac{\pi f^2}{k_r} \cdot \left[1 - \frac{2R_0}{c} \cdot \frac{1}{\left(1 - \frac{f_\tau^2 \lambda^2}{4v^2}\right)^{\frac{3}{2}}} \cdot \frac{k_r}{f_0} \cdot \left(\frac{f_\tau^2 \lambda^2}{4v^2} + \frac{1}{4} \left(\frac{f}{f_0}\right)^2 + \frac{f}{f_0} \right) \right]$$

$$(3.244)$$

The phase term due to equation (3.232) can now be linearised following the equation (3.244) and accordingly processed.

3.5.3 Calculation of the modified sweep rate

According to equation (3.244) the exponential term concerning the range frequency can be fractionised in a constant term, a linear term and in a quadratic term. The sweep rate also called chirp rate of the quadratic term can be considered as a modified chirp rate:

$$E(f, f_\tau) \cong \frac{4\pi R_0}{\lambda} \cdot \sqrt{1 - \frac{f_\tau^2 \lambda^2}{4v^2}} + \frac{4\pi R_0}{\lambda} \cdot \frac{f}{f_0} \cdot \frac{1}{\sqrt{1 - \frac{f_\tau^2 \lambda^2}{4v^2}}} + \frac{\pi f^2}{k_r'} \qquad (3.245)$$

with

$$\frac{1}{k_r'} = \frac{1}{k_r} \cdot \left[1 - \frac{k_r 2R_0}{c \cdot \left(1 - \frac{f_\tau^2 \lambda^2}{4v^2}\right)^{\frac{3}{2}}} \cdot \frac{1}{f_0} \cdot \left(\frac{f_\tau^2 \lambda^2}{4v^2} + \frac{1}{4} \cdot \left(\frac{f}{f_0}\right)^2 + \frac{f}{f_0} \right) \right] \tag{3.246}$$

approximatively is valid because of $f/f_0 \ll 0$:

$$\frac{1}{k_r'} = \frac{1}{k_r} \cdot \left[1 - \frac{k_r 2R_0 \cdot \frac{f_\tau^2 \lambda^2}{4v^2}}{c \cdot f_0 \cdot \left(1 - \frac{f_\tau^2 \lambda^2}{4v^2}\right)^{\frac{3}{2}}} \right] \tag{3.247}$$

and it follows for:

$$k_r' = k_r \cdot \left[1 - \frac{k_r 2R_0 \cdot \frac{f_\tau^2 \lambda^2}{4v^2}}{c \cdot f_0 \cdot \left(1 - \frac{f_\tau^2 \lambda^2}{4v^2}\right)^{\frac{3}{2}}} \right]^{-1} \tag{3.248}$$

The modified chir -rate will play an important role in the process of SAR processing.

3.5.4 Description of the modified point-target spectrum

The starting point equation (3.231) changes with the modified chirp rate to:

$$R_T(f, f_\tau, R_0, \tau_0) = \sigma(R_0, \tau_0) \cdot \text{rect}\left(\frac{f}{B_r}\right) \cdot \text{rect}\left[\frac{f_\tau - f_c}{B_{az}}\right] \cdot \frac{1}{\sqrt{k_r}}$$

$$\cdot \sqrt{\frac{c \cdot R_0}{2}} \cdot \frac{1}{v} \cdot \frac{f + f_0}{\left[(f + f_0)^2 - \frac{f_\tau^2 \cdot c^2}{4 \cdot v^2}\right]^{3/4}} \cdot e^{-j \cdot 2\pi f_\tau \tau_0}$$

$$\cdot e^{-j4\pi \frac{R_0}{\lambda} \cdot \left(\sqrt{1 - \frac{f_\tau^2 \lambda^2}{4v^2}} + \frac{f}{f_0} \cdot \frac{1}{\sqrt{1 - \frac{f_\tau^2 \lambda^2}{4v^2}}} \right)} \cdot e^{-j \frac{\pi f^2}{k_r'}} \tag{3.249}$$

With the description of the point-target response in the 2D-frequency domain according to equation (3.249) we solve the inverse Fourier integral:

$$R_T(t, f_\tau, R_0, \tau_0) = \int_{-\infty}^{\infty} R_T(f, f_\tau, R_0, \tau_0) \cdot e^{j2\pi ft} \cdot df$$

$$R_T(t, f_\tau, R_0, \tau_0) = \int_{-\infty}^{\infty} \sigma(R_0, \tau_0) \cdot \text{rect}\left(\frac{f}{B_r}\right) \cdot \text{rect}\left(\frac{f_\tau - f_c}{B_{az}}\right) \cdot \frac{1}{\sqrt{k_r}}$$

$$\cdot \sqrt{\frac{c \cdot R_0}{2}} \cdot \frac{1}{v} \cdot \frac{f + f_0}{\left[(f + f_0)^2 - \dfrac{f_\tau^2 \cdot c^2}{4 \cdot v^2}\right]^{3/4}} \cdot e^{-j \cdot 2\pi f_\tau \tau_0}$$

$$\cdot \exp\left[-j4\pi \frac{R_0}{\lambda} \cdot \left(\sqrt{1 - \frac{f_\tau^2 \lambda^2}{4v^2}} + \frac{f}{f_0} \cdot \frac{1}{\sqrt{1 - \dfrac{f_\tau^2 \lambda^2}{4v^2}}}\right)\right]$$

$$\cdot \exp\left(-j\frac{\pi f^2}{k_r'}\right) \cdot e^{j2\pi ft} \cdot df \qquad (3.250)$$

3.5.5 Evaluation of the point-target spectrum in the range-doppler domain

Again we use the principle of the stationary phase to solve the integral. Let us simplify at the beginning as follows:

$$\Phi(f) = +4\pi \frac{R_0}{\lambda} \cdot \left(\sqrt{1 - \frac{f_\tau^2 \lambda^2}{4v^2}} + \frac{f}{f_0} \cdot \frac{1}{\sqrt{1 - \dfrac{f_\tau^2 \lambda^2}{4v^2}}}\right) + \frac{\pi f^2}{k_r'} - 2\pi ft \qquad (3.251)$$

3.5.5.1 Usage of method of stationary phase

According to the principle of stationary phase we begin:

$$\Phi(f) \cong \Phi(f^*) + \frac{1}{2} \cdot \ddot{\Phi}(f^*) \cdot (f - f^*)^2$$

$$I = \int\limits_{-\infty}^{\infty} \exp\left[1 - j4\pi\frac{R_0}{\lambda} \cdot \left(\sqrt{1 - \frac{f_\tau^2\lambda^2}{4v^2}} + \frac{f}{f_0} \cdot \frac{1}{\sqrt{1 - \frac{f_\tau^2\lambda^2}{4v^2}}}\right)\right]$$
$$\cdot \exp\left(-j\frac{\pi f^2}{k_r'}\right) \cdot e^{j2\pi ft} \cdot df \tag{3.252}$$

$$I = \int\limits_{-\infty}^{\infty} e^{-\Phi(f)} \cdot df \tag{3.253}$$

$$I = \int\limits_{-\infty}^{\infty} e^{-\Phi(f^*)} \cdot e^{-\frac{1}{2}\cdot\ddot{\Phi}(f^*)\cdot(f-f^*)^2} \cdot df \tag{3.254}$$

Corresponding method to chapter 2 SAR basics offers the following solution:

$$I = e^{-\Phi(f^*)} \cdot \frac{\sqrt{2} \cdot \sqrt{\pi}}{\sqrt{\ddot{\Phi}(f^*)}} \cdot e^{-\frac{\pi}{4}} \tag{3.255}$$

According to equation (3.255) we need the respective phase derivative, the stationary point and the corresponding phase derivatives in the stationary point.

Let us remember equation (3.251)

$$\Phi(f) = +4\pi\frac{R_0}{\lambda} \cdot \left(\sqrt{1 - \frac{f_\tau^2\lambda^2}{4v^2}} + \frac{f}{f_0} \cdot \frac{1}{\sqrt{1 - \frac{f_\tau^2\lambda^2}{4v^2}}}\right) + \frac{\pi f^2}{k_r'} - 2\pi ft$$

The first phase derivative is:

$$\dot{\Phi}(f) = \frac{4\pi R_0}{c \cdot \sqrt{1 - \frac{f_\tau^2\lambda^2}{4v^2}}} + \frac{2\pi f}{k_r'} - 2\pi t \tag{3.256}$$

The second phase derivative is:

$$\ddot{\Phi}(f) = \frac{2\pi}{k_r'} \tag{3.257}$$

The stationary point results in:

$$\dot{\Phi}(f^*) = \frac{4\pi R_0}{c \cdot \sqrt{1 - \dfrac{f_\tau^2 \lambda^2}{4v^2}}} + \frac{2\pi f^*}{k_r'} - 2\pi t \stackrel{!}{=} 0 \tag{3.258}$$

$$f^* = k_r' \cdot 2\pi t - \frac{4\pi R_0}{c \cdot \sqrt{1 - \dfrac{f_\tau^2 \lambda^2}{4v^2}}} \cdot \frac{1}{2\pi} \tag{3.259}$$

$$f^* = k_r' \cdot t - \frac{2R_0}{c} \cdot \frac{k_r'}{\sqrt{1 - \dfrac{f_\tau^2 \lambda^2}{4v^2}}} \tag{3.260}$$

Inserting the stationary point equation (3.260) into the phase function equation (3-251) results in:

$$\Phi(f^*) = +4\pi \frac{R_0}{\lambda} \cdot \sqrt{1 - \frac{f_\tau^2 \lambda^2}{4v^2}} - 2\pi f f^* t + f^* \cdot \frac{4\pi R_0}{\lambda \cdot f_0 \sqrt{1 - \dfrac{f_\tau^2 \lambda^2}{4v^2}}} + j \frac{\pi f^{*2}}{k_r'} \tag{3.261}$$

$$\Phi(f^*) = +4\pi \frac{R_0}{\lambda} \cdot \sqrt{1 - \frac{f_\tau^2 \lambda^2}{4v^2}} - 2\pi k_r' t^2 + \frac{4\pi k_r' R_0}{c \cdot \sqrt{1 - \dfrac{f_\tau^2 \lambda^2}{4v^2}}}$$

$$+ \frac{4\pi R_0}{\lambda \cdot f_0 \cdot \sqrt{1 - \dfrac{f_\tau^2 \lambda^2}{4v^2}}} \cdot \left(k_r' t - \frac{2 R_0 k_r'}{c \cdot \sqrt{1 - \dfrac{f_\tau^2 \lambda^2}{4v^2}}} \right)$$

$$+ j \frac{\pi}{k_r'} \left(k_r' t - \frac{2 R_0 k_r'}{c \cdot \sqrt{1 - \dfrac{f_\tau^2 \lambda^2}{4v^2}}} \right)^2 \tag{3.262}$$

$$\Phi(f^*) = +4\pi \frac{R_0}{\lambda} \cdot \sqrt{1 - \frac{f_\tau^2 \lambda^2}{4v^2}} - \pi k_r' \cdot \left(t - \frac{2 R_0}{c \cdot \sqrt{1 - \dfrac{f_\tau^2 \lambda^2}{4v^2}}} \right)^2 \tag{3.263}$$

Inserting the stationary phase equation (3.263) into the integral I equation (3.255) finally results in:

$$I = e^{-j4\pi\frac{R_0}{\lambda}\cdot\sqrt{1-\frac{f_\tau^2\lambda^2}{4v^2}}} \cdot \exp\left[j\pi k_r' \cdot \left(t - \frac{2R_0}{c\cdot\sqrt{1-\frac{f_\tau^2\lambda^2}{4v^2}}}\right)^2\right] \cdot \sqrt{k_r'} \cdot e^{-j\frac{\pi}{4}}$$

$$(3.264)$$

Consideration of the rectangular term in range direction:

$$\text{rect}\left[\frac{f^*}{B_r}\right] = \text{rect}\left[\frac{k_r'}{B_r}\left(t - \frac{2R_0}{c} \cdot \frac{1}{\sqrt{1-\frac{f_\tau^2\lambda^2}{4v^2}}}\right)\right] \tag{3.265}$$

$$\text{rect}\left[\frac{f^*}{B_r}\right] = \text{rect}\left[\frac{k_r'}{k_r \cdot T}\left(t - \frac{2R_0}{c} \cdot \frac{1}{\sqrt{1-\frac{f_\tau^2\lambda^2}{4v^2}}}\right)\right] \tag{3.266}$$

with

$$\frac{k_r'}{k_r} = \left[1 - \frac{k_r 2R_0 \cdot \frac{f_\tau^2\lambda^2}{4v^2}}{c\cdot f_0 \cdot \left(1-\frac{f_\tau^2\lambda^2}{4v^2}\right)^{\frac{3}{2}}}\right]^{-1} . \tag{3.267}$$

3.5.6 The point-target spectrum in the range-doppler domain

Inserting the simplified integral solution equation (3.264) and the modified rectangular function equation (3.266) into the inverse Fourier integral of the approximated point-target spectrum in the 2-dimensional frequency domain equation (3.250) leads to the wanted result, the range-doppler point target (point-target spectrum at the range-doppler domain):

$$R_T(t, f_\tau, R_0, \tau_0) = \sigma(R_0, \tau_0) \cdot \text{rect}\left[\frac{f_\tau - f_c}{B_{az}}\right] \cdot \sqrt{\frac{c \cdot R_0}{2}} \cdot \frac{1}{v} \cdot e^{-j \cdot 2\pi f_\tau \tau_0}$$

$$\cdot \frac{f + f_0}{\left[(f + f_0)^2 - \frac{f_\tau^2 \cdot c^2}{4v^2}\right]^{3/4}} \cdot e^{-j\frac{\pi}{4}}$$

$$\cdot e^{-j\frac{4\pi R_0}{\lambda} \cdot \sqrt{1 - \frac{f_\tau^2 \lambda^2}{4v^2}}} \cdot e^{j\pi k_r' \cdot \left(t - \frac{2R_0}{c \cdot \sqrt{1 - \frac{f_\tau^2 \lambda^2}{4v^2}}}\right)^2}$$

$$\cdot \text{rect}\left[\frac{k_r'}{k_r \cdot T} \cdot \left(t - \frac{2R_0}{c \cdot \sqrt{1 - \frac{f_\tau^2 \lambda^2}{4v^2}}}\right)\right] \tag{3.268}$$

The essential characteristics expressed in equation (2.368) will be discussed at length in the following chapters.

3.6 Summary and interpretation of the results

In chapter 2.2 SAR systems the principles of the procedure are already described that enable to process a SAR-raw data scene.

The task consists in finding a way to describe the exact SAR-sensor movement to put us in the position to originate a focussed image from a 2-dimensional time- respectively distance-variant (range migration) inverse-filtering of the respective raw data.

In chapter 3 SAR-signal processing different facts e.g. the range and azimuth compression of the processing procedures are discussed under different points of view.

A comprehensive description, however, allows the system theoretical description of the SAR sensor respectively its transfer function as a 2-dimensional impulse response – the point-target response.

However, the description of the 2-dimensional conditions, in particular after the transformation into different time- and frequency domains complicates the understanding of essential correlations. For this reason, here is a summary of the most important results.

Transmitting signal

$$s_T(t) = \text{rect}\left(\frac{t}{T}\right) \cdot e^{j\pi k_r t^2}$$

The chirp as a transmitting signal has got especially good characteristics concerning the resolution of illuminated areas. Using this kind of signal in addition with special filtering, the so-called "matched filter", the SNR is substantially improved.

Receiving signal

$$r_T(t, \tau, \tau_0, R_0) = \sigma(\tau_0, R_0) \cdot s_T(t - t_v(\tau, R_0, \tau_0)) \cdot e^{-j2\pi f_0 \cdot t_v(\tau, R_0, \tau_0)} \cdot \text{rect}\left(\frac{\tau - \tau_c}{\Delta\tau}\right)$$

The receiving signal represents the backscattered, distance-dependent and time-delayed signal of the transmitted wave, weighted by the backscattering coefficient.

Point-target response at the 2-dimensional time domain

$$g_{ra_T}(t, \tau, R_0, \tau_0) = \frac{\sigma(R_0, \tau_0) \cdot e^{jp_0} \cdot T}{2} \cdot e^{j2\pi \cdot f_{dc} \cdot (\tau - \tau_0)}$$
$$\cdot \Lambda\left(\frac{t - t_0}{T}\right) \cdot \text{si}\left[\pi B_r(t - t_0) \cdot \Lambda\left(\frac{t - t_0}{T}\right)\right]$$
$$\cdot \Lambda\left(\frac{\tau - \tau_0}{\Delta\tau}\right) \cdot \text{si}\left[\pi B_{az}(\tau - \tau_0) \cdot \Lambda\left(\frac{\tau - \tau_0}{\Delta\tau}\right)\right]$$

After optimal scene processing, e.g. range and azimuth compression, the expected result will be this 2-dimensional pseudo si-function.

Transmitter-chirp spectrum in the frequency domain

$$S_T(f) = \frac{1}{\sqrt{k_r}} \cdot \text{rect}\left(\frac{f}{B_r}\right) \cdot e^{-j\frac{\pi f^2}{k_r}} \cdot e^{j\frac{\pi}{4}}$$

The usage of the transmitter-chirp spectrum is very common and explains the self-reciprocity of the chirp signals.

Point-target spectrum in the 2-dimensional frequency domain

$$R_T(f, f_\tau, R_0, \tau_0) = S_T(f) \cdot \sigma(R_0, \tau_0) \cdot \text{rect}\left[\frac{f_\tau - f_c}{B_{az}}\right] \cdot \sqrt{\frac{c \cdot R_0}{2}} \cdot \frac{1}{v}$$
$$\cdot \frac{f + f_0}{\left[(f + f_0)^2 - \frac{f_\tau^2 \cdot c^2}{4 \cdot v^2}\right]^{3/4}} \cdot e^{-j \cdot \pi/4} \cdot e^{-j \cdot 2\pi f_\tau \tau_0}$$
$$\cdot e^{-j4\pi \frac{R_0}{c} \cdot \sqrt{(f + f_0)^2 - \frac{f_\tau^2 c^2}{4v^2}}}$$

This point-target spectrum contains all required system parameters to allow exact processing. It describes in particular the time/location-variant presentation of the range migration by the implicit connection of the range and azimuth frequency in the non-linear square-root term of the distance-dependent phase term. As already described, this spectrum is the common starting point of different signal-processing algorithms.

Modified sweep rate

$$k_r' = k_r \cdot \left[1 - \frac{k_r 2 R_0 \cdot \frac{f_\tau^2 \lambda^2}{4v^2}}{c \cdot f_0 \cdot \left(1 - \frac{f_\tau^2 \lambda^2}{4v^2}\right)^{\frac{3}{2}}} \right]^{-1}$$

The so-called modified sweep rate reveals the mistake inevitably done by the necessary approximation to calculate the corresponding point-target responses. Chapter 4 Signal processing algorithms reflects this topic in detail.

Point-target spectrum in the range-doppler domain

$$R_T(t, f_\tau, R_0, \tau_0) = \sigma(R_0, \tau_0) \cdot \text{rect}\left[\frac{f_\tau - f_c}{B_{az}}\right] \cdot \sqrt{\frac{c \cdot R_0}{2}} \cdot \frac{1}{v} \cdot e^{-j \cdot 2\pi f_\tau \tau_0}$$

$$\cdot \frac{f + f_0}{\left[(f + f_0)^2 - \frac{f_\tau^2 \cdot c^2}{4v^2}\right]^{3/4}} \cdot e^{-j\frac{\pi}{4}}$$

$$\cdot e^{-j\frac{4\pi R_0}{\lambda} \cdot \sqrt{1 - \frac{f_\tau^2 \lambda^2}{4v^2}}} \cdot e^{j\pi k_r' \cdot \left(t - \frac{2R_0}{c \cdot \sqrt{1 - \frac{f_\tau^2 \lambda^2}{4v^2}}}\right)^2}$$

$$\cdot \text{rect}\left[\frac{k_r'}{k_r \cdot T} \cdot \left(t - \frac{2R_0}{c \cdot \sqrt{1 - \frac{f_\tau^2 \lambda^2}{4v^2}}}\right)\right]$$

Just as the point-target spectrum in the 2-dimensional frequency domain, also the point-target spectrum in the range-doppler domain describes all system parameters necessary for SAR processing. However, the effect of the time/location-variant range migration is to obtain here as well as in the distance-dependent phase functions and also in the rectangular-formed envelope.

This point-target spectrum could be derivated from the common formulation of the point-target response in the 2D-frequency domain, thus is, considering the modified chirp rate, an identical description and equal to the calculations from [RAN92].

Target of the SAR processing is the focussing of entire (2-dimensional) raw-data scenes. Therefore, in order to find out the whole scene-transmitting functions, the super-position integral over all target points has to be calculated (see Chapter 4).

With a view to greater clarity let us discuss first some basic processing methods towards the respective point-target transfer function.

Signal processing algorithms

4.1 Exact point-target compression

The final result of the spectral derivation in chapter 3.4 Point-target spectrum in 2-dimensional frequency domain is the 2-dimensional point-target spectrum.

$$
R_T(f, f_\tau, R_0, \tau_0) = S_T(f) \cdot \sigma(R_0, \tau_0) \cdot \mathrm{rect}\left[\frac{f_\tau - f_c}{B_{az}}\right] \cdot \sqrt{\frac{c \cdot R_0}{2}} \cdot \frac{1}{v}
$$

$$
\cdot \frac{f + f_0}{\left[(f + f_0)^2 - \frac{f_\tau^2 \cdot c^2}{4v^2}\right]^{3/4}} \cdot e^{-j \cdot \frac{\pi}{4}} \cdot e^{-j \cdot 2\pi f_\tau \tau_0}
$$

$$
\cdot e^{-j \cdot 4\pi \frac{R_0}{c} \sqrt{(f+f_0)^2 - \frac{f_\tau^2 c^2}{4v^2}}}
\tag{4.1}
$$

This term represents the 2-dimensional Fourier transformation of the point-target response (Image 4.1). As already discussed in chapter 3 SAR-signal processing, this formula contains all necessary parameters of the SAR sensor for processing. Proceeding from this 2-dimensional point-target spectrum the possibility of an exact point-target compression in the frequency domain is derivated.

The point-target response in Image 4.1 is described in the 2-dimensional time domain. In range direction it characterises the transmitting chirp. In azimuth-direction it characterises the azimuth-chirp depending on the distance, thus embodies the "brightness" of the corresponding amplitude of a 2-dimensional chirp. As already partially discussed in chapter 3 SAR-signal processing, the exact compression of point targets in the frequency domain and at the location (τ_0, R_0) are explained as follows.

4.1.1 Range compression

After the 2-dimensional Fourier transformation the range-compression, as discussed in chapter 3.2 Geometric resolution, is performed by matched filtering in the frequency domain:

$$
H_{K1}(f) = S_T(f)^*
\tag{4.2}
$$

Image 4.1: Point-target response of a SAR sensor; ERS-1 parameter

$S_T(f)$ is the range-chirp spectrum (transmitted signal).

After the range compression the compressed point-target spectrum is described as follows:

$$G_{T1}(f, f_\tau, R_0, \tau_0) = \sigma(R_0, \tau_0) \cdot \Phi_{SS_T}(f) \cdot \text{rect}\left[\frac{f_\tau - f_c}{B_{az}}\right] \cdot \sqrt{\frac{c \cdot R_0}{2}} \cdot \frac{1}{v}$$

$$\cdot \frac{f + f_0}{\left[(f + f_0)^2 - \frac{f_\tau^2 \cdot c^2}{4v^2}\right]^{3/4}} \cdot e^{-j\cdot\frac{\pi}{4}} \cdot e^{-j\cdot 2\pi f_\tau \tau_0}$$

$$\cdot e^{-j\cdot 4\pi\frac{R_0}{c}\sqrt{(f+f_0)^2 - \frac{f_\tau^2 c^2}{4v^2}}} \tag{4.3}$$

4.1.2 Compensation of the amplitude

The next step describes the elimination of the "constant" amplitude terms by inverse multiplication.

$r_T(t, \tau, R_0, \tau_0)$

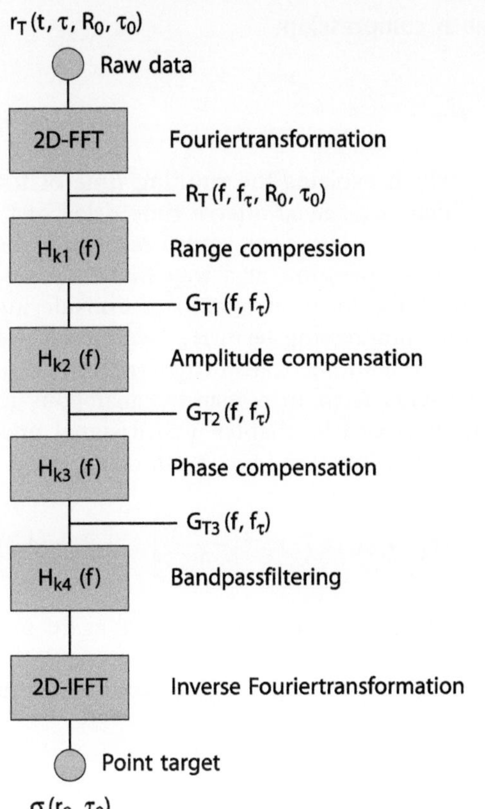

Image 4.2: Exact point-processing

$\sigma(r_0, \tau_0)$

$$H_{K2}(f, f_\tau) = \sqrt{\frac{2}{c \cdot R_0}} \cdot v \cdot \frac{\left[(f + f_0)^2 - \dfrac{f_\tau^2 \cdot c^2}{4v^2} \right]^{\frac{3}{4}}}{f + f_0} \tag{4.4}$$

After elimination of the amplitude term we get as an intermediate result:

$$G_{T2}(f, f_\tau, R_0, \tau_0) = \sigma(R_0, \tau_0) \cdot \Phi_{SS_T}(f) \cdot \text{rect}\left[\frac{f_\tau - f_c}{B_{az}} \right]$$

$$\cdot e^{-j\frac{\pi}{4}} \cdot e^{-j \cdot 2\pi f_\tau \tau_0} \cdot e^{-j \cdot 4\pi \frac{R_0}{c} \sqrt{(f+f_0)^2 - \frac{f_\tau^2 c^2}{4v^2}}} \tag{4.5}$$

4.1.3 Phase compensation: azimuth compression

$$H_{K3}(f, f_\tau) = e^{+j\cdot 4\pi\frac{R_0}{c}\sqrt{(f+f_0)^2 - \frac{f_\tau^2 c^2}{4v^2}}} \cdot e^{j\frac{\pi}{4}} e^{-j2\pi\cdot\frac{2\cdot R_0}{c}\cdot(f+f_0)^2} \tag{4.6}$$

Please note the term $e^{-j2\pi\cdot\frac{R\cdot R_0}{c}\cdot(f+f_0)}$ which explains the running time of the signal in the band pass domain which is received after a time delay of t_v. The term allows the start-stop approximation used whilst recording the data to become part of the azimuth compression in a way that the compressed point target can be found at the location $t_0 = 2R_0/c$. Considering the non-linear phase function of the compressing term H_{K3} we merely see the complex conjugated part of the remaining transformation term G_{T2}. Thus, the azimuth compression presents itself in a manner analogous to the range compression (as already discussed in chapter 3 SAR-signal processing) and is described as a correlation function in azimuth direction.

$$G_{T3}(f, f_\tau, R_0, \tau_0) = \sigma(R_0, \tau_0) \cdot \Phi_{SS_T}(f) \cdot \Phi_{AA_T}(f_\tau) \cdot \mathrm{rect}\left[\frac{f_\tau - f_c}{B_{az}}\right] \tag{4.7}$$

Thus, the point-target spectrum contains approximatively (once again under the assumption that the distance trend follows a parabolic term) the autocorrelation function of a chirp signal, both in range direction and in azimuth direction.

4.1.4 Band pass filtering

As last step has to be extracted the part being of interest from the total spectral range by band pass filter:

$$H_{K4}(f, f_\tau) = \mathrm{rect}\left[\frac{f_\tau - f_c}{B_{az}}\right] \tag{4.8}$$

The point-target result can be described as follows:

$$\boxed{\sigma(R_0, \tau_0) = \sigma(R_0, \tau_0) \cdot \varphi_{SS_T}(t) \cdot \varphi_{AA_T}(\tau)} \tag{4.9}$$

After the band pass filtering and the 2-dimensional inverse Fourier transformation appears the expected backscattering coefficient weighted with the corresponding autocorrelation functions.

Thus, the exact point-target compression preferably is a 2-dimensional autocorrelation function which is pretty similar to a si-function. As the optimal compression's result for the point-target response of a SAR sensor (Image 4.1) Image 4.3 illustrates the 2d-autocorrelation function (please have a look also at chapter 3.3 Azimuth compression and Image 4.16).

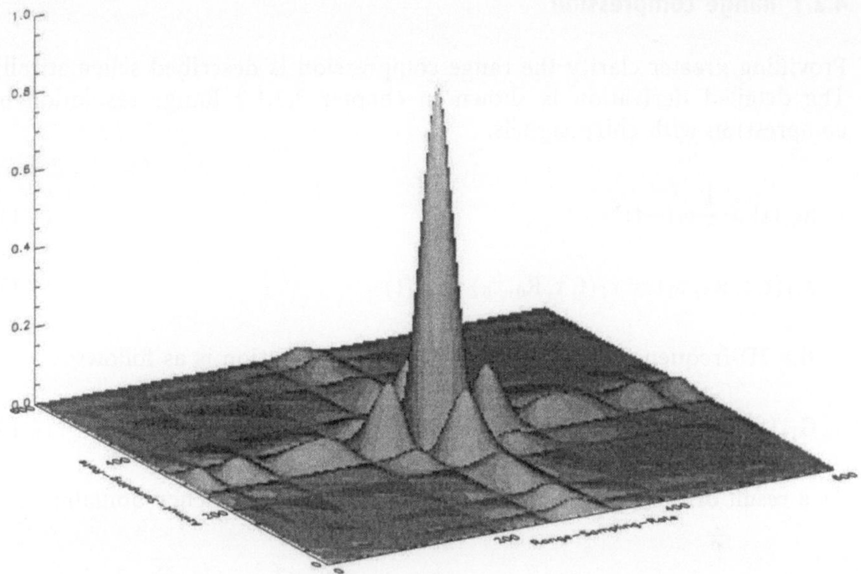

Image 4.3: Cutout of a range- and azimuth-compressed point target

4.2 The range-doppler processor

The range-doppler processor is an approximation of the exact compression algorithm, but can according to the so-called Secondary Range Compression (SRC) be considered as a very good approximation. The naming of the range-doppler processor specifies in particular the transformation domains the corresponding compression is performed in, especially "range" for the range-time domain (respectively range-location domain) and "doppler" for the azimuth-frequency domain, the so-called doppler domain.

Starting point of the consideration again is the result of the 2-dimensional point-target spectrum, the 2d-Fourier transformation of the echo signal:

$$
\begin{aligned}
R_T(f, f_\tau, R_0, \tau_0) = {} & S_T(f) \cdot \sigma(R_0, \tau_0) \cdot \mathrm{rect}\left[\frac{f_\tau - f_c}{B_{az}}\right] \cdot \sqrt{\frac{c \cdot R_0}{2}} \cdot \frac{1}{v} \\
& \cdot \frac{f + f_0}{\left[(f + f_0)^2 - \dfrac{f_\tau^2 \cdot c^2}{4v^2}\right]^{3/4}} \cdot e^{-j \cdot \frac{\pi}{4}} \cdot e^{-j \cdot 2\pi f_\tau \tau_0} \\
& \cdot e^{-j \cdot 4\pi \frac{R_0}{c}\sqrt{(f+f_0)^2 - \frac{f_\tau^2 c^2}{4v^2}}}
\end{aligned}
\tag{4.10}
$$

4.2.1 Range compression

Providing greater clarity the range compression is described schematically. The detailed derivation is shown in chapter 3.2.1.2 Range resolution by compression with chirp signals.

$$h_{K1}(t) = \frac{1}{2} s_T(-t)^*$$ (4.11)

$$g_{T1}(t, \tau, R_0, \tau_0) = r_T(t, \tau, R_0, \tau_0) \cdot h_{K1}(t)$$ (4.12)

In the 2D-frequency domain the convolution operation is as follows:

$$G_{T1}(f, f_\tau, R_0, \tau_0) = R_T(f, f_\tau, R_0, \tau_0) \cdot H_{K1}(f)$$ (4.13)

As a result of the range compression we get in the frequency domain:

$$G_{T1}(f, f_\tau, R_0, \tau_0) = \sigma(R_0, \tau_0) \cdot \Phi_{SS_T}(f) \cdot \text{rect}\left[\frac{f_\tau - f_c}{B_{az}}\right] \cdot \sqrt{\frac{c \cdot R_0}{2}} \cdot \frac{1}{v}$$
$$\cdot \frac{f + f_0}{\left[(f + f_0)^2 - \dfrac{f_\tau^2 \cdot c^2}{4v^2}\right]^{3/4}} \cdot e^{-j\cdot\frac{\pi}{4}} \cdot e^{-j\cdot 2\pi f_\tau \tau_0}$$
$$\cdot e^{-j\cdot 4\pi \frac{R_0}{c} \sqrt{(f+f_0)^2 - \frac{f_\tau^2 c^2}{4v^2}}}$$ (4.14)

4.2.2 Compensation of the amplitude

The exact amplitude compensation

$$H_{K2}(f, f_\tau) = \sqrt{\frac{2}{c \cdot R_0}} \cdot v \cdot \frac{\left[(f + f_0)^2 - \dfrac{f_\tau^2 \cdot c^2}{4v^2}\right]^{\frac{3}{4}}}{f + f_0}$$ (4.15)

is approximated by

$$\frac{\left[(f + f_0)^2 - \dfrac{f_\tau^2 \cdot c^2}{4v^2}\right]^{\frac{3}{4}}}{f + f_0} \cong \frac{\left[(f + f_0)^2\right]^{\frac{3}{4}}}{f + f_0} = \sqrt{f + f_0}$$ (4.16)

to a simplified, constant amplitude term:

$$H_{K2}(f) \cong \sqrt{\frac{2}{c \cdot R_0}} \cdot v \cdot \sqrt{f + f_0} \cong \sqrt{\frac{2f_0}{c \cdot R_0}} \cdot v = \text{const.} \tag{4.17}$$

4.2.3 Phase compensation: azimuth compression

The task of the azimuth compression, after all, is compensating the known phase term:

$$\Phi(f, f_\tau, R_0) = -4\pi \frac{R_0}{c} \sqrt{(f + f_0)^2 - \frac{f_\tau^2 c^2}{4v^2}} \tag{4.18}$$

Again, note the start-stop approximation and the performance in the real band pass domain. Thus, we receive as compression transfer function:

$$\Phi_K(f, f_\tau, R_0) = 4\pi \frac{R_0}{c} \sqrt{(f + f_0)^2 - \frac{f_\tau^2 c^2}{4v^2}} - 2\pi \cdot \frac{2R_0}{c} \cdot (f + f_0) \tag{4.19}$$

Further:

$$\Phi_K(f, f_\tau, R_0) = 4\pi \frac{R_0 \cdot (f + f_0)}{c} \cdot \left[\sqrt{1 - \frac{f_\tau^2 c^2}{4v^2 \cdot (f + f_0)^2}} - 1 \right] \tag{4.20}$$

Equation (4.20) is now replaced by an approximation:

$$\Phi_K(f, f_\tau, R_0) = \Phi_{K0}(f_\tau, R_0) + \Phi_{K1}(f_\tau, R_0) \cdot f + \Phi_{K2}(f_\tau, R_0) \cdot f^2 + \ldots \tag{4.21}$$

At this, Φ_{K0} can be regarded as azimuth-correlation term needed for focussing. The linear term (in range-frequency direction) Φ_{K1} characterizes the range migration (which has to be compensated). The quadratic term in range direction Φ_{K2} contains sort of a pulse dispersion (phase distortion) of the transmitting pulse, which declines when increasing the squint angle [RAN94].

The corresponding factors are determined as follows. The respective necessary compression functions are determined as appropriately conjugated phase functions:

Azimuth-correlation term:

$$\boxed{\Phi_{K0}(f_\tau, R_0) = 4\pi \frac{R_0 \cdot f_0}{c} \cdot \left[\sqrt{1 - \frac{f_\tau^2 c^2}{4v^2 \cdot f_0^2}} - 1 \right]} \tag{4.22}$$

Range migration:

$$\Phi_{K1}(f_\tau, R_0) = \frac{\partial \Phi_K(f, f_\tau, R_0)}{\partial f} = 4\pi \frac{R_0 \cdot (f + f_0)}{c} \cdot \left[\sqrt{(f + f_0)^2 - \frac{f_\tau^2 c^2}{4v^2}} \right]^{-1} - 4\pi \frac{R_0}{c}$$

(4.23)

Considering at this point f=0 results in:

$$\boxed{\Phi_{K1}(f_\tau, R_0) = 4\pi \frac{R_0}{c} \cdot \left[\left(\sqrt{1 - \frac{f_\tau^2 c^2}{4v^2 f_0^2}} \right)^{-1} - 1 \right]}$$

(4.24)

The range migration, also called range-cell migration (RCM), additionally contains the information of the correct point position in range and is interpreted alternatively to equation (4.24) as the running time of the point-target response as function of the azimuth frequency:

$$\Phi_{K1}(f_\tau, R_0) = 2\pi \cdot \left[\frac{2R_0}{c \cdot \sqrt{1 - \frac{f_\tau^2 c^2}{4v^2 \cdot f_0^2}}} - \frac{2R_0}{c} \right] = 2\pi \cdot [\tau_m(f_\tau, R_0) - t_0(R_0)]$$

(4.25)

Now we can recognize the point-target position in the term t_0 and the so-called range-curvature factor as τ_m [DAV94]:

$$\tau_m(f_\tau, R_0) = \frac{2R_0}{c \cdot \sqrt{1 - \frac{f_\tau^2 c^2}{4v^2 \cdot f_0^2}}}$$

(4.26)

The whole range-migration term equation (4.25) depends on the corresponding range distance – and thus is range-dependent. This range dependency unfortunately is not an independent parameter in the 2-dimensional frequency domain. If some backscattering signals are reflected and recorded from different targets the corresponding transformations will overlap, too. Single point targets can well be processed exactly, however, concerning receiving-signal overlappings only an approximative compression in the 2-dimensional frequency range is possible [RAN94].

The interpretation of the range migration as a running-time term of the point-target response equation (4.26) describes the range migration as position displacement of the signal envelope. It is a term of the azimuth frequency and the range distance and can be determined for each point.

As a result we receive the so-called range parabolas (see also chapter 3.3 Azimuth compression), which themselves feature different gradients in range direction.

4.2.4 Secondary range compression

So far, the quadratic factor of eq (4.21) has been neglected. To derivate the secondary range compression (SRC), the quadratic part of the approximation has to be considered more closely:

$$\Phi_{K2}(f, f_\tau, R_0) = \ddot{\Phi}(f, f_\tau, R_0) = \frac{\partial}{\partial f}\left\{4\pi\frac{R_0}{c} \cdot \left[\left(1 - \frac{f_\tau^2 c^2}{4v^2 \cdot f_0^2}\right)^{-1/2} - 1\right]\right\}$$

(4.27)

$$\Phi_{K2}(f, f_\tau, R_0)|_{f=0} = -\frac{2 \cdot \pi R_0 \cdot f_\tau^2 c^2}{c \cdot \left[1 - \dfrac{f_\tau^2 c^2}{4v^2 \cdot f_0^2}\right]^{3/2} \cdot 4v^2 \cdot f_0^3}$$

(4.28)

At the point $f = 0$ is:

$$\Phi_{K2}(f_\tau, R_0) = -\frac{\pi R_0 \cdot c}{2v^2 \cdot f_0^3} \cdot \frac{f_\tau^2}{\left[1 - \dfrac{f_\tau^2 c^2}{4v^2 \cdot f_0^2}\right]^{3/2}}$$

(4.29)

This quadratic term is not compensated without considering the SRC during SAR processing. It depends very much on the azimuth frequency whereas concerning the distance shows only a linear correlation. In particular the strong doppler dependency (azimuth frequency) results in defocussing effects at high squint angles. Correcting this phase dispersion is called secondary range compression (SRC).

The phase dispersion is not a term of the modulation rates k_n, it is in fact a geometrically founded term, which can be reduced to the loss of the orthogonality between range- and azimuth-signal components particularly at big squint angles [RAN94]. Although equation (4.29) is not a function of the range-sweep rate k_r, correcting this phase dipersion can be done by adapting the range-sweep rate.

Let us now have a look at a resulting chirp rate k_r consisting of the original chirp rate plus the derivated phase dispersion:

$$\frac{\pi}{k_r'} = \frac{\pi}{k_r} + \Phi_{K2}(f_\tau, R_0)$$

(4.30)

$$\frac{1}{k_r'} = \frac{1}{k_r} + \frac{\Phi_{K2}(f_\tau, R_0)}{\pi}$$

$$\frac{1}{k_r'} = \frac{1}{k_r}\left[1 - \frac{R_0 \cdot c}{2v^2 \cdot f^3} \cdot \frac{k_r \cdot f_\tau^2}{\left[1 - \frac{f_\tau^2 c^2}{4v^2 \cdot f_0^2}\right]^{3/2}}\right] \tag{4.31}$$

$$\boxed{\frac{1}{k_r'} = \frac{1}{k_r}\left[1 - \frac{2R_0}{f_0 \cdot c} \cdot \frac{k_r \cdot \frac{f_\tau^2 \cdot \lambda^2}{4v^2}}{\left[1 - \frac{f_\tau^2 \lambda^2}{4v^2}\right]^{3/2}}\right]} \tag{4.32}$$

If the chirp rate is adapted corresponding to equation (4.32) during the range-compression, in other words the phase trend is pre-distorted corresponding to the expected phase dispersion, the addressed geometric effects will be corrected and a secondary range compression will be realised.

After derivation of the range-doppler processor out of the 2-dimensional frequency domain we get the following block diagram (concerning the point-target compression). Proceeding from the received raw data, the number of operations described in Image 4.4 the different phase terms are egalised to gain a processed SAR scene. The range-doppler processor with secondary range compression is a pretty good approximation towards the exact processor!

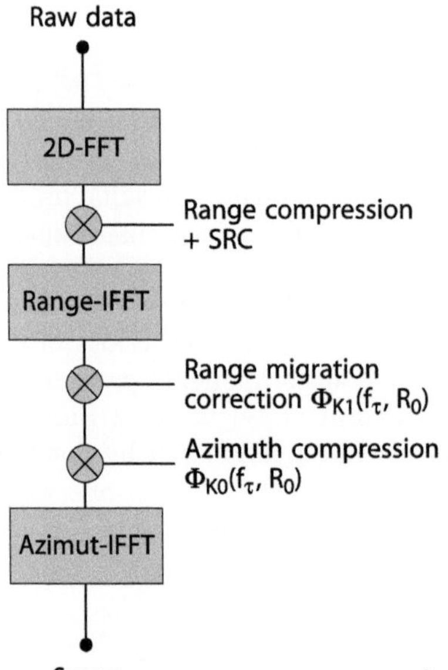

Image 4.4: Range-doppler processor

4.3 The SIFT processor

In the previous chapters we discussed the range and azimuth compression particularly concerning point targets. Now we will have a close look at the problem of processing whole scenes based on the example of the scaled inverse Fourier transformation processor (SIFT processor). For this purpose let us have a look at the series of Images on page 14, Image 1.13. The task of a SAR processor can clearly be recognised, the task of calculating a focussed scene (Image 1.13) out the received raw data (Image 1.12).

The range compression can be performed without any problems as the transmitting signal and its necessary parameters needed for processing are very well known. Implementing the azimuth compression is more difficult because of the different range parabolas under the influence of different point targets, in other words due to the range-based variance. Last but not least because of this reason, SAR processors differ substantially from each other because of their different way of implementing the azimuth-compression.

Both compressions can be performed either in the frequency domain or in the time domain. The developed novel SIFT processor performs the range compression in the frequency domain as a matched-filter operation.

4.3.1 Spectrum of the whole scene and first steps towards compression

To illustrate the ideas, which led to the development of the Scaled-Inverse-Fourier-Transform Processor (SIFT), let us briefly remember the necessary steps towards the starting point of signal processing. After that let us start with the topic of 2-dimensional range variant signal processing with all its mathematical steps and system theoretical ideas.

4.3.1.1 Derivation of the whole scene spectrum

To calculate the 2-dimensional point-target spectrum, the received backscattered signal is described with the range-dependent time delay:

$$t_v(\tau) = \frac{2R(\tau)}{c} = t_v(\tau, R_0, \tau_0)$$

Thus we received:

$$r_T(t, \tau, R_0, \tau_0) = \sigma(R_0, \tau_0) \cdot s_T(t - t_v(\tau, R_0, \tau_0)) \cdot e^{-j2\pi f_0 t_v(\tau, R_0, \tau_0)} \cdot \mathrm{rect}\left(\frac{\tau - \tau_c}{\Delta\tau}\right)$$

Illustrating the spectrum we had to fourier-transform twice:

$$R_T(f, f_\tau, R_0, \tau_0) = \mathscr{F}_t \mathscr{F}_\tau \{r_T(t, \tau, R_0, \tau_0)\}$$

$$R_T(f, f_\tau, R_0, \tau_0) = \int\limits_{-\infty}^{\infty} \int\limits_{-\infty}^{\infty} \sigma(R_0, \tau_0) \cdot s_T(t - t_v(\tau, R_0, \tau_0)) \cdot e^{-j2\pi f_0 t_v(\tau, R_0, \tau_0)}$$

$$\cdot \, \text{rect}\left(\frac{\tau - \tau_c}{\Delta \tau}\right) \cdot e^{-j2\pi f t} \cdot e^{-j2\pi f_\tau \tau} \cdot dt \cdot d\tau$$

After solving the integral, the 2-dimensional point-target spectrum of the receiving signal of a point target could be formulated at the location (τ_0, R_0) in the equivalent low pass domain as:

$$R_T(f, f_\tau, R_0, \tau_0) = S_T(f) \cdot \sigma(R_0, \tau_0) \cdot \text{rect}\left[\frac{f_\tau - f_c}{B_{az}}\right] \cdot \sqrt{\frac{c \cdot R_0}{2}} \cdot \frac{1}{v}$$

$$\cdot \, \frac{f + f_0}{\left[(f + f_0)^2 - \dfrac{f_\tau^2 \cdot c^2}{4 \cdot v^2}\right]^{3/4}} \cdot e^{-j \cdot \pi/4} \cdot e^{-j \cdot 2\pi f_\tau \tau_0}$$

$$\cdot \, e^{-j4\pi \frac{R_0}{c} \sqrt{(f+f_0)^2 - \frac{f_\tau^2 c^2}{4v^2}}}$$

Regarding the whole scene, the point-target spectrum must be integrated via all of the illuminated back scattering targets:

$$\sigma(R, \tau) = \int\limits_{-\infty}^{\infty} \int\limits_{-\infty}^{\infty} \sigma(R_0, \tau_0) \cdot \delta(R - R_0, \tau - \tau_0) \cdot dR_0 \cdot d\tau_0 \tag{4.33}$$

Enlarging the SAR system:

$$\sigma(R_0, \tau_0) \cdot \delta(R - R_0, \tau - \tau_0) \quad \text{O—[SAR scene]—O} \quad R_T(f, f_\tau, R_0, \tau_0)$$

by the integral illustrated in equation (4.33), the result for the whole SAR scene is:

$$\sigma(R, \tau) \quad \text{O—[SAR scene]—O} \quad S_T(f, f_\tau) = \int\limits_{-\infty}^{\infty} \int\limits_{-\infty}^{\infty} R_T(f, f_\tau, R_0, \tau_0) \cdot dR_0 \cdot d\tau_0$$

The spectrum of an entire SAR scene in the equivalent low passdomain is represented by:

$$S_T(f, f_\tau) = S_T(f) \cdot \text{rect}\left[\frac{f_\tau - f_c}{B_{az}}\right] \cdot \sqrt{\frac{c}{2}} \cdot \frac{1}{v} \cdot \frac{f + f_0}{\left[(f + f_0)^2 - \frac{f_\tau^2 \cdot c^2}{4v^2}\right]^{3/4}} \cdot e^{-j \cdot \pi/4}$$

$$\cdot \int\limits_{-\infty}^{\infty} \int\limits_{-\infty}^{\infty} \sqrt{R_0} \cdot \sigma(R_0, \tau_0) e^{-j \cdot 2\pi f_\tau \tau_0} \cdot e^{-j4\pi\frac{R_0}{c} \cdot \sqrt{(f+f_0)^2 - \frac{f_\tau^2 c^2}{4v^2}}} \cdot dR_0 \cdot d\tau_0$$

$$(4.34)$$

Traditional SAR processors working in the frequency-domain use equation (4.34) as a starting point, to perform an azimuth compression with the help of the so-called "Stolt-interpolation" [STO78]. We will start via a scaled Fourier transformation. First of all, equation (4.34) has to be transformed.

We have a look at the integral:

$$I(f, f_\tau) = \int\limits_{-\infty}^{\infty} \int\limits_{-\infty}^{\infty} \sqrt{R_0} \cdot \sigma(R_0, \tau_0) \cdot e^{-j2\pi f_\tau \tau_0} \cdot e^{-j4\pi\frac{R_0}{c} \cdot \sqrt{(f+f_0)^2 - \frac{f_\tau^2 c^2}{4v^2}}} \cdot dR_0 \cdot d\tau_0$$

$$(4.35)$$

If the absolute distance R_0 is substituted by the average distance R_m, which varies by the change in distance r, the factor sqrt(R_0) can be preferred within the integral (4.35), because the change in distance r is neglegibly small in contrast to the average distance R_m. Equation (4.34) reduces itself by $R_0 = R_m + r$ and $dR_0 = dr$ to:

$$I(f, f_\tau) = \sqrt{R_m} \cdot \int\limits_{-\infty}^{\infty} \int\limits_{-\infty}^{\infty} \sigma(\{R_m + r\}, \tau_0) \cdot e^{-j2\pi f_\tau \tau_0} \cdot e^{-j4\pi\frac{\{R_m + r\}}{c} \cdot \sqrt{(f+f_0)^2 - \frac{f_\tau^2 c^2}{4v^2}}}$$

$$\cdot dr \cdot d\tau_0 \qquad (4.36)$$

The integration via τ_0 is equivalent to a Fourier-transformated integral which can be used to discribe equation (4.36) as follows:

$$I(f, f_\tau) = \sqrt{R_m} \cdot e^{-j4\pi\frac{R_m}{c} \cdot \sqrt{8f+f_0)^2 - \frac{f_\tau^2 c^2}{4v^2}}} \cdot \int\limits_{-\infty}^{\infty} \sigma(\{R_m + r\}, f_\tau)$$

$$\cdot e^{-j4\pi\frac{r}{c} \cdot \sqrt{(f+f_0)^2 - \frac{f_\tau^2 c^2}{4v^2}}} dr \qquad (4.37)$$

To simplify we set:

$$f_r(f, f_\tau) = \frac{2}{c} \cdot \sqrt{(f + f_0)^2 - \frac{f_\tau^2 c^2}{4v^2}} \tag{4.38}$$

$$I(f, f_\tau) = \sqrt{R_m} \cdot e^{-j2\pi R_m \cdot f_r(f, f_\tau)} \cdot \int_{-\infty}^{\infty} \sigma(\{R_m + r\}, f_\tau) \cdot e^{-j2\pi f_r(f, f_\tau)r} \cdot dr \tag{4.39}$$

After the Fourier transformation and after using the displacement proposition of the Fourier transformation, the result is:

$$\begin{aligned} I(f, f_\tau) &= \sqrt{R_m} \cdot e^{-j2\pi R_m \cdot (f, f_\tau)} \cdot \sigma(f_r(f, f_\tau), f_\tau) \cdot e^{+j2\pi R_m f_r(f, f_\tau)} \\ &= \sqrt{R_m} \cdot \sigma(f_r(f, f_\tau), f_\tau) \end{aligned} \tag{4.40}$$

The whole spectrum of a SAR scene can now be described as follows:

$$\begin{aligned} S_T(f, f_\tau) = S_T(f) \cdot \mathrm{rect}\left[\frac{f_\tau - f_c}{B_{az}}\right] \cdot \sqrt{\frac{c}{2} \cdot \frac{1}{v}} \cdot \frac{f + f_0}{\left[(f + f_0)^2 - \frac{f_\tau^2 \cdot c^2}{4v^2}\right]^{3/4}} \cdot e^{-j\cdot\pi/4} \\ \cdot \sqrt{R_m} \cdot \sigma\left(\frac{2}{c} \cdot \sqrt{(f + f_0)^2 - \frac{f_\tau^2 \cdot c^2}{4v^2}}, f_\tau\right) \end{aligned} \tag{4.41}$$

This contains the spectrum of the complete backscattering coefficient:

$$\sigma\left(\frac{2}{c} \cdot \sqrt{(f + f_0)^2 - \frac{f_\tau^2 \cdot c^2}{4v^2}}, f_\tau\right) = \sigma(f_r(f, f_\tau), f_\tau) \tag{4.42}$$

with

$$f_r(f, f_\tau) = \frac{2}{c} \cdot \sqrt{(f + f_0)^2 - \frac{f_\tau^2 \cdot c^2}{4v^2}}$$

Within the spectrum according to equation (4.42) we discover in contrast to equation (4.36) that there is a special significance to the total backscattering coefficient respectively its spectrum as it can not be transformed back in the range time domain by taking the conventional way because of its dependency on both the range frequency f and the azimuth frequency f_τ.

To keep the coming considerations on a level easy to understand, let us introduce now a first compression term.

4.3.1.2 First compression-transfer function

The task of the first compression-transfer function on one hand is the range compression and the other the amplitude correction. For this purpose, the total spectrum of the SAR scene has to be multiplied with the complex conjugate spectrum of the transmitting chirp and additionally with an amplitude-correction term. The resulting compression-transfer function is:

$$H_{k1_T}(f, f_\tau, R_m) = S_T(f)^* \cdot \frac{\left[(f + f_0)^2 - \dfrac{f_\tau^2 c^2}{4v^2}\right]^{3/4}}{f + f_0} \cdot e^{j\frac{\pi}{4}} \cdot v \cdot \sqrt{\frac{2}{c}} \cdot \frac{1}{\sqrt{R_m}} \qquad (4.43)$$

After the first compression we receive the following result:

$$\begin{aligned} G_{1_T}(f, f_\tau) &= S_T(f, f_\tau) \cdot H_{k1_T}(f, f_\tau, R_m) \\ &\approx \sigma(f_r(f, f_\tau), f_\tau) \cdot \mathrm{rect}\left(\frac{f_\tau - f_c}{B_{az}}\right) \cdot \mathrm{rect}\left(\frac{f}{B_r}\right) \end{aligned} \qquad (4.44)$$

Please note:

$$S_T(f) \cdot S_T(f)^* \cong \mathrm{rect}\left(\frac{f}{B_r}\right) \qquad \text{(see chapter 2.1.2.3)}$$

Equation (4.44) reveals that the spectrum of the total backscattering coefficient equation (4.42) will be kept after the first compression function without any further amplitude-weighting functions. The remaining operation leading to a correctly compressed SAR image is the azimuth compression.

4.3.2 Basic ideas for further processing

The spectrum of a point target depends directly on its position. Thus, the compression function for focussing this point target depends on the location of this very point target. A point target with the shortest distance R_{01} has to be compressed with another transfer function as one with the shortest distance R_{02}. If all point targets were focussed with the same transfer function, only those point targets within the correct distance of the transfer function would be transferred properly. All other point targets would only be incomplete if not focussed at all. The receiving signal in the SAR is an overlay of many closely adjacent point targets of a footprint (the "print" of the earth's surface illuminated by the antenna's lobe). Each point target of the SAR scene shows an own shortest distance R_0 and thus creates (equation (4.34)) a location-dependent share in the 2-dimensional total spectrum. To compress all point targets accurately, it is necessary to treat all these point targets individually. Practically, this is a difficult task if not impossible at all. The usual approximation towards this problem consists of several more or less complex steps depending on how accurate the result is supposed to be. For an arbitrarily exact elimination of the range migration whilst maintaining simultaneous, location-variant signal processing, the following basic ideas will be discussed.

If we regard the whole backscattering spectrum after the first compression function equation (4.44), we set the task (considering the azimuth compression) to re-transform the spectrum of the total backscattering coefficient $\sigma(f_r(f, f_\tau), f_\tau)$ both in the range-time domain and in the azimuth-time domain. Let us regard the radar-backscattering coefficient in the 2-dimensional frequency domain:

$$\sigma(f_r(f, f_\tau), f_\tau) = \sigma\left(\frac{2}{c} \cdot \sqrt{(f + f_0)^2 - \frac{f_\tau^2 \cdot c^2}{4v^2}}, f_\tau\right) \tag{4.45}$$

Here is

$$f_r(f, f_\tau) = \frac{2}{c} \cdot \sqrt{(f + f_0)^2 - \frac{f_\tau^2 \cdot c^2}{4v^2}} \tag{4.46}$$

a non-linear distortion of the range-frequency axis. The frequency f_r in equation (4.46) depends on the actual range frequency f and the azimuth frequency f_τ. The following correlations have already been described:

Range-migration parabola due to different, in particular azimuth-dependent time delay.

Different range-migration parabolas due to different range-point target distances to the SAR carrier.

These correlations, performed merely in the frequency domain, are reflected in equation (4.46). This non-linear, scaled term depends on both the range frequency and the azimuth frequency and thus characterises the range migration. The task of SAR processing is, when derivated in the 2-dimensional frequency domain, to transform the in equation (4.45) illustrated non-linear, scaled backscattering spectral performance into the range-distance domain, azimuth-time domain, in a way that the non-linear scaling is eliminated. In this case, all distortion effects would be corrected and the wanted information, the genuine radar-backscattering coefficient, would be calculated in the corresponding coordinates. As already discussed, equation (4.45) can not be transformed into the range-time domain using the traditional Fourier-back transformation because of equation (4.46). Finding a way that puts us in the position to perform this kind of transformation, let us have a look at the following approach:

Is it possible to find an approximation of the scaled spectrum equation (4.45) in a way that the non-linear distortion acts as a linearised deformation?!

$$\sigma(f_r(f, f_\tau), f_\tau) = \sigma\left(\frac{2}{c} \cdot \sqrt{(f + f_0)^2 - \frac{f_\tau^2 \cdot c^2}{4v^2}}, f_\tau\right) = \sigma(a \cdot f + b \cdot f_0, f_\tau)$$

$$(4.47)$$

And is it possible at the same time to find a Fourier back-transformation being able to fourier-transform inversely and re-scalingly?!

$$h(t, \tau) = \mathfrak{I}_{t(a)}^{-1} \, \mathscr{F}_\tau^{-1}\{H(a \cdot f + b \cdot f_0, f_\tau)\}$$

$$(4.48)$$

Then we could calculate step by step a linearised term $\sigma(af+bf_0, f_\tau)$ from $\sigma(f_r(f, f_\tau), f_\tau)$ via the approximation following equation (4.47) and the wanted time term respectively SAR image $\sigma(t, \tau)$ created by the corresponding back-transformation equation (4.48). In equation (4.47) an approximation is performed which puts the total backscattering coefficient $\sigma(f_r(f, f_\tau), f_\tau)$ to a scaled form in the frequency range $(\sigma(af + bf_0, f_\tau))$.

In fact, both assumptions can be realized and thus a new concept on SAR processing is reality.

4.3.3 Derivation of the "Inverse-Scaled Fourier Transformation"

An essential part of the SIFT-SAR processor is the inverse scaled Fourier transformation. In the following chapter an algorithm is described that re-transforms a scaled signal in the frequency domain in such a way that it is re-scaled in the time domain. In order to perform a comparison let us at first describe the traditional Fourier transformation.

4.3.3.1 Traditional fourier back-transformation

Given is an arbitrary signal $G(f)$ in the Fourier domain containing a peculiarity, a scaling according to equation (4.49) with a scaling factor a:

$$G(f) = S(a \cdot f) \tag{4.49}$$

The back-transformation into the time domain is performed by executing the following operations:

a) Multiplication by the factor $e^{j2\pi ft}$ and following integration via f:

$$g(t) = \int_{-\infty}^{\infty} S(a \cdot f) \cdot e^{j2\pi ft} \cdot df \tag{4.50}$$

b) Substitution of $f' = a \cdot f$:

$$g(t) = \int_{-\infty}^{\infty} S(f') \cdot e^{j2\pi \cdot \left(\frac{f'}{a}\right) \cdot t} \cdot d\left(\frac{f'}{a}\right) = \frac{1}{|a|} \cdot s\left(\frac{t}{a}\right) \tag{4.51}$$

According to equation (4.51) a scaled signal in the frequency domain shows a reciprocally corresponding time axis due to the analogy theorem of the Fourier transformation in the time domain.

SAR signal theory poses the problem to transform a scaled band pass signal in the frequency domain to the time domain. At the same time the scaling of the time axis has to be eliminated (Chapter 4.3.2 Basic ideas). To change the time scaling, Papoulis [PAP68] derivates a calculation rule that transforms a scaled time signal in the time domain into a "re-scaled" time signal. To solve the present problem, however, an algorithm has to be developed that egalises the scaling created in the frequency domain with the help of a corresponding back-transformation.

4.3.3.2 Scaled fourier transformation

4.3.3.2.1 Mathematical formulation

Again, let us have a look at the initial term in the frequency domain:

$$G(f) = S(a \cdot f) \tag{4.52}$$

If we now, other than described in chapter 4.3.3.1 Traditional Fourier transformation, multiply by factor $e^{j2\pi aft}$, and then integrate via f, the result is:

$$g(t) = \int_{-\infty}^{\infty} S(a \cdot f) \cdot e^{j2\pi aft} \cdot df \tag{4.53}$$

After the substitution $f' = a \cdot f$ results:

$$g(t) = \int_{-\infty}^{\infty} S(f') \cdot e^{j2\pi f't} \cdot d\left(\frac{f'}{a}\right) = \frac{1}{|a|} \cdot s(t) \tag{4.54}$$

Equation (4.54) says that a multiplication by a corresponding phase term and the following integration via f, a scaled signal in the frequency domain can indeed be re-scaled into the time domain by a special transformation.

Now the above described signal-theoretical derivation is system-theoretically converted as the indicated algorithm has to be implemented on a computer.

4.3.3.2.2 *Systemtheoretical formulation*

Initial point is the correlation:

$$\frac{1}{|a|} \cdot s(t) = \int\limits_{-\infty}^{\infty} S(a \cdot f) \cdot e^{j2\pi aft} \cdot df = \Im^{-1}\{S(a \cdot f)\} \tag{4.55}$$

If the quadratic supplement of the phase comes into operation, like:

$$e^{j2\pi aft} = e^{j\pi at^2} \cdot e^{j\pi af^2} \cdot e^{-j\pi a(t-f)^2} \tag{4.56}$$

then:

$$\frac{1}{|a|} \cdot s(t) = \int\limits_{-\infty}^{\infty} S(a \cdot f) \cdot e^{j\pi at^2} \cdot e^{j\pi af^2} \cdot e^{-j\pi a(t-f)^2} \cdot df \tag{4.57}$$

$$\frac{1}{|a|} \cdot s(t) = e^{j\pi at^2} \cdot \int\limits_{-\infty}^{\infty} S(a \cdot f) \cdot e^{j\pi af^2} \cdot e^{-j\pi a(t-f)^2} \cdot df \tag{4.58}$$

and as final result:

$$\frac{1}{|b|} \cdot s(t) = e^{j\pi at^2} \cdot \left[\left\{S(a \cdot f) \cdot e^{j\pi af^2}\right\} \cdot e^{-j\pi atf^2}\right] \tag{4.59}$$

The amazing final result according to equation (4.59) allows the re-transformation of a scaled signal from the frequency domain into the time domain and the elimination of the scaling factor at the same time!

For the whole transformation, called "Scaled Inverse Fourier Transform" (SIFT), only standard operations from communications engineering are used, by name the convolution and the multiplication both in the frequency domain and in the time domain.

The illustration in Image 4.4 requires a consideration unit by unit because of a standardisation:

$$e^{j2\pi a \cdot [f_e t_e] \cdot \left(\frac{f}{f_e} \cdot \frac{t}{t_e}\right)} = e^{j2\pi a \cdot [f_e t_e] \cdot \left(\frac{t}{[t_e]}\right)^2} \cdot e^{j\pi a \cdot [f_e t_e] \cdot \left(\frac{f}{[f_e]}\right)^2} \cdot e^{-j\pi a \cdot [f_e t_e] \cdot \left(\frac{t}{[t_e]} - \frac{f}{[f_e]}\right)^2}$$

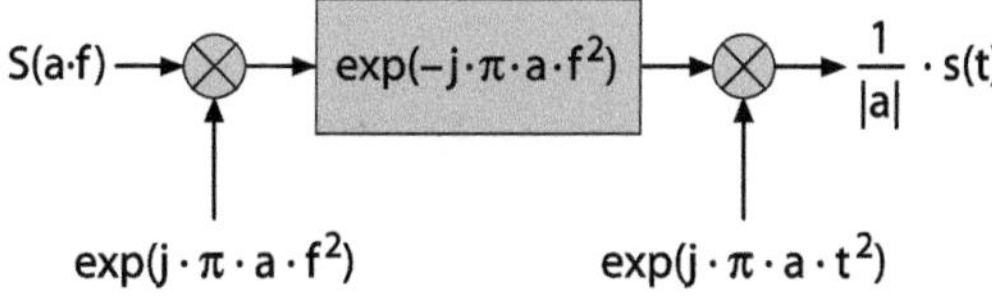

Image 4.4: Scaled inverse Fourier transformation

For this purpose, the corresponding standardisation factors of the units in square brackets are given the name index e. Inserting the terms for time axis and frequency axis results in unit-free phase terms.

4.3.4 The phase approximation

In chapter 4.3.2 Basic ideas, besides of the inverse scaled Fourier transformation in addition a phase approximation is required putting us in the position to use a SIFT algorithm in the SAR-raw data respectively the SAR data after the first step of compression. For this purpose, equation (4.46) has to be replaced by an approximation, here called phase approximation.

4.3.4.1 Derivation of the approximation

In the following, the employed approximation is derivated for the total backscattering spectrum $\sigma(f_r(f, f_\tau), f_\tau)$.
Therefore we calculate the Taylor series for equation (4.46):

$$f_r(f, f_\tau) = \frac{2}{c} \cdot \sqrt{(f + f_0)^2 - \frac{f_\tau^2 \cdot c^2}{4v^2}} \tag{4.46}$$

$$s(x) = s(x_0) + \dot{s}(x_0) \cdot (x - x_0) + \frac{1}{2}\ddot{s}(x_0) \cdot (x - x_0)^2 + \dots \tag{4.60}$$

At this, we neglect the quadratical and higher order terms of the approximation:

$$f_r(f, f_\tau) = \frac{2}{c} \cdot \sqrt{(f + f_0)^2 - \frac{f_\tau^2 \cdot c^2}{4v^2}} = \frac{2}{c} \cdot f_0 \cdot \sqrt{\left(1 + \frac{f}{f_0}\right)^2 - \frac{f_\tau^2 \cdot c^2}{4v^2 \cdot f_0}} \tag{4.61}$$

The first part of the Taylor series is calculated as follows:

$$f_r(f = 0, f_\tau) = \frac{2}{c} \cdot f_0 \cdot \sqrt{1 - \frac{f_\tau^2 \cdot c^2}{4v^2 \cdot f_0}} = \frac{2}{c} \cdot f_0 \cdot \sqrt{1 - \frac{f_\tau^2 \cdot \lambda^2}{4v^2}} \tag{4.62}$$

The first derivative within the Taylor series is:

$$\dot{f}_r(f = 0, f_\tau) = \frac{2}{c} \cdot f_0 \cdot \frac{\left(1 + \frac{f}{f_0}\right) \cdot \frac{1}{f_0}}{\sqrt{\left(1 + \frac{f}{f_0}\right)^2 - \frac{f_\tau^2 \cdot c^2}{4v^2 \cdot f_0}}} = \frac{2}{c} \cdot \frac{1}{\sqrt{1 - \frac{f_\tau^2 \cdot \lambda^2}{4v^2}}} \tag{4.63}$$

The non-linear spectrum can now be described as follows:

$$f_r(f, f_\tau) = \frac{2}{c} \cdot \sqrt{(f + f_0)^2 - \frac{f_\tau^2 \cdot c^2}{4v^2}} \cong \frac{2}{c} \cdot f_0 \cdot \sqrt{1 - \frac{f_\tau^2 \cdot \lambda^2}{4v^2}} + \frac{2}{c} \cdot \frac{f}{\sqrt{1 - \frac{f_\tau^2 \cdot \lambda^2}{4v^2}}}$$

(4.64)

Now equation (4.46) is equivalent to a linear-scaled spectrum with an additional constant frequency part, and has to be transformed via the developed scaled-inverse Fourier transformation as demanded in the beginning. Therefore we write:

$$f_r(f, f_\tau) = \frac{2}{c} \cdot \sqrt{(f + f_0)^2 - \frac{f_\tau^2 \cdot c^2}{4v^2}} \cong a(f_\tau) \cdot f + b(f_\tau) \cdot f_0$$

(4.65)

The illustration of the backscattering spectrum is modified to:

$$\sigma\left(\frac{2}{c} \cdot \sqrt{(f + f_0)^2 - \frac{f_\tau^2 \cdot c^2}{4v^2}}, f_\tau\right) = \sigma(f_r(f, f_\tau), f_\tau) \cong \sigma(a(f_\tau) \cdot f + b(f_\tau) \cdot f_0, f_\tau)$$

(4.66)

The two approximation factors are:

$$b(f_\tau) = \frac{2}{c} \cdot f_0 \cdot \sqrt{1 - \frac{f_\tau^2 \cdot \lambda^2}{4v^2}}$$

(4.67)

$$a(f_\tau) = \frac{2}{c} \cdot \frac{f}{\sqrt{1 - \frac{f_\tau^2 \cdot \lambda^2}{4v^2}}}$$

(4.68)

4.3.4.2 Approximation error

Images 4.5 and 4.6 illustrate different error functions as a result of the phase approximation. The relative error is calculated by subtracting the approximation of the solution from the exact term. Image 4.5 illustrates the approximation error without doppler displacement in consequence of a squint angle. Corresponding to the approximation according to equation (4.66) the parabolic error in azimuth direction as well as the linear range error can easily be detected. Its absolute maximum value is about 4.17E-11.

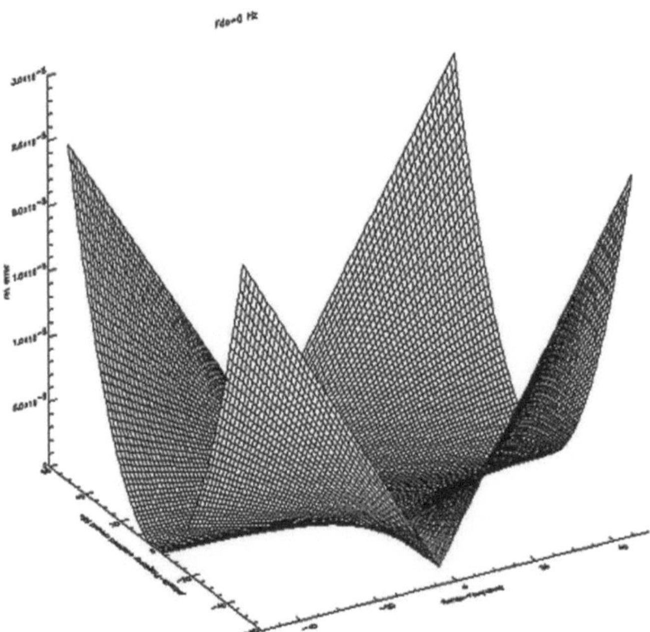

Image 4.5: Relative approximation error for a doppler-centroid frequency of 0 Hz

If the phase approximation is analysed with the help of squint angles, on one hand we will get a more serious error according to the squint angle and on the other hand a dissenting performance of the error function. Decisive is only the maximum of the absolute error that results in Image 4.6 with a value of 3.55 E-7.

An improvement of the mentioned approximation results can be achieved as already discussed concerning the range-doppler processor, by the so-called secondary range compression. This is discussed in the separate sub chapter 4.6 Significance of the secondary range compression.

4.3.5 Consideration of the backscattering spectrum in the band pass domain

To derivate the SIFT-SAR processor correctly, we have to refrain from the simplified consideration of the signals, namely in the equivalent low-pass domain. SAR data indeed are band pass signals and thus have to be processed in the band pass domain resp. processed as band pass signals.

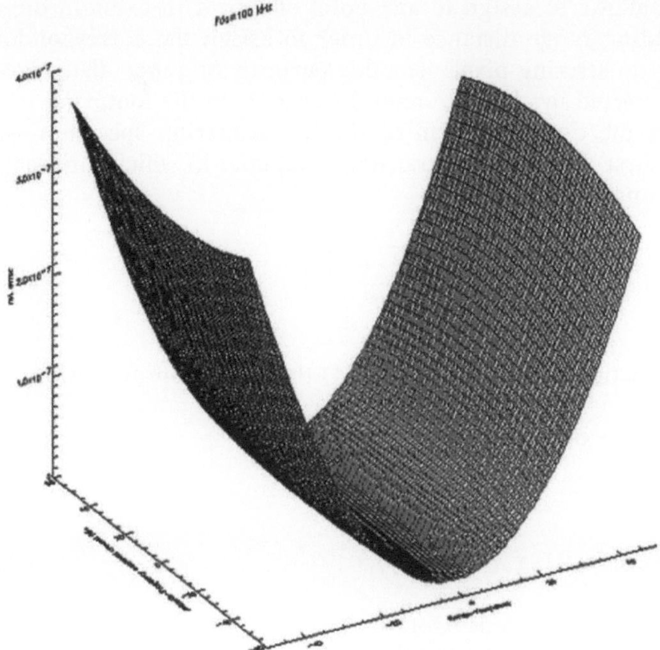

Image 4.6: Relative approximation error with doppler-centroid frequency (100 kHz)

4.3.5.1 Consideration of the zero-suppression

To compress the backscattered spectrum correctly, we need to reflect on the origin of the raw data matrix of the SAR system.

In reality, the received signals can be detected and recorded only after a running time t_v. For sensible reasons the raw data are recorded also from the initial point t_v on, so as not to waste memory space when receiving non-backscattering echoes. According to the displacement theorem a displacement in the time domain by $-t_v$ is equivalent to a multiplication in the frequency domain by $\exp(-j2\pi ft)$. Thus, the present equation (4.44) has to be extended by a phase term.

Corresponding to the delay of band pass signals, according to the above performed reflections, all in all results in the following term:

$$G_1(f, f_\tau) = G_{1_T}(f, f_\tau) \cdot e^{-j2\pi(f+f_0)t_v} = \sigma(f_r(f, f_\tau), f_\tau) \tag{4.69}$$

4.3.5.2 Transformation into the range-distance domain

If the transforming term described in equation (4.69) is processed respectively scaled into the time domain, a signal in the azimuth-time domain and in the range-time domain is created. However, the desired term we need is one that allows to assign to any point of time τ in azimuth direction a corresponding range distance in order to assign the corresponding distances directly to striking pixels. For this purpose the range distance already is to be corrected by the minimum distance R_0 to the footprint.

Initial point is the derivated term of the backscattering spectrum (see chapter 4.3.1.2 First compression-transform function), which originates after the first compression:

$$\sigma(f_r(f, f_\tau), f_\tau) = \sigma\left(\frac{2}{c} \cdot \sqrt{(f + f_0)^2 - \frac{f_\tau^2 \cdot c^2}{4v^2}}, f_\tau\right) \tag{4.70}$$

With the approximation from equation (4.3.4) the phase approximation

$$f_r(f, f_\tau) \cong a(f_\tau) \cdot f + b(f_\tau) \cdot f_0 \tag{4.71}$$

is valid:

$$\sigma(f_r(f, f_\tau), f_\tau) \cong \sigma(\{a(f_\tau) \cdot f + b(f_\tau) \cdot f_0\}, f_\tau) \tag{4.72}$$

As conclusion of the whole processor we have to look for a function that depends in azimuth direction on the flight time and in range direction on the distance $\sigma(r, \tau)$:

Valid is:

$$\sigma(\{a(f_\tau) \cdot f + b(f_\tau) \cdot f_0\}, f_\tau) = \mathscr{F}_{\tau \to f_\tau}\left\{\int_{-\infty}^{\infty} \sigma(r, \tau) \cdot e^{-j2\pi \cdot r \cdot (a(f_\tau) \cdot f + b(f_\tau) \cdot f_0)} \cdot dr\right\} \tag{4.73}$$

To simplify things, we consider only the 1-dimensional case with the f_r-depending part.

With:

$$\sigma(\{a(f_\tau) \cdot f + b(f_\tau) \cdot f_0\}, f_\tau) = F_\sigma(a \cdot f + b \cdot f_0) \tag{4.74}$$

$$f_\sigma(r) = \mathfrak{S}^{-1}\{F_\sigma(f)\} = \int_{-\infty}^{\infty} F_\sigma(f) \cdot e^{j2\pi f \cdot r} \cdot df = \int_{-\infty}^{\infty} F_\sigma(u) \cdot e^{j2\pi u \cdot r} \cdot du \tag{4.75}$$

and the substitution $u = a \cdot f$, $du = df \cdot a$ is:

$$f_\sigma(r) = a \cdot \int_{-\infty}^{\infty} F_\sigma(a \cdot f) \cdot e^{j2\pi a \cdot f \cdot r} \cdot df \tag{4.76}$$

Further substitutions are $f = x + \dfrac{b}{a} \cdot f_0$, $df = dx$. We then get:

$$f_\sigma(r) = a \cdot \int\limits_{-\infty}^{\infty} F_\sigma(a \cdot x + b \cdot f_0) \cdot e^{j2\pi(a \cdot x + b \cdot f_0) \cdot r} \cdot dx \tag{4.77}$$

$$f_\sigma(r) = a \cdot e^{j2\pi \cdot b \cdot f_0 \cdot r} \cdot \int\limits_{-\infty}^{\infty} F_\sigma(a \cdot x + b \cdot f_0) \cdot e^{j2\pi \cdot a \cdot r \cdot x} \cdot dx \tag{4.78}$$

The back substitution $f = x$ reveals now the 1-dimensional intermediate result:

$$f_\sigma(r) = a \cdot e^{j2\pi \cdot b \cdot f_0 \cdot r} \cdot \int\limits_{-\infty}^{\infty} F_\sigma(a \cdot f + b \cdot f_0) \cdot e^{j2\pi \cdot a \cdot r \cdot f} \cdot df \tag{4.79}$$

The wanted transformation instruction is described as follows:

$$\sigma(r, f_\tau) = a(f_\tau) \cdot e^{j2\pi \cdot r \cdot b(f_\tau) \cdot f_0} \cdot \int\limits_{-\infty}^{\infty} \sigma(\{a(f_\tau) \cdot f + b(f_\tau) \cdot f_0\}, f_\tau) \cdot e^{j2\pi \cdot a(f_\tau) \cdot f \cdot r} \cdot df$$

$$\tag{4.80}$$

With the reflections in chapter 4.3.5.1 Zero suppression, the described instruction according to equation (4.80) has to be extended respectively "shifted" by the minimum range distance R_0:

$$\sigma(\{r + r_0\}, f_\tau) = a(f_\tau) \cdot e^{j2\pi \cdot \{r + r_0\} \cdot b(f_\tau) \cdot f_0}$$

$$\cdot \int\limits_{-\infty}^{\infty} \sigma(\{a(f_\tau) \cdot f + b(f_\tau) \cdot f_0\}, f_\tau) \cdot e^{j2\pi \cdot a(f_\tau) \cdot f \cdot \{r + r_0\}} \cdot df$$

$$\tag{4.81}$$

The result in equation (4.81) reveals the wanted total backscattered coefficient in the range-time-azimuth frequency domain and has only to be transformed into azimuth direction to receive the SAR image.

4.3.6 Representation of the complete processor

All necessary requirements to realise the whole processor are discussed in the previous chapters. Using these requirements, the SAR processor now can be developed and illustrated in a block diagram.

The SAR-raw-data matrix on hand in the first compression step are compressed and standardised (according to chapter 4.3.1 Total scene spectrum and first compression step). That results in the total backscattering spectrum in the equivalent low-pass domain:

$$G_{1_T}(f, f_\tau) \approx \sigma(f_r(f, f_\tau), f_\tau) \cdot \text{rect}\left(\frac{f_\tau - f_c}{B_{az}}\right) \cdot \text{rect}\left(\frac{f}{B_r}\right) \tag{4.82}$$

With the approximation from chapter 4.3.4 Phase approximation the result of the first compression term is described as follows:

$$G_1(f, f_\tau) = \sigma(f_r(f, f_\tau), f_\tau) = \sigma(\{a(f_\tau) \cdot f + b(f_\tau) \cdot f_0\}, f_\tau) \tag{4.83}$$

In fact, the zero-suppressed band pass signal equation (4.83) in the equivalent low-pass domain has to be processed. So equation (4.8) changes according to equation (4.69) to:

$$G_1(f, f_\tau) = G_{1_T}(f, f_\tau) \cdot e^{-j2\pi(f+f_0)\cdot t_v} = \sigma(\{a(f_\tau) \cdot f + b(f_\tau) \cdot f_0\}, f_\tau) \tag{4.84}$$

To explain the required steps for the azimuth compression let us have a look at equation (4.81):

$$\sigma(\{r + r_0\}, f_\tau) = a(f_\tau) \cdot e^{-j2\pi\cdot\{r+r_0\}\cdot b(f_\tau)\cdot f_0}$$
$$\cdot \int_{-\infty}^{\infty} \sigma(\{a(f_\tau) \cdot f + b(f_\tau) \cdot f_0\}, f_\tau) \cdot e^{-j2\pi\cdot a(f_\tau)\cdot f\cdot\{r+r_0\}} \cdot df$$
$$\tag{4.85}$$

The total backscattering spectrum in the band pass domain is recognisable. Equation (4.81) is rewritten with the help of equation (4.83) to:

$$\sigma(\{r + r_0\}, f_\tau) = a(f_\tau) \cdot e^{j2\pi\cdot\{r+r_0\}\cdot b(f_\tau)\cdot f_0}$$
$$\cdot \int_{-\infty}^{\infty} G_{1_T}(f, f_\tau) \cdot e^{-j2\pi(f+f_0)t_v} \cdot e^{-j2\pi\cdot a(f_\tau)\cdot f\cdot\{r+r_0\}} \cdot df \tag{4.86}$$

$$\sigma(\{r + r_0\}, f_\tau) = a(f_\tau) \cdot e^{j2\pi\cdot\{r+r_0\}\cdot b(f_\tau)\cdot f_0} \cdot e^{-j2\pi f_0 t_v}$$
$$\cdot \int_{-\infty}^{\infty} G_{1_T}(f, f_\tau) \cdot e^{-j2(t_v-a(f_\tau)\cdot r_0)} \cdot e^{j2\pi\cdot a(f_\tau)\cdot f\cdot r} \cdot df \tag{4.87}$$

With $t_v = 2r_0/c$ equation (4.87) is the final result:

$$\sigma(\{r + r_0\}, f_\tau) = a(f_\tau) \cdot e^{j2\pi \cdot r_0 \cdot (b(f_\tau) - 2/c) \cdot f_0} \cdot e^{j2\pi f_0 r b(f_\tau)}$$
$$\cdot \int_{-\infty}^{\infty} G_{1_T}(f, f_\tau) \cdot e^{-j2\pi f(2r_0/c - a(f_\tau) \cdot r_0)} \cdot e^{j2\pi \cdot a(f_\tau) \cdot f \cdot r} \cdot df \tag{4.88}$$

Considering a possible time variance of the zero suppression $t_v = t_v(\tau) = \frac{2}{c} \cdot R(\tau)$ results in:

$$\sigma(\{r + r_0\}, f_\tau) = a(f_\tau) \cdot e^{j2\pi \cdot (r + r_0) \cdot (b(f_\tau)) \cdot f_0} \cdot e^{j2\pi f_0 r t_v(\tau)}$$
$$\cdot \int_{-\infty}^{\infty} G_{1_T}(f, f_\tau) \cdot e^{-j2\pi(t_v(\tau) - a(f_\tau) \cdot r_0)} \cdot e^{j2\pi \cdot a(f_\tau) \cdot f \cdot r} \cdot df \tag{4.89}$$

Equation (4.88) contains all parameters necessary for the compression. To simplify things, the following abbreviations are introduced:

$$a_0 = a(f_\tau) \cdot \frac{c}{2} = \frac{1}{\sqrt{1 - \dfrac{f_\tau^2 \cdot \lambda^2}{4v^2}}} \tag{4.90}$$

$$b_0 = b(f_\tau) \cdot \frac{c}{2} = \sqrt{1 - \frac{f_\tau^2 \cdot \lambda^2}{4v^2}} \tag{4.91}$$

$$b_1 = \frac{2}{\lambda} \cdot b_0 \tag{4.92}$$

$$G_2(f, f_\tau) = G_{1_T}(f, f_\tau) \cdot H_{k2}(f, f_\tau) \tag{4.93}$$

$$H_{k2}(f, f_\tau) = e^{-j2\pi f(2r_0/c - a(f_\tau) \cdot r_0)} = e^{-j4\pi f \frac{r_0 \cdot (1 - a_0)}{c}} \tag{4.94}$$

respectively without approximation for t_v :

$$H_{k2}(f, f_\tau) = e^{-j2\pi f(t_v(\tau) - a(f_\tau) \cdot r_0)}$$

$$H_{k3}(r, f_\tau) = a(f_\tau) \cdot e^{-j4\pi \cdot \frac{r_0}{\lambda}(1 - b_0)} \cdot e^{j2\pi \cdot r \cdot b_1} \tag{4.95}$$

resp. without approximation for t_v:

$$H_{k3}(r, f_\tau) = a(f_\tau) \cdot e^{j2\pi f_0 \cdot (b(f_\tau) \cdot r_0 - t_v(\tau))} \cdot e^{j2\pi \cdot f_0 \cdot r \cdot b(f_\tau)}$$

Considering the simplifications, the processing algorithm results in:

$$\sigma(\{r + r_0\}, f_\tau) = H_{k3}(r, f_\tau) \cdot \int_{-\infty}^{\infty} G_2(f, f_\tau) \cdot e^{j2\pi \cdot a(f_\tau) \cdot f \cdot r} \cdot df \qquad (4.96)$$

For the in equation (4.96) described integral:

$$I_{SIFT}(f, f_\tau) = \int_{-\infty}^{\infty} G_2(f, f_\tau) \cdot e^{j2\pi \cdot a(f_\tau) \cdot f \cdot r} \cdot df \qquad (4.97)$$

we use the derivated SIFT algorithm (see chapter 4.3.3 Derivation of the inverse scaled Fourier transformation) and the processing of a complete SAR scene is finished.

The Secondary Range Compression (SRC) illustrated in Image 4.8 Complete SIFT processing is described in the separate chapter 4.6 Significance of the Secondary Range Compression.

4.4 Scaling algorithms

As we already made use of scaling techniques with chirps when derivating the SIFT processor, we will discuss the field of cognate processing methods used in SAR processing and also e.g. when performing fast Fourier transformations [OPP89].

4.4.1 Introduction

The principle task of the SAR image processing still remains the inverse Fourier transformation of the backscattering spectrum in the 2-dimensional time domain. The reflection of the 2-dimensional frequency domain implements the image distortion, expressed by the non-linear frequency-axis distortions in the frequency domain which has to be corrected after or during the retransformation.

Concerning the derivation of the chirp scaling processor we will not use a scaled inverse Fourier transformation but choose the way via the traditional inverse Fourier transformation to correct afterwards the remaining scaling properties in the time domain. This means we separate the inverse Fourier transformation from the actual scaling.

Initial point of the considerations again is the 2-dimensional, distorted non-linear backscattering spectrum illustrated in chapter 4.3 The SIFT processor:

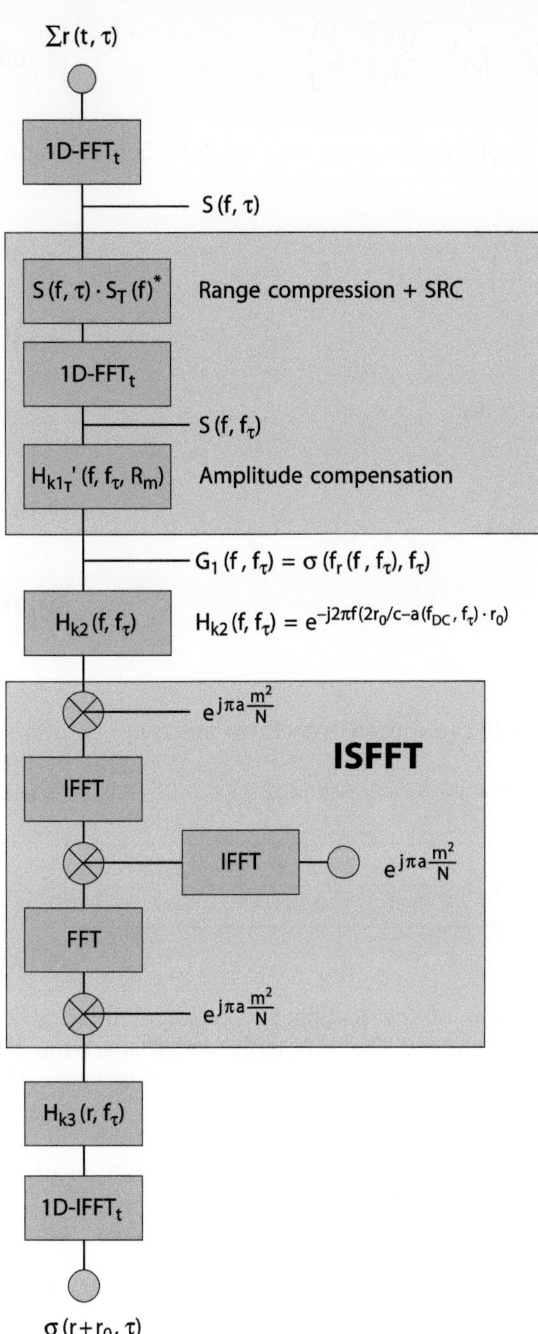

Image 4.8: Complete SIFT processing

$$\sigma(f_r(f, f_\tau), f_\tau) = \sigma\left(\frac{2}{c} \cdot \sqrt{(f + f_0)^2 - \frac{f_\tau^2 \cdot c^2}{4v^2}}, f_\tau\right) \tag{4.100}$$

We again approximate this spectrum the same way we did when derivating the SIFT processor:

$$\sigma\left(\frac{2}{c} \cdot \sqrt{(f + f_0)^2 - \frac{f_\tau^2 \cdot c^2}{4v^2}}, f_\tau\right) = \sigma(f_r(f, f_\tau), f_\tau) \cong \sigma(a(f_\tau) \cdot f + b(f_\tau) \cdot f_0, f_\tau) \tag{4.101}$$

The two approximation factors are:

$$b(f_\tau) = \frac{2}{c} \cdot f_0 \cdot \sqrt{1 - \frac{f_\tau^2 \cdot \lambda^2}{4v^2}} \tag{4.102}$$

$$a(f_\tau) = \frac{2}{c} \cdot \frac{f}{\sqrt{1 - \frac{f_\tau^2 \cdot \lambda^2}{4v^2}}} \tag{4.103}$$

After the approximation we now can fourier-transform inverse:

$$\mathscr{F}^{-1}\{\sigma(f_r(f, f_\tau), f_\tau)\} \cong \mathscr{F}^{-1}\{\sigma(a(f_\tau) \cdot f + b(f_\tau) \cdot f_0, f_\tau)\} \tag{4.104}$$

$$\boxed{\mathscr{F}^{-1}\{\sigma(a(f_\tau) \cdot f + b(f_\tau) \cdot f_0, f_\tau)\} = \frac{1}{|a(f_\tau)|} \cdot \sigma\left(\frac{r}{a(f_\tau)}\right) \cdot e^{-j2\pi b(f_\tau) \cdot f_0 \cdot \frac{r}{a(f_\tau)}}} \tag{4.105}$$

In equation (4.105) the retention of the scaling properties of the Fourier transformation can be realised clearly. The originally scaled spectrum remains scaled even after the inverse Fourier transformation, however, the multiplicative frequency scaling changes corresponding to the now temporal performance to a reciprocal multiplicative time scaling.

To simplify things, we picture the inverse Fourier transformation of the total backscattering spectrum as follows:

$$\frac{1}{|a(f_\tau)|} \cdot \sigma\left(\frac{r}{a(f_\tau)}\right) \cdot e^{-j2\pi b(f_\tau) \cdot f_0 \cdot \frac{r}{a(f_\tau)}} = \frac{1}{|a(f_\tau)|} \cdot n\left(\frac{r}{a(f_\tau)}\right) \tag{4.106}$$

After calculating the first step towards the processed SAR scene, in the next step the remaining scaling is to eliminate. That means we have to look for a scaling technique that allows the correction of the time-scaling factor a.

$$\frac{1}{|a(f_\tau)|} \cdot n\left(\frac{r}{a(f_\tau)}\right) \xrightarrow{\text{scaling}} n(r) \tag{4.107}$$

4.4.2 Frequency scaling

A corresponding scaling in the frequency domain according to equation (4.108) is already indicated by Papoulis [PAP68]:

$$N(a(f_\tau) \cdot f) = N(a \cdot f) \xrightarrow{\text{scaling}} N(f) \tag{4.108}$$

By combining convolution operations and multiplication operations with corresponding signals a re-scaling of the frequency axis is achieved. For this purpose please have a look at Image 4.9.

The scaling algorithm illustrated in Image 4.9 uses different chirps for re-scaling of the frequency domain. The four chirps used feature different chirp rates. According to Papoulis [PAP68] these four chirp rates can be determined with the help of three fundamental equations:

$$\frac{\gamma}{\alpha} = -\frac{1}{a} \tag{4.109}$$

$$\gamma + \alpha = \beta \tag{4.110}$$

$$\varepsilon = \gamma \cdot \frac{\beta}{\alpha} \tag{4.111}$$

The next step towards development of the chirp scaling algorithm is the displacement of the in Image 4.9 illustrated last multiplication chirp from the output of the algorithm to the input. We have to consider the modification of the chirp rate. The displaced chirp must use a corresponding chirp rate whilst the displacement to the input of a sequence takes place. The chirp rate is selected on the basis of the prevailing scaling at the input.

Chirp at the output:

$$N(f) = N_0(f) \cdot \frac{\alpha}{\pi} \cdot e^{-j \cdot \varepsilon \cdot f^2} \tag{4.112}$$

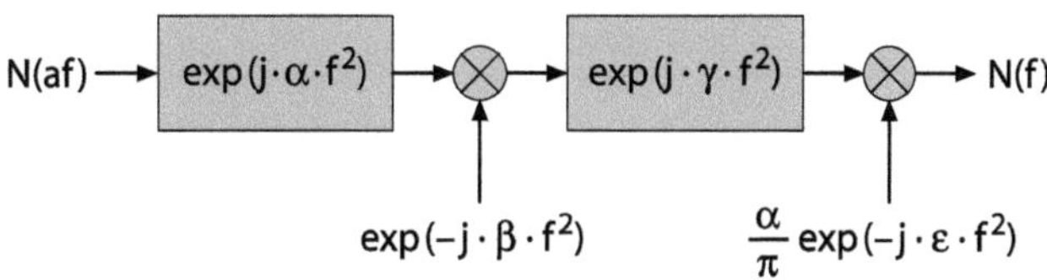

Image 4.9: Frequency scaling according to Papoulis

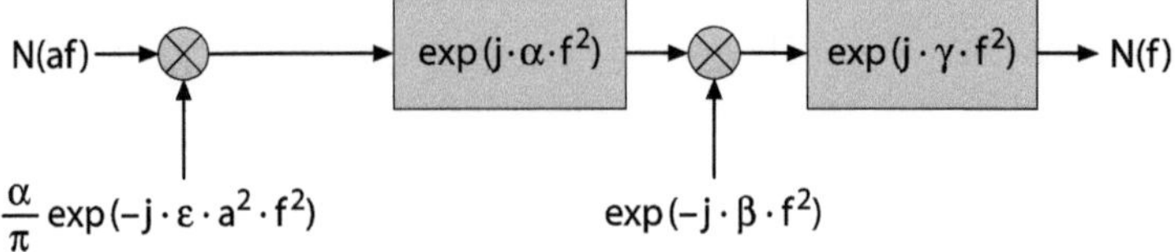

Image 4.10: Modified frequency scaling

Chirp at the input:

$$N(a(f_\tau)\cdot f) = N_0(a(f_\tau)\cdot f)\cdot\frac{\alpha}{\pi}\cdot e^{-j\cdot\varepsilon a(f_\tau)^2\cdot f^2} \tag{4.113}$$

The modified scaling instructions change to:

To determine the corresponding scaling instructions required according to equation (4.107) in the time domain, we transform the modified frequency scaling back to the time domain. For this purpose, we use a traditional Fourier transformation.

To derivate the chirp-scaling algorithm the current chirp rates are calculated and adapted to the corresponding SAR-signal characteristics.

4.4.3 Time scaling and chirp scaling

As already discussed above, in order to determine the corresponding chirp rates within the frequency-scaling routine, three determination functions are available to calculate four parameters, the chirp rates. That is why it is possible to choose one of the four parameters freely. For the derivation of the chirp-scaling algorithm it is sensible to set the chirp rate of the first multiplication chirp equal to the chirp rate of the actually used transmitting signal. Then the first convolution is performed implicitly by the use of the transmitting chirp.

$$-\pi\cdot\frac{f^2}{k_r}\overset{!}{=} -\varepsilon\cdot a^2\cdot f^2 \tag{4.114}$$

The chirp rate k_r has to be interpreted dependent on the range distance (so far, the chirp rates have been time-dependent parameters). With equation (4.114) the further chirp rates can be determined by inserting in equations (4.109, 4.110, 4.111) and result in:

$$\alpha = \frac{\pi}{k_r}\cdot(1-a)^{-1} \tag{4.115}$$

$$\beta = -\frac{\pi}{a\cdot k_r} \tag{4.116}$$

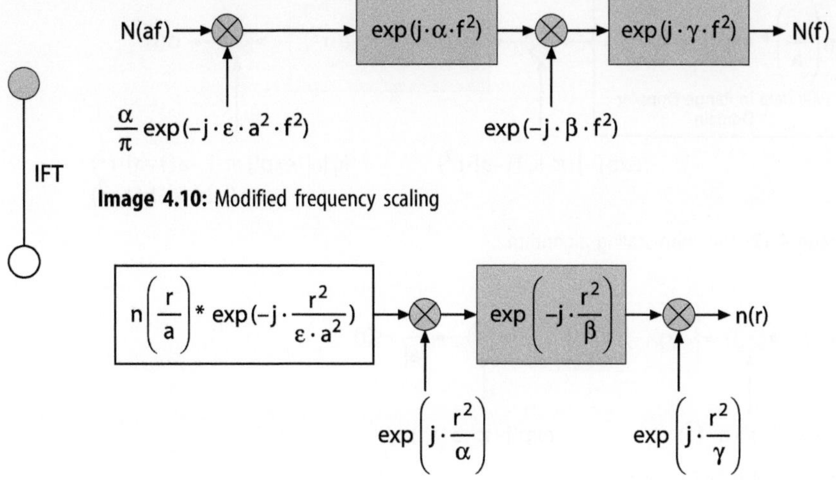

Image 4.10: Modified frequency scaling

Image 4.11: Modified time scaling

$$\gamma = -\frac{\pi}{k_r} \cdot \frac{1}{a \cdot (1 - a)} \tag{4.117}$$

Inserting equations (4.114–4.117) into Image 4.11 results in the wanted algorithm that can re-scale the 2-dimensional total backscattering coefficient originated by the inverse Fourier transformation. The stated scaling sequence is called chirp scaling [RUN92].

However, please note the modified phaseterm of the last multiplication and remember the simplified time-domain illustration of the inverse Fourier transformation of the distorted total backscattering spectrum equation (4.106):

$$\frac{1}{|a(f_\tau)|} \cdot \sigma\left(\frac{r}{a(f_\tau)}\right) \cdot e^{-j2\pi b(f_\tau) \cdot f_0 \cdot \frac{r}{a(f_\tau)}} = \frac{1}{|a(f_\tau)|} \cdot n\left(\frac{r}{a(f_\tau)}\right) \tag{4.106}$$

Image 4.11 (* means convolution) shows the modified time scaling delivering as a result n(r). We are looking for $\sigma(r)$, though. Thus, in addition we have to multiply our previous result by the conjugated phase term according to equation (4.106):

$$\sigma(r) = e^{j2\pi b(f_\tau) \cdot f_0 \cdot r} \cdot n(r) \tag{4.118}$$

In Image 4.12 The chirp-scaling algorithm this phase term is implemented as final multiplication.

Like the range-doppler processor, we could derivate the chirp-scaling algorithm from the common initial point, by name the 2-dimensional point-target spectrum. The diverse illustration of the different processor types reveals different advantages and disadvantages, which are discussed in the following chapter.

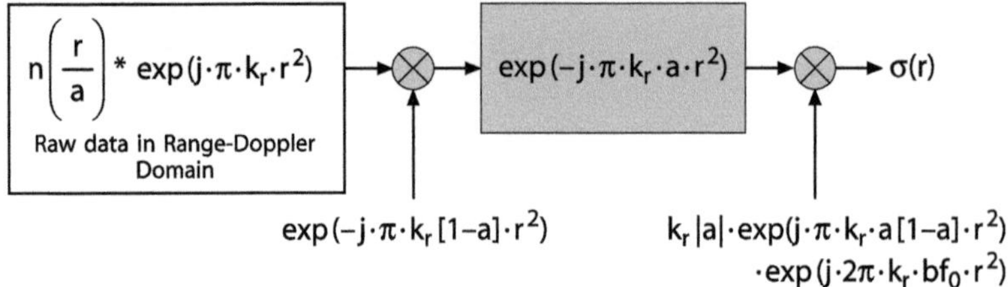

Image 4.12: The chirp-scaling algorithm

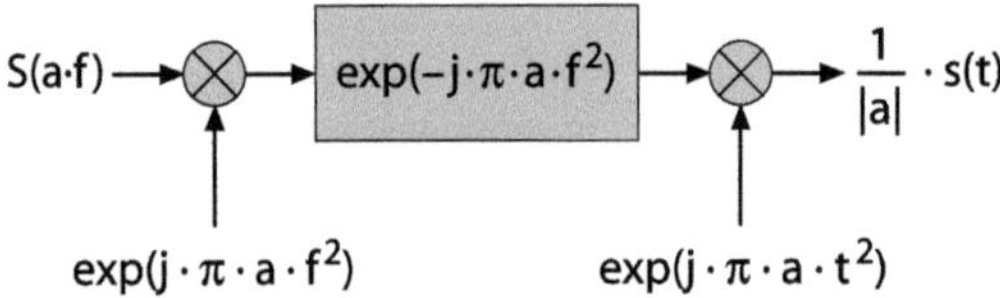

Image 4.13 a: SIFT scaling

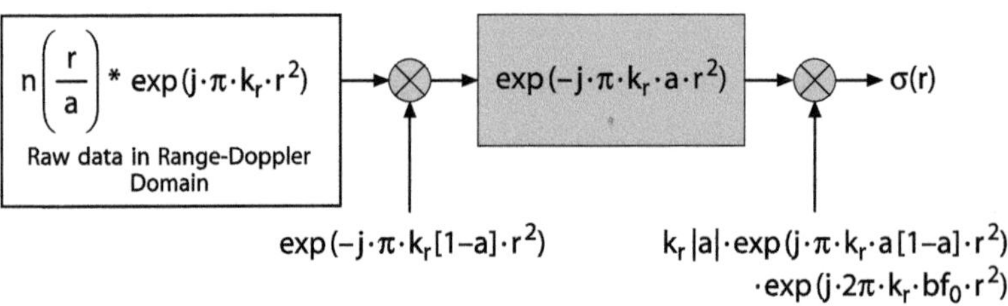

Image 4.14: Comparison SIFT-chirp scaling
Image 4.13 b: Chirp scaling

4.5 Connection between the different processors

We have a look only at the criteria of the SIFT processor and the chirp-scaling processor. Both algorithms obviously make use of the scaling characteristics of signals with quadratic phase trend (chirps).

The essential characteristics can clearly be realised in Image 4.14.

As a matter of principle, the functionality is similar with both sequences. The scaling is achieved by a succession of chirp convolutions and chirp multiplications.

We would like to remind you that only to keep things simple, the described sequences are displayed 1-dimensionally in equation (4.105). In fact, the simplified input and output phase functions are 2-dimensional terms derivated out of the 2-dimensional frequency domain.

Regarding the description of both algorithms, the different levels the corresponding scalings are performed in can be realised.

The SIFT processor works in the 2-dimensional frequency domain whereas the chirp-scaling processor in the range-, time- and azimuth-frequency domain. The advantage of the chirp-scaling processor is easy to see. Using the chirp-scaling, in opposite to the SIFT processor, a Fourier transformation can be leaped and thus calculating time saved.

The disadvantages of the chirp-scaling processor are also clear:

It requires a linear frequency-modulated input signal, "chirps", for successful scaling (first convolution of the input signal)

The chirp rates of the scaling chirps have to be modified permanently during the scaling sequence.

4.6 Description of the secondary range compression

This chapter describes the significance of the so-called Secondary Range Compression for the three derived SAR signal-processing algorithms. A geometric reflection had already been shown by [RAN94]. He describes the Secondary Range Compression (SRC) as a term due to the loss of the orthogonality between range components and azimuth components.

Let us choose the signal theoretical way that describes the SRC as compensation of an approximation error, that error that results in the non-linear, distorted system-transfer function (2-dimensional radar backscattering coefficient).

All derived processor types here are proceeded from a 2-dimensional description of the point-target spectrum. Thus, the mentioned approximation of the non-linear spectrum is equally performed in all processors in the frequency domain and has to result in identical error-description functions at equal output-phase functions.

The common initial point of the processor was equation (3.227):

$$R_T(f, f_\tau, R_0, \tau_0) = S_T(f) \cdot \sigma(R_0, \tau_0) \cdot \text{rect}\left[\frac{f_\tau - f_c}{B_{az}}\right] \cdot \sqrt{\frac{c \cdot R_0}{2}} \cdot \frac{1}{v}$$

$$\cdot \frac{f + f_0}{\left[(f + f_0)^2 - \frac{f_\tau^2 \cdot c^2}{4v^2}\right]^{3/4}} \cdot e^{-j \cdot \pi/4} \cdot e^{-j \cdot 2\pi f_\tau \tau_0}$$

$$\cdot e^{-j4\pi \frac{R_0}{c} \cdot \sqrt{(f + f_0)^2 - \frac{f_\tau^2 c^2}{4v^2}}}$$

The description of the 2-dimensional Fourier spectrum in the range-time-azimuth-frequency domain requires an approximation of the pattern (Chapter 3.5 Point-target spectrum in the range-doppler domain eq. (3.237))

$$\sqrt{1 + x} \cong 1 + \frac{1}{2} \cdot x - \frac{1}{8} \cdot x^2$$

We had to determine a modified chirp rate for this approximation equation (3.247):

$$\frac{1}{k_r'} = \frac{1}{k_r} \cdot \left[1 - \frac{k_r 2R_0 \cdot \dfrac{f_\tau^2 \lambda^2}{4v^2}}{c \cdot f_0 \cdot \left(1 - \dfrac{f_\tau^2 \lambda^2}{4v^2}\right)^{\frac{3}{2}}} \right] \tag{4.119}$$

The stated modified chirp rate was created due to that approximation required for the back-transformation of the 2D-point target spectrum in the range-time-azimuth-frequency domain.

The range-doppler processor used (in opposite to the derivation out of the common initial point, the 2-dimensional point-target spectrum) for the Taylor series the known 2-dimensional phase term (see chapter 4.2 The range-doppler processor equation (4.20)):

$$\Phi(f, f_\tau, R_0) = -j \cdot 4\pi \frac{R_0 \cdot (f + f_0)}{c} \cdot \left[\sqrt{1 - \frac{f_\tau^2 c^2}{4v^2 \cdot (f + f_0)^2}} - 1 \right]$$

in order to calculate after performing the corresponding conjugated phase function:

$$\Phi_K(f, f_\tau, R_0) = j \cdot 4\pi \frac{R_0 \cdot (f + f_0)}{c} \cdot \left[\sqrt{1 - \frac{f_\tau^2 c^2}{4v^2 \cdot (f + f_0)^2}} - 1 \right]$$

the SRC as second derivative equation (4.31):

$$\frac{1}{k_r'} = \frac{1}{k_r} \cdot \left[1 - \frac{R_0 \cdot c}{2 \cdot v^2 \cdot f_0^3} \cdot \frac{k_r \cdot f_\tau^2}{\left[1 - \dfrac{f_\tau^2 c^2}{4v^2 \cdot f_0^2} \right]^{3/2}} \right] \tag{4.120}$$

Both results equations (4.119, 4.120) of the modified chirp rates are identical. On one hand we can assume that the described possibility to derivate the range-doppler processor is calculated correctly with the help of the 2D-point-target spectrum (see also [RAN94]). On the other hand the 2D-frequency domain description obviously is a more common, respectively more exact solution than the usually used range-time-azimuth-frequency domain for the range-doppler processor (see also [RAN94]). Proceeding from the 2D-frequency domain, the transformation to the range-time-azimuth-frequency domain took place by an approximation towards the Fourier transformation.

Still, the comparison between the two processors (SIFT- and range-doppler processor in the 2D-frequency domain), which were both derivated deliberately with the help of identical phase derivatives (equations (4.18, 4.46)), leads also to identical modified chirp rates.

A further comparison between the SIFT processor and the chirp-scaling processor (both were derived from the 2-dimensional spectrum) must result in identical modified chirp rates for the Secondary Range Compression as identical approximations of a common initial point were again used.

Thus, the Secondary Range Compression has to be implemented in all described types of processors (e.g. during the range compression).

4.7 Quality of SAR processors

The quality of an SAR processor plays an important role concerning the quality of the SAR-scene products. A number of testing and calibrating methods to eveluate the corresponding criteria are known [ZIN93]. But we only want to focus on two essential tests:

the point-target response/system-transform function
the phase preservance.

A SAR processor should fulfill the theoretically derived functions, that means to be in the position to generate out of a simulated point-target function the expected 2-dimensional pseudo si-function with the help of the derivated matched-filter operation. Doing so, the functionality is proven (as also done in the progress of this work). The first developments of a SIFT processor can be dated back to the year 1966 [HEI96]. Comparing the further descriptions with those made in [HEI96] reveals substantial improvements both in the field of point-target descriptions and for regarding the phase preservance.

If a SAR processor is intended to be used to process interferometric data, that means he is supposed to process interferometrically recorded SAR raw data in a way that interferograms can be generated from the compressed scenes, so the SAR processor has to meet the requirements of the so-called phase preservance. Phase preservance is the treatment of the original phase information of the raw data in a way that after the complete processing no modification of this phase is visible. If the requirements of the phase preservance were not met then already whilst processing the necessary single scenes phase errors would be created and would falsify every single following working step.

Now let us examine the SIFT processor on its phase preservance. In general, the phase preservance of SAR processors is proven by the formation of interferograms of its own. Concerning the point targets it is proceeded as follows:

a SAR-scene is processed in a distance R_0
the same scene is displaced and processed again in $R_1 = R_0 +$ displacement
the processed displaced scene is re-displaced
both scenes are processed by conjugated complex multiplications to an interferogram of its own.

In that created interferogram a statement concerning both the phase preservance and the range variance can be made by watching the phase. In the beginning let us examine in the above described way a SIFT processor with a point target.

In Image 4.15 we can see the result of a point-target compression. The real part with the abscissa as range direction and the ordinate as azimuth direction is illustrated. It does indeed meet our expectations. In particular the processing of the that way generated 'raw data', in other words the focussing of Image 4.15 after describing the input data beforehand in a conjugated complex way, correspond with the derivated and in Image 4.3 illustrated ideal point-target response. The result of the point-target response, generated from the conjugated point-target raw data, must be verified in Image 4.16 Simulated point target. A comparison with the ideal, calculated response (Image 4.3, page 155) reveals an almost optimal congruence. The functions created with the SAR processor are calculated without exception only for ERS-1 parameters.

As already described, this point target is calculated with an interferogram of its own by using a displaced point target (100 Pixel = 790 m; ERS-1 parameter). The resulting interferogram and the corresponding phase difference is illustrated in Images 4.17 and 4.18. Please note the missing

Image 4.15: Point-target processing, real part, ERS-1 parameter

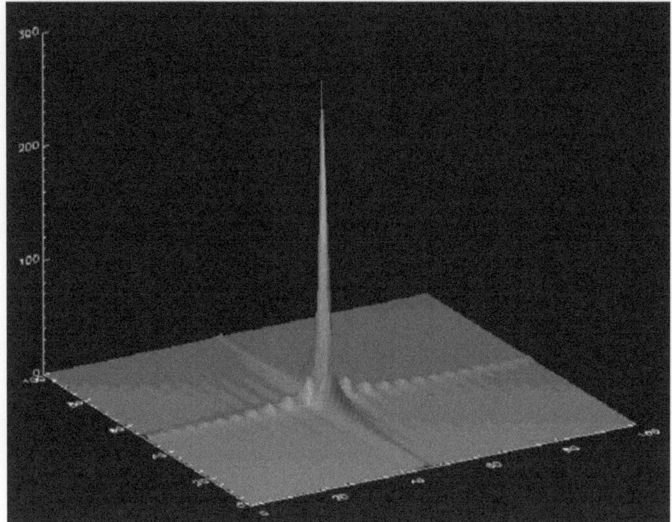

Image 4.16: Simulated point target response

Image 4.17: Interferogram of the displaced and processed point target results

Image 4.18: Phase difference of the displaced and processed point target results

fringe formation (phase skipping). If fringes are visible now the processor has caused non-corrigible phase errors.

However, the phase relation to be determined gets more concrete by comparing the measurement phase with the theoretically expected phase characteristics in view of the modified range distance. This comparison is qualitatively illustrated in Image 4.19.

Image 4.19 describes the direct comparison between the measured phase and the expected phase. The expected phase can be determined according to chapter 2.2.2 Doppler effect and frequency shift as follows:

$$k_{az} = -\frac{2v_a^2}{\lambda \cdot R(\tau_c)} \tag{4.121}$$

The wanted parabola as result of the corresponding spacer difference is calculated as follows:

$$\Delta k_{az} = k_{at1} - k_{az2} = -\frac{2v_a^2}{\lambda} \cdot \left(\frac{1}{R(\tau_1)} - \frac{1}{R(\tau_2)} \right) \tag{4.122}$$

Both parabolas are qualitatively equal. The displacement in terms of an offset of the calculated parabola is inserted deliberately because of a better

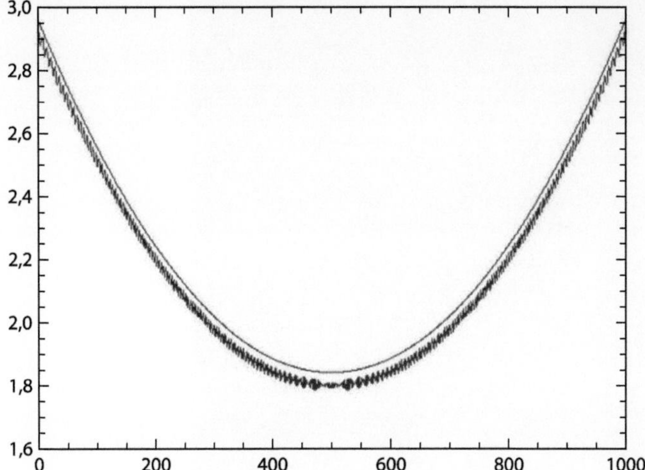

Image 4.19: Phase difference in azimuth direction (the target phase is illustrated in a displaced way)

comparability. A comparison in quantitative terms will not be made as on the one hand this contemplation has already been done in [HEI96] and on the other hand a much more complex test will follow.

If only single points are used to verify a SAR processor, the occurring input-signal energies are limited to the point target. Furthermore (provided that the processor works properly), by using only single points particularly with regard to the planar-shaped targets a verification can not be done.

On this account, an interferogram of a real SAR scene has to be generated. For this purpose another method has to apply as the above mentioned one:

a SAR scene (location: Oberpfaffenhofen) is processed within the correct, shortest distance R_0

the same scene is displaced and processed again, but now the shortest distance R_0 is not enlarged corresponding to the input-array displacement, but reduced in a way that the processed phase characteristics in the displaced input field is equivalent to the phase characteristics chosen whilst focussing correctly, in other words: $R_1 = R_0$-displacement

the processed displaced scene is re-displaced

both scenes are processed by a complex conjugate multiplication to an interferogram of its own.

The following images visualise the generated test and tell us something about the located errors.

At the beginning, the test area (Image 4.20) is introduced to get an idea of the data record in use. In Image 4.21 the interferometric phase of its own is displayed that appears at first glance as a low phase noise that easily could be considered as not noteworthy.

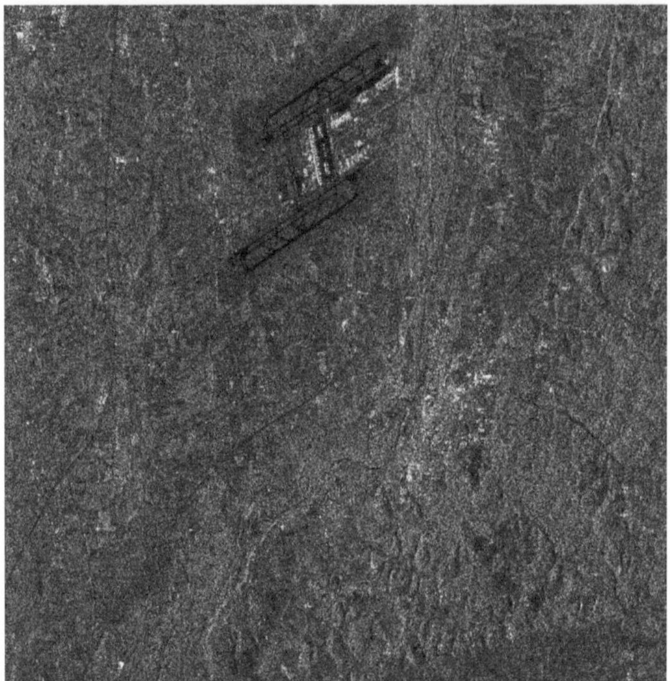

Image 4.20: ERS-1 recording of Oberpfaffenhofen, Germany

However, the following Image 4.22 reveals a segment of Image 4.21. In this image we realise that the appearing phase peaks very well cause a disturbance noticable as corresponding level errors in a possible interferogram.

Besides these observable peaks appearing at first glance as phase errors, no systematic phase errors let alone phase skippings can be detected, though. Thus, the tested processor shows only single phase peaks as erroneous which require further examination.

To achieve a clearer comparability we only use plots of entire 2-dimensional arrays. Image 4.23 illustrates the phase characteristics of the interferogram of its own as a plot and thus corresponds with Image 4.21. The mentioned peaks are clearly recognisable. Let us have a look at their origins.

If a SAR scene shows only low radar backscattering coefficients then those poorly reflected parts of the scene appear only as dark areas in the SAR image. In those areas not able to receive real characteristic signals of the sensor, due to the missing signal information, no correct phase information can be extracted. The phase information generated with the help of the interferogram of its own can lead to phase errors within the dark areas of the SAR image which can be identified as peaks in the phase image.

Image 4.21: Phase characteristics of the interferogram

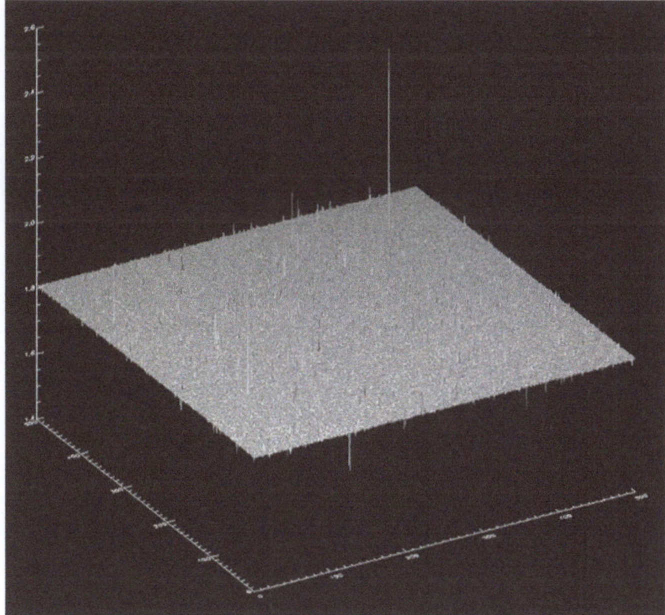

Image 4.22: Part of the phase characteristic of the eigen-interferogram

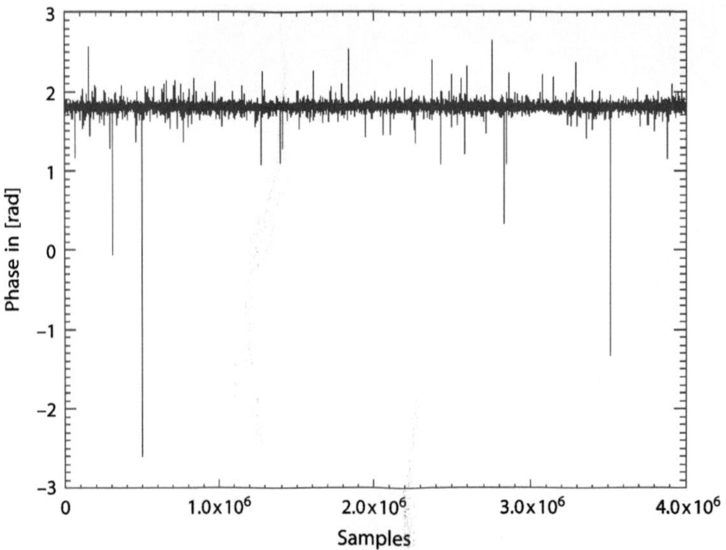

Image 4.23: Phase characteristics of the interferogram

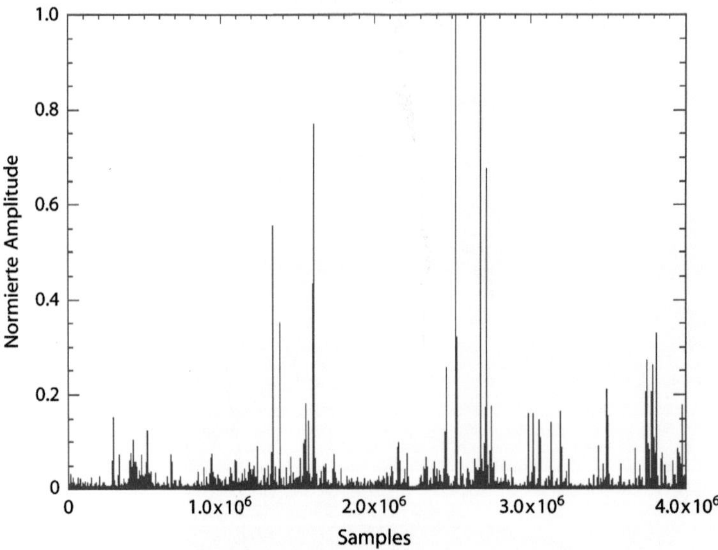

Image 4.24: Standardised interferogram (absolute)

Now we can assume that eliminating the dark areas in the SAR image could lead to a reduction of the phase peaks. This would prove our reflections.

First of all we have a look at the interferogram in Image 4.24 which corresponds to the brightness of the SAR scene. As the small amplitudes are interesting for us we choose an inverse plot, Image 4.25.

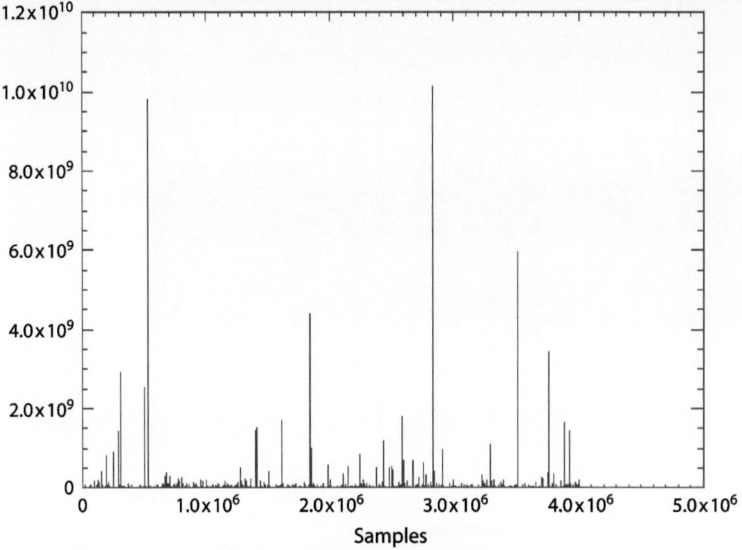

Image 4.25: Inverse standardised interferogram (absolute)

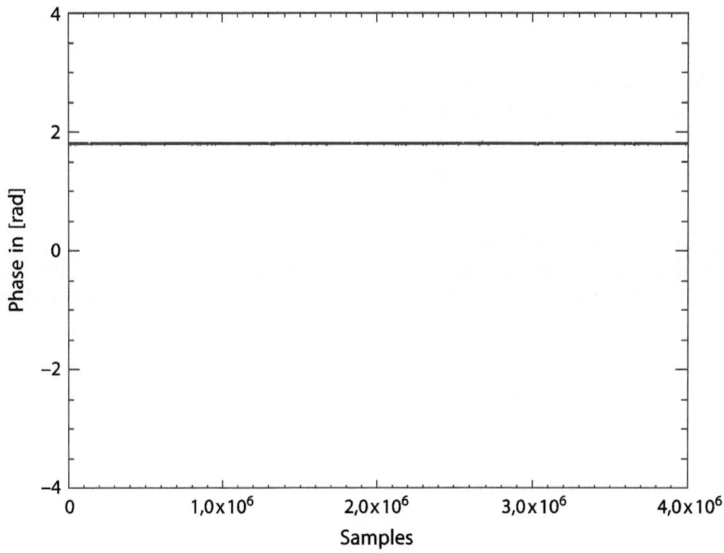

Image 4.26: Verified phase characteristic of the interferogram of its own

The poorly reflecting spots are clearly visible in Image 4.25 as peaks in the interferogram.

To eliminate those poorly reflecting areas, both the average value and the variance of the value of the processed SAR scene are calculated. As borderline of that so-called dark spot a concrete deviation of the average

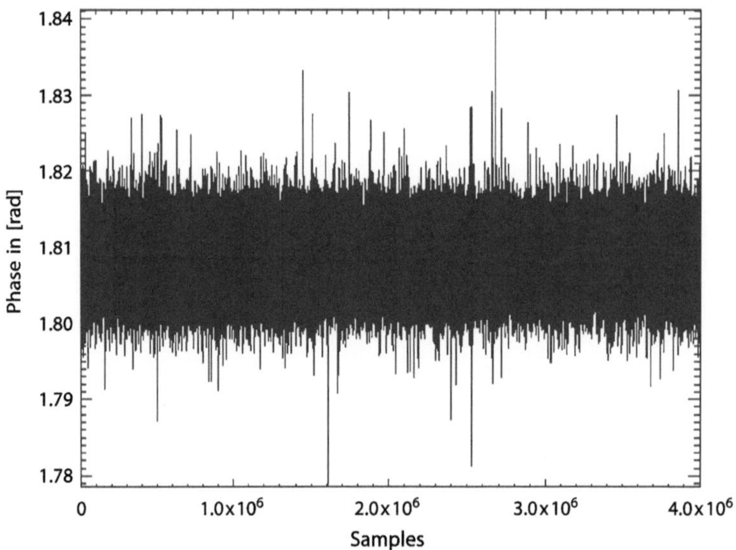

Image 4.27: Verified phase response of the interferogram of its own

value is defined, the 100-sigma distance. As a limit not the usual 3-sigma border is used but a substantially more demanding criteria. Thus, peaks must be up to thirty times below the average value. Finally we can speak now of peaks or poorly reflecting targets.

In the SAR scene all the spots smaller than defined in the above mentioned 100-sigma distance are marked. In the next step all those marked pixels in the phase array that correspond to the marked areas in the SAR scene are set to calculated average value of the phase response.

Image 4.26 illustrates the results of the phase response. To make the image more comparable to Image 4.23 we did not change the axis. Image 4.26 clearly shows a substantial improvement. Even the disturbing peaks in the phase response are eliminated. The phase errors initially assumed as erroneous processor properties are due to insufficient backscattering characteristics of the observed area.

Image 4.27 shows the verified phase characteristics in a more detailed way. With its help we can detect a phase noise respectively phase error of the processor of about 0.95% based upon '2π'. Another test (not illustrated here) used the mentioned 3-sigma and reduced the phase noise to a level of 0.16% or $0.5°$. Already an accuracy of $3°$ is called precision SAR Processing' [RAN94].

Thus, the described novel SIFT processor meets the requirements concerning the phase preservance and, thanks to its accuracy, can be assigned to the precise processors.

Interferometry

5.1 Introduction

Besides the determination of geophysical parameters there is an essential area of application of the radar with synthetic aperture, the multi-dimensional mapping of areas. Extracting in particular height information of the corresponding data of a certain area is interesting.

In the field of earth science the exact knowledge of the topography is an important condition to perform and analyse measurements of different kind. For instance, at the framework of earth observing measurements, minimal movements of the earth's crust along tectonic plates are registered. Volcanic eruptions can be predicted by observing topographic height characteristics [SCHW95]. In the meantime, the supply of digital altitude information has become a necessity for applications in the fields of land use planning and environmental observation. It also serves as data base for geo information systems (GIS).

Digital height models play also an important role in the fields of processing, analysing and interpretating of corresponding remote sensing data. Especially in the field of geocoding of different input-data bases (Landsat, Spot) digital-height models of the terrain are required.

The traditional methods to generate digital-height models basing upon different methods of data recording and modified data processing (tacheometry, digitalising of topographic maps, photogrammetric methods) will soon be obsolete. The application of the more accurate interferometric is more and more advancing. In the future the traditional methods will only be used for special applications.

5.2 Theory of SAR interferometry

[GRA74] pointed out for the first time the possibility to achieve knowledge concerning the topography of a terrain with the help of the interferometric processing of SAR data. The method of the SAR interferometry, already discussed at the beginning of this work, among others enables us according to [SCHW95] to take advantage of the possibility of measuring the absolute ground level height:

geocoding the recorded data and thus comparing with satellite-based recording systems
the radiometric calibration of SAR images
creation of data bases in geo-information systems.

The method of SAR interferometry (also called SARIF or InSAR) is based
on the coherence of the radar's signal, the recording of two images of the
same area from different perspectives and the utilisation of the phase information of the SAR images.

Considering Image 5.1, the conventional SAR method (only sensor 2)
point 1 can not be distinguished from point 2 because of $r_{S2P1} = r_{S2P2}$. Thus,
there is no detectable phase difference of both echo signals.

Applying a second sensor this ambiguity is solved (because $r_{S1P1} \neq r_{S1P2}$).
The phase of the echo signal of point 2 on the sensor is:

$$\phi_1 = \frac{2\pi \cdot 2r_{S1P2}}{\lambda} + \phi_0 \tag{5.1}$$

On sensor 2 the following phase is measured:

$$\phi_2 = \frac{2\pi \cdot 2r_{S2P2}}{\lambda} + \phi_0 \tag{5.2}$$

Now forming the phase difference of both signals results in the interferometric phase:

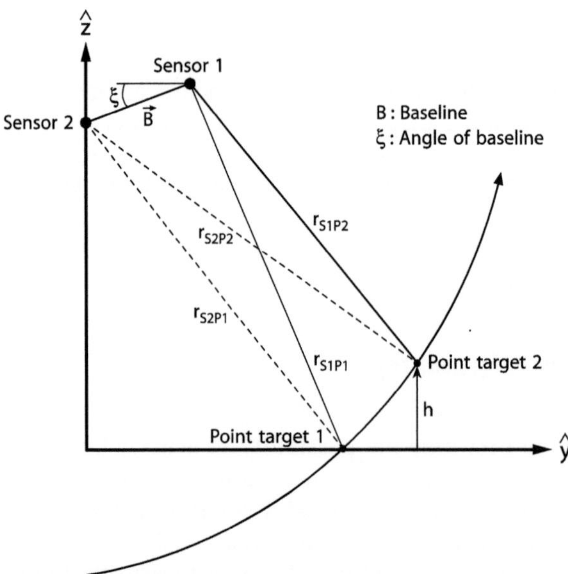

Image 5.1: To the principle of the SAR interferometry

$$\delta_\phi = \frac{4\pi}{\lambda} \cdot \delta_r \tag{5.3}$$

The height can be calculated according to [LOF] as follows:

$$h = H - r \cdot \left[\sqrt{1 - a^2} \cdot \cos(\xi) - a \cdot \sin(\xi) \right] \tag{5.4}$$

whereas

$$a = \frac{\left(r + \frac{\lambda}{4\pi} \cdot \phi \right)^2 - r^2 - B^2}{2 \cdot B \cdot r} \tag{5.5}$$

In practise, ϕ_0 is not identical at two overflights as the recording conditions have changed. For this reason (among others) the extracted interferogram calculated by the conjugated complex pixelwise multiplication of the two complex SAR images:

$$\underline{Z} = \underline{Z}_1(x, y) \cdot \underline{Z}_2(x, y)^* \tag{5.6}$$

in its phase, the interferometric phase:

$$\delta_\phi = \arctan \frac{\mathrm{Re}\{\underline{Z}(x, y)\}}{\mathrm{Im}\{\underline{Z}(x, y)\}} \tag{5.7}$$

is not accurate. An inaccuracy results in determining the evaluation according to equation (5.4) and (5.5). The signals echoes received with sensor 1 are not completely coherent to those received with sensor 2 which means there is no fixed, defined phase relation between the two signals. The degree of the coherence between two SAR signals s_1 and s_2 can be defined as follows [BAM93]:

$$\gamma = \frac{E\{s_1 \cdot s_2{}^*\}}{\sqrt{E\{|s_1|^2\} \cdot E\{|s_2|^2\}}} \tag{5.8}$$

whereas $E\{\ldots\}$ is the expected value. The value of this "correlation coefficient" γ can have values between zero (incoherent) and one (coherent) and serves as quality indicator of the interferometric phase. To get exact height information whilst extracting and processing SAR data, a high value on the phase preciseness has to be set.

The achievable phase preciseness of the interferogram is influenced by the following possible de-correlation effects:

change of the physical characteristics of the observed scene within the two overflights (for example change in vegetation, in the water's surface)

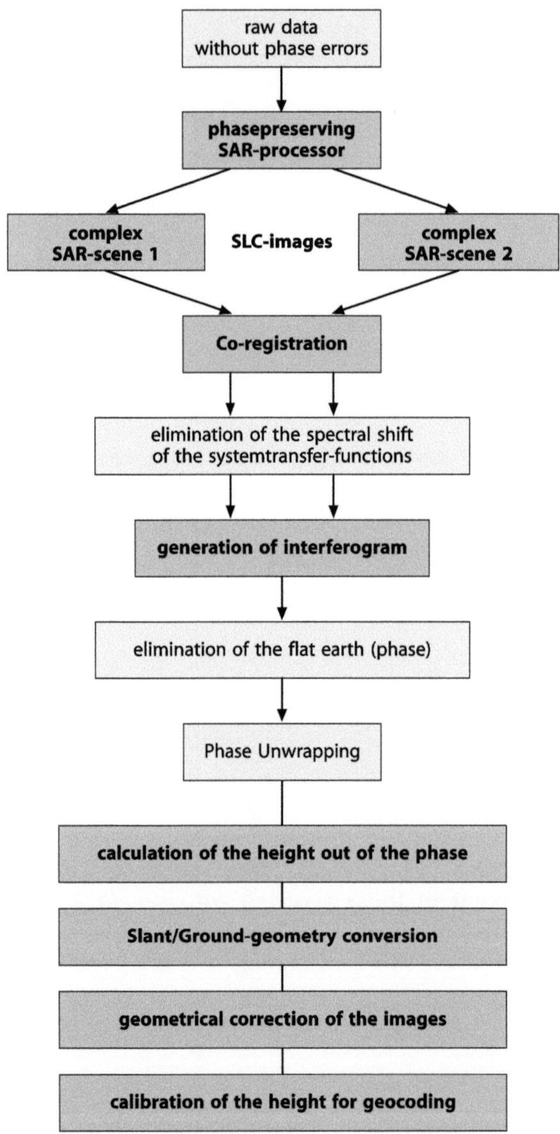

Image 5.1 a: Processing chain

running-time differences of the radar waves due to changing tropospheric or ionospheric conditions between two overflights
thermic noise of the radar system
rotation of the target concerning the perspective of the radar in non-parallel orbits
phase errors whilst recording or processing data
non-precise coregistration

spectral displacement of the system-transfer functions (due to different perspectives a dissolution cell has been considered. Caused in range by the (necessary) baseline, in azimuth by probable different antenna squint angles).

Examinations of those decorrelation effects concerning their appearance, their effects on the phase preciseness of the interferogram, their system theoretical descriptions are described among others in [ZEB92], [JUS94], [GEU95]. They will not be considered any further here. The necessary steps towards the received raw data to the geometrically corrected digital height model will be revealed in an easily comprehensible way (see Image 5.1).

Further on we illustrate some newly developed parts of the interferometry processor. After a detailed discussion the corresponding results are presented.

In particular the discussion on the newly developed parts differing from those earlier mentioned will be considered with the help of a further component.

In order to derivate the interferometric processing flow analogously, we will systematically discuss the developed algorithms sequentially following their particular purpose.

5.3 From the raw data to the digital evaluation model

The processing flow is described in an overview starting from the raw data to the geometric corrected digital evaluation model. The used images do not compulively corresponding with each other, in fact meaningful scenes or parts of them are used to demonstrate the corresponding processing effects.

DHM calculation from interferometric SAR raw data

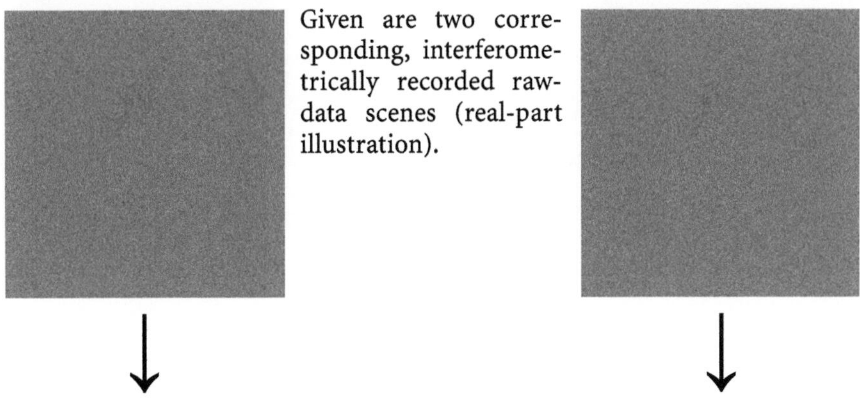

Given are two corresponding, interferometrically recorded raw-data scenes (real-part illustration).

Processing

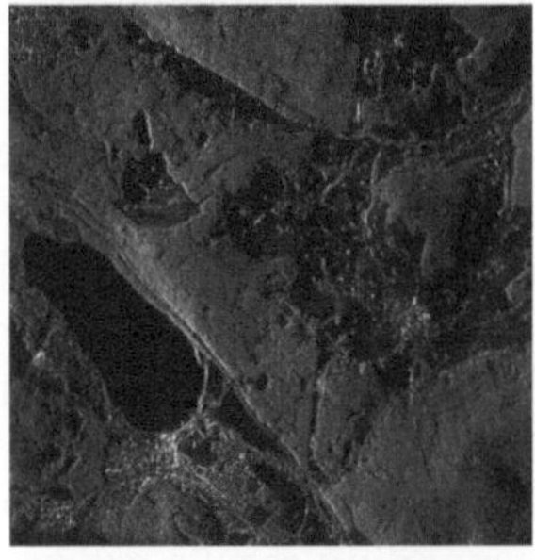

After the phase-preserving processing two complex SAR scenes originate, the so-called single-look complex (SLC) Images (real-part illustration). The displaced illustration serves to clarify the different recording positions.

$$\underline{z}_1(r, a) = a_1(r, a) \cdot e^{j \cdot \varphi_1(r,a)}$$

$$\underline{z}_2(r, a) = a_2(r, a) \cdot e^{j \cdot \varphi_2(r,a)}$$

Coregistration

The coregistration corrects the geometric recording distortions due to the different recording positions of both receiving antennas (real-part illustration)

Calculation of the interferogram

$$\underline{z}(r, a) = \underline{z}_1(r, a) \cdot \underline{z}_2(r, a)^* = a(r, a) \cdot e^{j\varphi(r,a)}$$

with

$$\varphi(r, a) = \varphi_1(r, a) - \varphi_2(r, a)$$

After complex conjugated multiplication of both complex single images the complex interferogram originates containing the current height information.

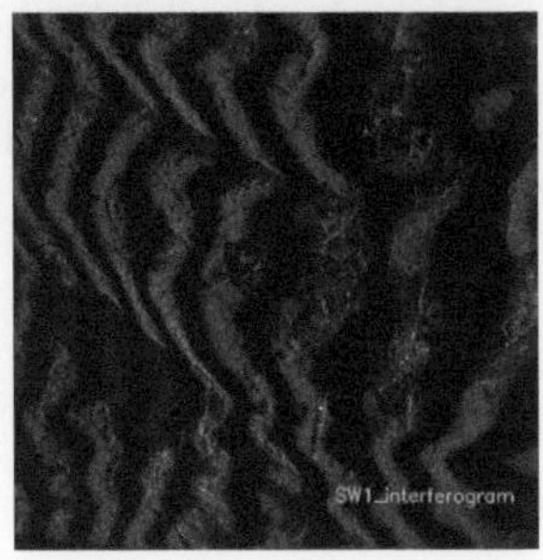

Displayed is the real part of the complex interferogram. The fringe formation (phase ambiguity) is already clearly recognisable in the interferogram.

Calculation of the interferometric phase

$$\varphi_m(r, a) = \arctan_2\left(\frac{\mathrm{Im}\{\underline{z}(r, a)\}}{\mathrm{Re}\{\underline{z}(r, a)\}}\right)$$

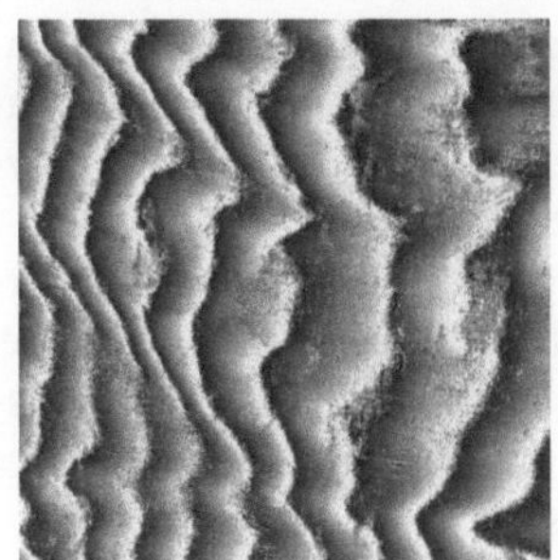

To utilize the height information out of the complex interferogram with the help of the creation of the arcus tangent function, the phase is calculated.

The phase skippings due to the not illustratable evaluation dynamics are clearly visible.

Calculation of the unambiguous phase

$$\varphi_m(r, a) \Rightarrow \varphi_h(r, a)$$

The problem of the phase ambiguity is solved with the help of the so-called phase-unwrapping algorithm. Different methods are discussed in [KRÄ98] and will not be considered here.

The illustration contains clearly visible the "flat earth" due to the side-looking method (Side Looking Geometry – DO-SAR) as slant-offset term in the evaluation.

Geometric calculation of the slant height out of the unambiguous phase

$$h(r, a) = f(\varphi_h(r, a))$$

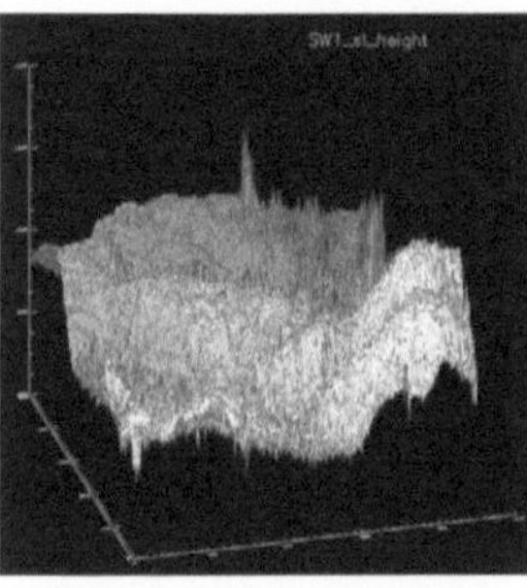

By using the derivations made in chapter 5.2 and [LOF] we can generate an height-information from the unambiguous phase-information. The required calculation needs as entry-information the so-called base line.

Without knowledge of the baseline, an exact calculation of the height is not possible. The estimation of the baseline is not discussed at this work. The baseline problematics are discussed for instance at [SMA93].

Geometric calculation of the ground height from the slant height

The substantial change in the illustration in opposite to the slant-height model allows an easier control of the changing ground distances due to the prevailing evaluation of the observed area.

The higher the recorded area the larger is the distance of the corresponding pixels (fore-shortening).

Geometric corrections (layovers, shadows) of the ground height

The last geometric correction concerns the shadows of the radar and layover areas before finally creating a digital terrain model.

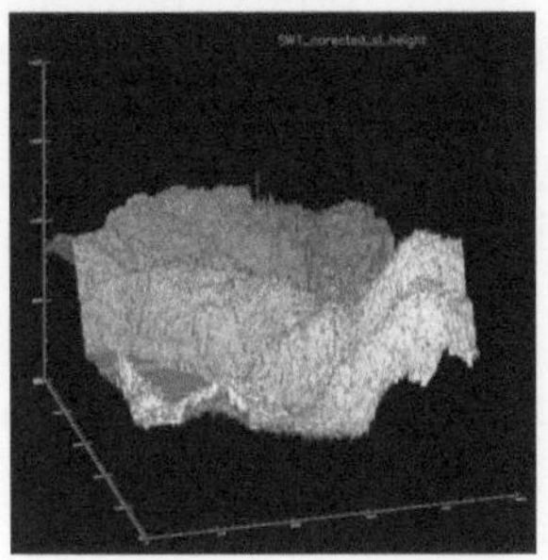
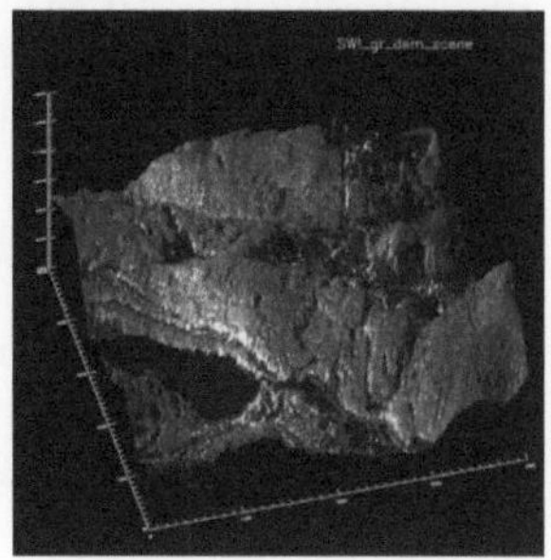

A new method for SAR processing (SIFT) is already presented in chapter 4. The problem concerning the phase unwrapping is already discussed in [KRÄ98]. The task of precise estimating of the base line (base line estimation) is processed at the moment [LOF]. Further on we will discuss mainly the missing details for the coregistration, for the slant/ground conversion, for the layover-, shadow- and foreshortening correction, which are necessary for the closed description of an interferometry processor. The last step reveals a method to automatically generate the elevation-pass points (plus geocoding).

5.4 Coregistration

5.4.1 Problem and requirements for the image registration

As already mentioned at the beginning of this work, the imaging geometry of both single SAR scenes due to the different satellite orbits (ERS) and thus the different recording positions (aircraft-SAR) the imaging geometry of both single SAR scenes is generally different from each other. The distorted and against each other shifted image data have to be adapted in a way that they overlap preferably well (high coherence). This method is called coregistration.

According to the results of [JUS94] concerning phase statistics and decorrelation of SAR images, inaccuracies within the coregistration lead not to a phase offset but to a phase variance.

This attracts attention to a relative displacement of both images by fractions α from a dissolution cell a decorrelation part for the correlation coefficient of:

$$|\gamma| = \mathrm{si}(\alpha) \tag{5.9}$$

The standard deviation of the interferogram phase dependent on the displacement α is illustrated in Image 5.3 for SNR = 10 dB and SNR = ∞:

When fixing the requirements towards exactness for the coregistration process it is to consider that even a perfectly coregistrated image and an identical transfer function (after eliminating by filtering the existing spec-

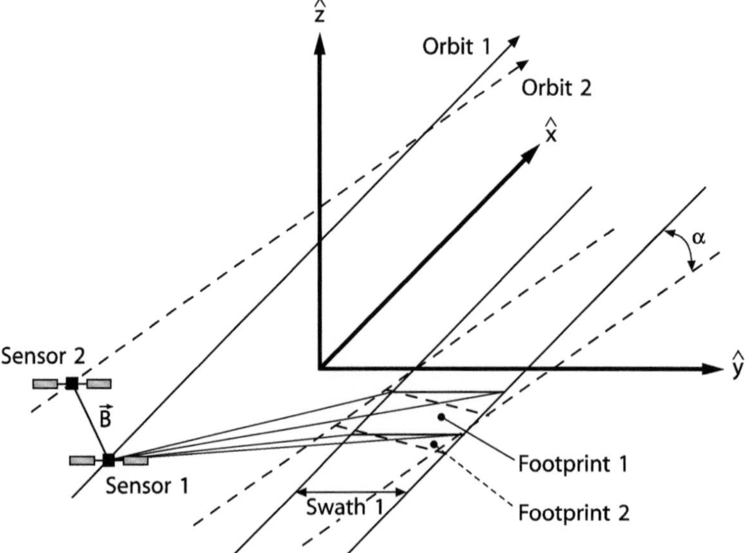

Image 5.2: Geometry to the coregistration

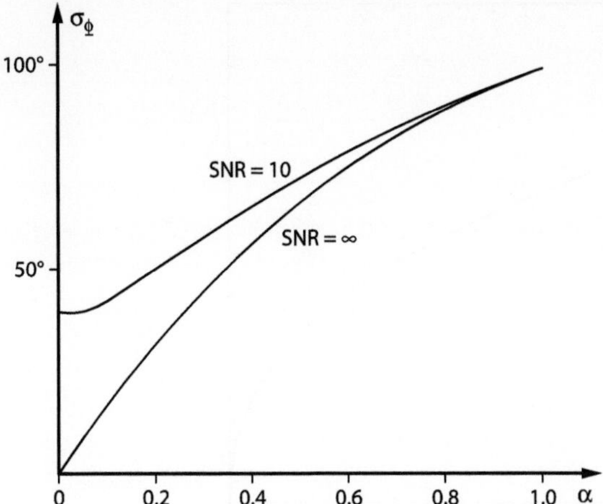

Image 5.3: Standard deviation of the phase depending on the displacement α in parts of a dissolution cell in azimuth direction according to [JUS94] (is valid also for the displacement in range direction)

tral displacement in range direction) cannot prevent the appearance of a phase deviance due to the thermic noise. The value of the correlation coefficient according to equation (5.8) is reduced (with identical SNR's in both overflights) to:

$$\gamma = \frac{1}{1 + \mathrm{SNR}^{-1}} \tag{5.10}$$

The standard deviation of the phase as term of the coherence is illustrated in Image 5.4.

To achieve a error standard deviation of $< 60°$ the average coherence has to be better than 0.8!

With typical SNR's of SAR images (10–20 dB) [GEU95] phase-deviations of about $15°$ to $40°$ already result. These values are trend-setting concerning the required exactness for the coregistration.

Image 5.3 says that at a phase standard deviation of $40°$ (for SNR = 10dB) already available due to the thermic noise, a displacement α of about ≤ 0.1 resolution elements affects no further aggravation of the phase preciseness.

The both complex SAR images must be coregistered with a preciseness better than 0.1 resolution element, so as not to increase the phase difference. A resolution element has (ERS-1 SAR) a size of about 9.6 m in range direction and 5.0 m in azimuth direction. In the SAR images provided, a pixel is in accordance with 7.9 m in range and 3.9 m in azimuth.

The exactness of the adaption of the two images after the coregistration should be better than about 1/8 pixel (0.1 resolution element is equivalent to 0.122 pixels in range and 0.128 pixels in azimuth).

The usual method to perform an image correction originates from the photogrammetry [SCHW95]. It requires three essential processing steps:

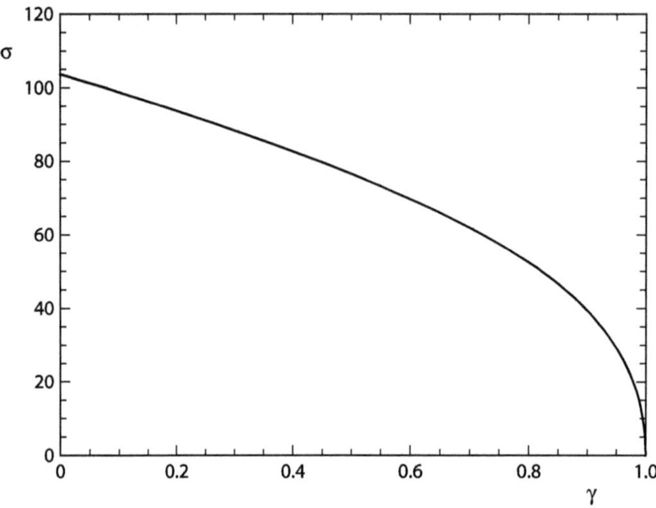

Image 5.4: Standard deviation of the phase error as term of the coherence [LOF]

the determination of the mapping function between both images
the new mapping of a slave of both images depending on the corrected distortion
a following resampling of the equalised image to the sampling distance of the first non-equalised image (master).

In the past years a series of different methods to determine the mapping term [SCHW95] were suggested. All of them based on the method of using the coherence of the SAR signal as a dimension for the quality of the adaption of both images (master, slave).

The accuracy of the adaption of the two images, as already discussed, has to be in the subpixel range. In practice, for this purpose a rough registration is performed first, which overlaps the two images in a way that the positions are deviating corresponding to their pass points (at best 1 to 10 pixels from each other per dimension).

Using the following precise registration we get the exactness in the subpixel-range whereas for the subpixel exact determination of the position of the image points we use interpolation methods.

1. Determination of the position of n pass points
   Determination of the position of the pass points (X_i, Y_i) in the slave image (image to be displaced)
   Usage of a capable method for determination of the position of the same (corresponding) pass points in the master image (x_i, y_i) (image serving as reference)

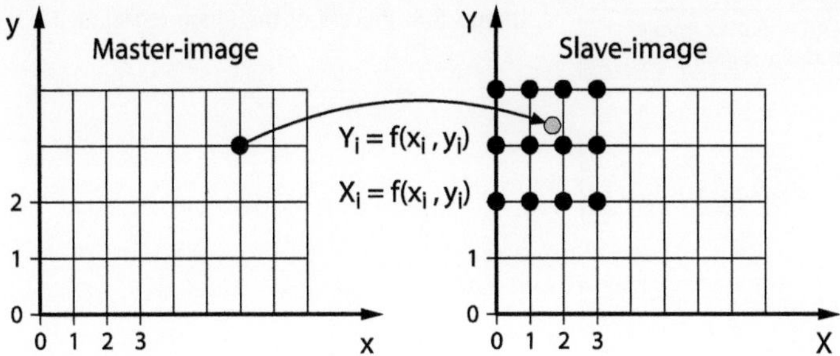

Image 5.5: Necessity of resampling

2. Estimation of a transformation function (mapping function) by using the known position of the pass points in a way that:

$$X_i = f(x_i, y_i) \tag{5.11}$$

and

$$Y_i = g(x_i, y_i) \tag{5.12}$$

3. Transformation
Usage of the mapping function: finding out all points positions of the slave image corresponding to the associated points in the master image, in other words transformation of the slave-image points to the master-image grid. That results in whole-numbered coordinate positions that have to be interpolated. This procedure of scanning for a regular grid is called resampling (see Image 5.5).

As already mentioned, in the beginning of the precise registration at least segments of an image have to be interpolated in order to achieve the sub-pixel preciseness in the coregistration.

Most coregistration methods known work by following the described principle. However, they notably differ from each other in the way of the correlations analysis (in the location respectively frequency range), in the transformation functions used and in the interpolation methods (achieving a subpixel preciseness and when resampling). A discussion of the different principles is described in detail in [KNE98].

The coregistration is performed in principle as explained in the beginning of chapter 5.4.

5.4.2 Rough registration

The rough registration of the two Images is performed according to pattern in Image 5.6. As for the determination of pass-point pairs (reference-

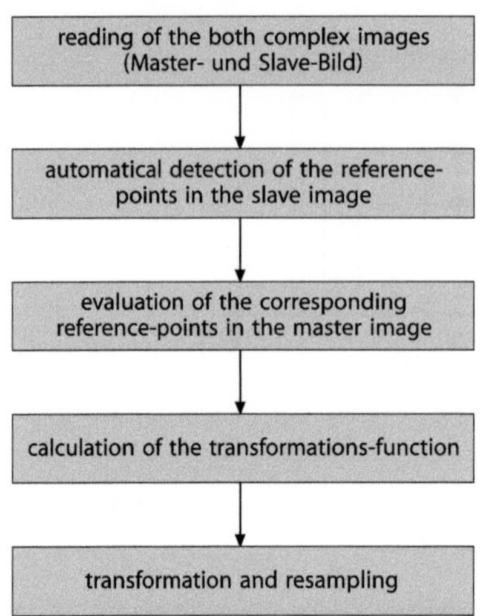

Image 5.6: Process of the rough registration

point pairs) a special correlation analysis with both images has to be performed we at first use a slave-image window to select reference points. The
pixel on top of the left side of the corresponding window is the reference
point (see Image 5.7).

For selecting the windows of the slave image certain eligibility criteria
apply. At first the reference points should be distributed in a geometrically
reasonable way around the image (e.g. they should not be all in a row or
in a line), to allow the generation of exact transformation functions.

As a further criterion we have a look at the average information content
of the window (the slave-image segment) as an image range with high information content can be found more easily in the master image (respectively can be differentiated from other image segments). An explanation
for the necessary information, entropy and others, a historically and

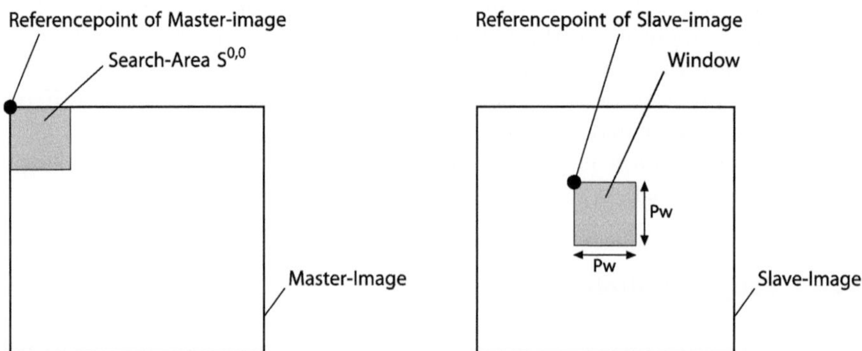

Image 5.7: Display detail and position of reference point

mathematically detailed description of the information theoretical approaches and terms can be retrieved from [ARN96].

To describe a set S_N, containing $N = 2^n$ elements, this information will suffice:

$$I = n = ld(N) \quad \text{(pseudo unit: bit)} \tag{5.13}$$

known as Hartley's information, when all elements occur with the same plausibility and their appearance is answered with "0" or "1". If we now accept sets appearing with different plausibility their elements with the same plausibility, though, we need to describe an element of the set S_{total} the information in which set is that element respectively which element within the set is meant.

Concerning the interesting display detail, each value (maximum p values at P pixel) basically represents a set with the element number 1. That means this is the information indicating which set the element is part of. This is the so-called Shannon's information (entropy):

$$I = -\sum_{k=1}^{n} p_k \cdot ld(p_k) \tag{5.14}$$

whereas p_k represents the plausibility of appearance of the value k (and thus also the set S_k).

The maximum value of the information entropy, $I = ld(n)$, occurs in the identical plausibility ($p_k = 1/n$). If there is an image containing a maximum of 256 grey scales (8-bit quantisation) and the histogram is showing that each grey scale occurs exactly x-times, an equipartition is presented, and the average information content is 8 bit (= maximum).

The necessary windows are automatically selected in a geometrically reasonable way and in addition due to the possibly highest average information content. The user has to specify the number and the wanted size of windows (whereupon it can be selected between 3, 4, 5, 8, 12, 16, and 25). (To keep things simple, a quadratic window is chosen. A differentiation of the windows' size in x-direction respectively y-direction would not reveal a noticable improvement for the continuing coregistration process).

Generally, a display detail of the slave image is displaced with the increment 1 pixel over the complete master image in order to find the corresponding master-image segment (Image 5.9). That means it is assumed that the user has no information on the position of the corresponding master-image segment. This assumption is reasonable with SAR images because of their characteristical structure (Speckle effect), similar display details without performing any further processing can only be found with high uncertainty (see Image 5.10).

Whilst proceeding with the registration process we have to examine the corresponding reference positions in the master image.

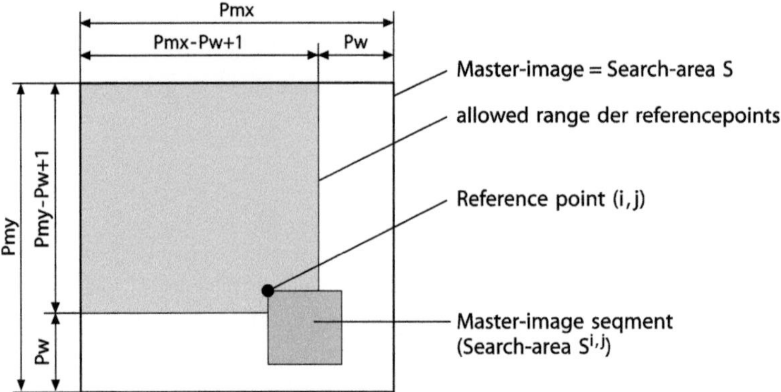

Image 5.9: Formation of the master-image segments

Image 5.10: SAR-image segment, ERS-1, Egypt

Surface Spline

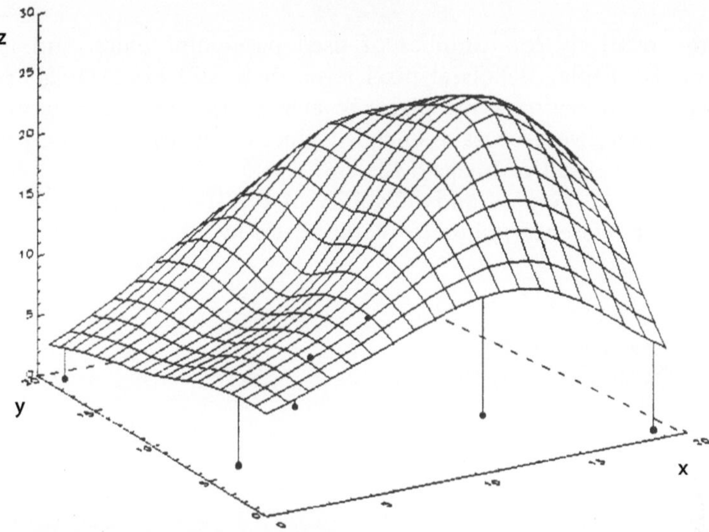

Image 5.11: Surface-spline interpolation

tion analysis explained in [KNE98]:
    calculation of the sum of the absolute value of the differences – as number of the similarity of the image segments
    direct calculation of the correlation coefficient in the time domain
    calculation of the correlation coefficient by multiplication in the frequency domain.

The calculation of the reference point positions is done because of the velocity advantages in opposite to the alternative method, the method of the sum of the abolute value of the differences.

After determination of all pass-point pairs the transformation functions can be arranged. According to the examination in [GEU95] and [KNE98] surface splines are used as transformation functions. They show in opposite to the traditional polynom transformations (bilinear interpolation) an improvement of almost 10% concerning the following measurable coherence.

To explain the principle let us have a look at Image 5.11. The used terms x and y illustrate the master-image coordinates, the fictive "elevation" z is equivalent to the slave-image coordinates. The final resampling is also done with surface splines because of the above mentioned reasons. The rough registration is finished.

5.4.3 Precise registration

Because of the relatively low number of used pass-point pairs, and the comparison of the display details at pixel level, there still exist coregistration inaccuracies concerning some pixel. To achieve the desired exactness in the subpixel range, we process a so-called precise registration. The process in principle is a result of Image 5.12.

To achieve a high exactness and reliability of the transformation function, quite a lot of reference points (in comparison to the size of the slave

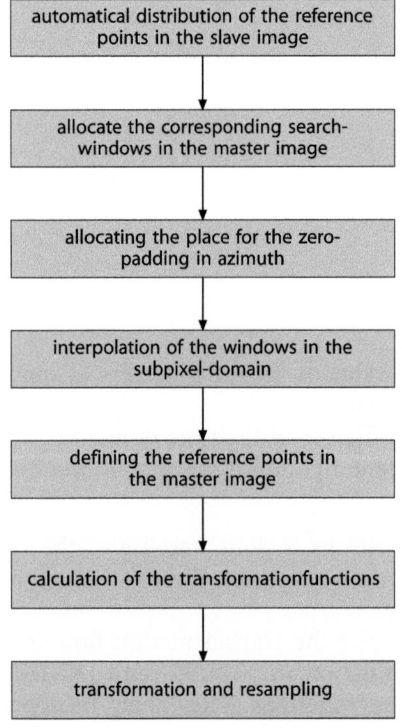

Image 5.12: Process of the Precise-registration

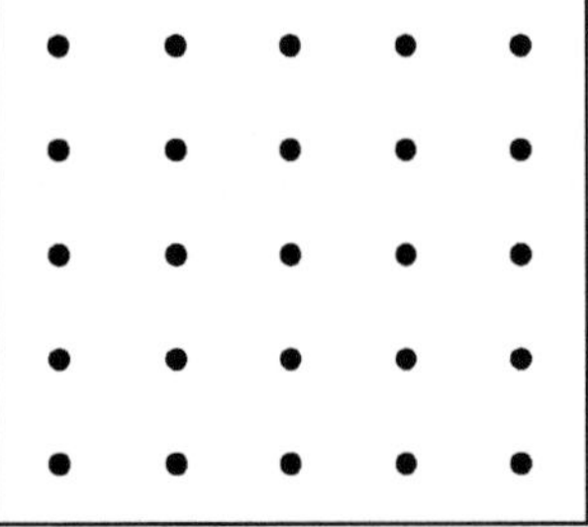

Image 5.13: Automatic generation of pass points

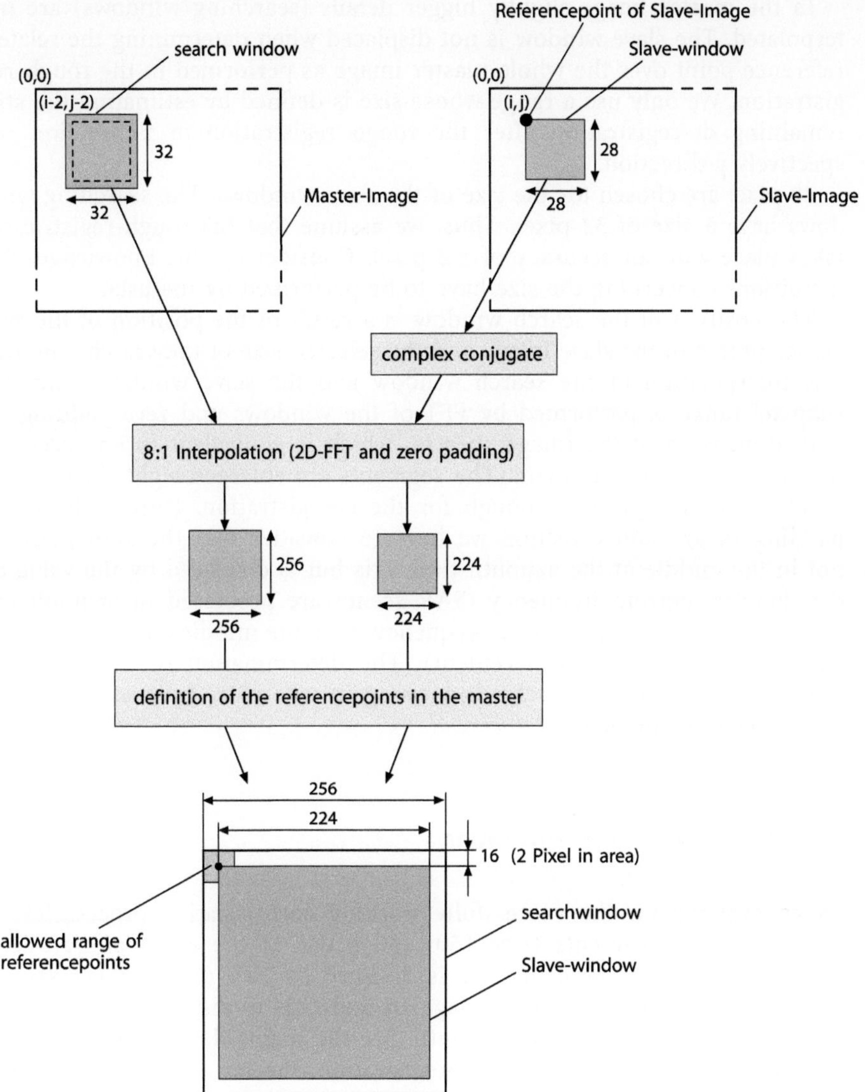

Image 5.14: Principle for determination of the pass-point pairs at the precise-registration

window) are distributed in the slave image. A selection due to the information content is not considered according to the great number of reference points and the thus relatively low displacement of a window.

Selecting the windows size requires that neither the slave image nor the master image has to be interpolated completely into the subpixel range.

Instead of this, we consider in the slave image only the segments specified by the position of the reference point and the selected size of window.

In the master image slightly bigger details (searching windows) are interpolated. The slave window is not displaced when determining the related reference point over the whole master image as performed in the rough registration. We only use a range whose size is defined by estimating the still remaining de-registration after the rough registration in x-direction respectively y-direction.

28 pixel are chosen for the size of the slave windows. The searching windows have a size of 32 pixel. Thus, we assume that the rough-registration takes place with an accuracy of ± 2 pixel. Considering this knowledge, the definitions concerning the size have to be performed by the user.

The position of the search window is a result of the position of the reference points in the slave image and the selected size of the search window. The interpolation of the search window and the slave windows into the subpixel range is performed by FFT of the windows and zero-padding in both dimension of the image spectra, which is equivalent to an excessive scanning in the time domain. The segments are enlarged eight times as an exactness of 1/8 pixel is enough for the coregistration. During the zero-padding in azimuth direction we have to consider that the zero point is not in the middle of the azimuth-time axis but is displaced by the value of the doppler-centroid frequency (SAR scenes are processed in azimuth direction in the doppler-centroid frequency, thus the middle frequency of the characteristic doppler displacement). The determination of the reference point positions in the search windows again take place by examination of the correlation coefficient.

5.4.4 Results of the coregistration

As an example for the successfully working coregistration process let us have a look at segments (size 150×150 pixel) of the given aircraft-SAR images (DO-SAR). The coherence is declined by 20% (additive Gauss-distributed noise) and the slave image in addition is displaced by about 2 pixel in range and 1 pixel in azimuth. See the manipulated SAR images illustrating now input data of the coregistration (master image and slave image) in Images 5.15 and 5.16.

The rough registration proceeds by spatial correlation analysis via 8 search windows each with a size of 15×15 pixel. The following precise registration works with 15 windows and a window size of 32×32 pixel.

In image 5.17, the coregistrated result is decribed less meaningful, though. The performed calculation for five randomly selected windows for the current range displacements and azimuth displacements result in a registration preciseness of 1/8 pixel and in a substantial improvement of the coherence of partially more than 200% (see table). For this purpose we adhere to the phase failures below 1% with the aid of the surface-spline interpolation.

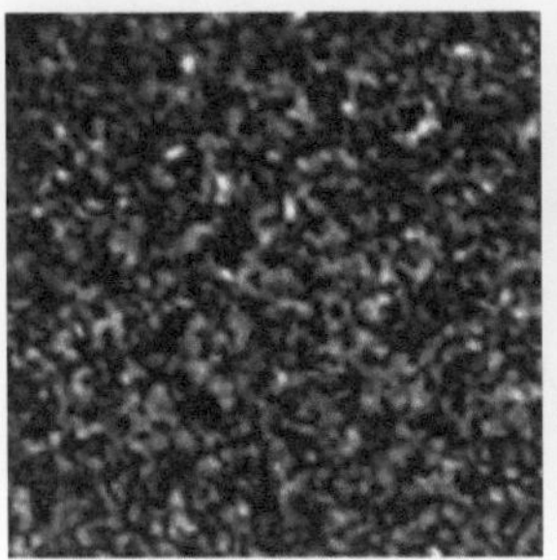  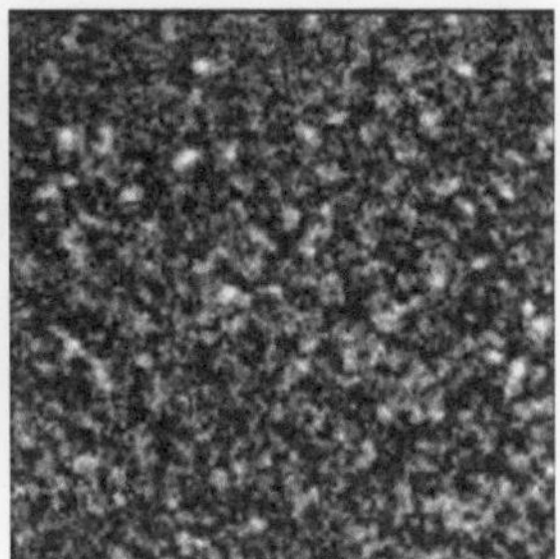

Image 5.15: Master image   **Image 5.16:** Slave image   **Image 5.17:** Coregistrated slave image

Value of the coherence Window no.	1	2	3	4	5
Before coregistration	0.2108	0.2919	0.2457	0.2307	0.2519
After rough registration	0.6144	0.6671	0.5292	0.5628	0.5858
After precise registration	0.6423	0.6993	0.5452	0.5981	0.6093

The introduced method is not only an extremely powerful tool to solve the coregistration problem, the coordination of distorted data arrays is elementary in the field of remote-sensing methods e.g. for processing of the geo-referencing problem.

Above all, the application of geo-information systems (GIS) due to different recording sources and their different recording positions confront us time after time with the task of referencing local non-linear distorted data sources against each other instead of allowing the corresponding usage of the GIS, namely evaluation of different data-fusion techniques.

In the course of this work we will use again this explained coregistration method during the so-called geocoding. Then not the local distortion (distortion of two levels) is corrected but calculated the correction of generated evaluation models.

But first we will discuss preliminary problems concerning the geometric transformation and corresponding modifications of the evaluation model.

5.5 Geometrical conversions and correction

5.5.1 Orthometric projection and foreshortening

In chapter 5.2 (Image 1.8) and chapter 5.3 further steps for generation of the interferogram after the performed SAR processing and the described

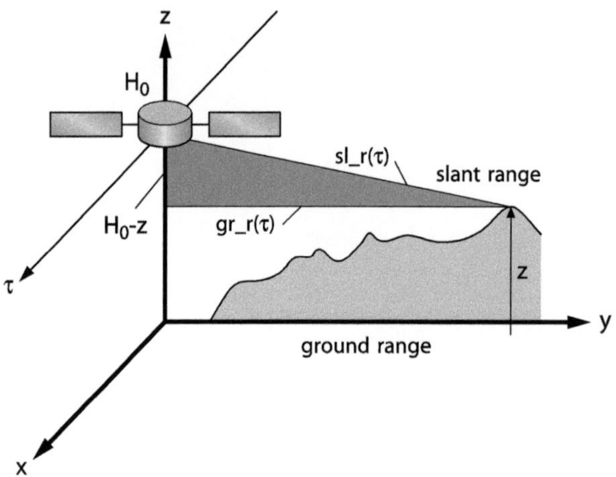

Image 5.18: Slant/ground-recording geometry

coregistration apply. To them count among others the "elimination of the spectral displacement of the system-transforming function", "the elimination of the flat earth" and the "phase unwrapping". This listed processing steps are derived in [KRÄ98] and are not considered here.

The conversion of the unambiguous phase due to phase unwrapping into the corresponding altitude information is explained in chapter 5.1 and illustrated in detail in [LOF].

In the course of this work the respectively necessary correction methods are illustrated and their realisation discussed afterwards.

In consequence of the recording geometry, the elevation informations coming from the interferogram are created in the slant-range domain, thus not in a rectangular coordinate system but in a slant-distance system. Image 5.18 *Slant/ground-recording geometry* clarifies this correlation.

Initial point of the conversion therefore are the interferometric altitude data in the so-called slant distance (slant distance sl_r). The calculation of the ground distance gr_r takes place by inclusion of the known flying altitude H_0 of the SAR sensor and the known elevation of the object (quasi online GPS measurement on board of the sensor) z via:

$$gr\_r = \sqrt{sl\_r^2 - (H_0 - z)^2} \tag{5.16}$$

As already mentioned in chapter 2, distortions in the altitude illustration are created by the side-looking recording method which has to be realised and eliminated during the slant/ground conversion. That requires among others the so-called foreshortening effect that reproduces mountain sides shortened when they are tilted towards the sensor. Corresponding tilts in the opposite direction from the sensor result in an extension of the reproduction. Regarding Image 5.19, these effects can be seen clearly. During re-

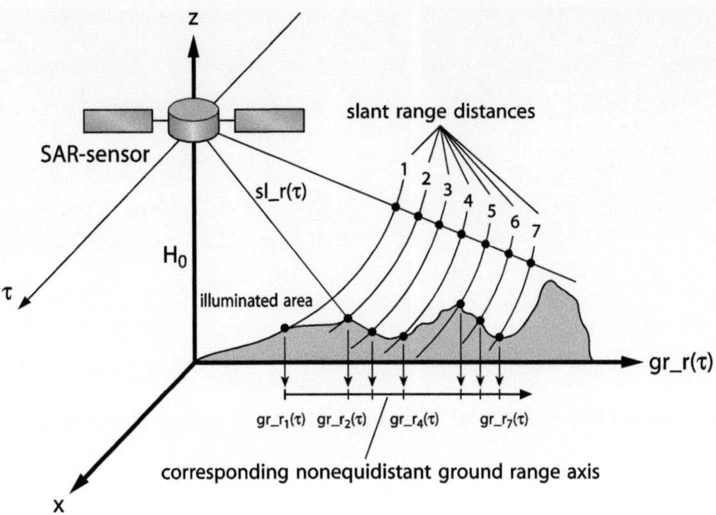

Image 5.19: Foreshortening geometry

gistration of the interesting surface, original ground-elevation values are recorded as equidistant slant-elevation values by the sensor. The equidistance is only valid concerning the slant distance (slant-range distance) towards the sensor.

The complete SAR-signal processing chain works within the slant-range coordinates system until the geo-referencing (calculation of the heights) begins. To change back into the wanted original ground-coordinates system the corresponding recording geometry has to be reproduced. That means we calculate a ground axis according to equation (5.16) and get for each known height value its current position in the searched ground-coordinates system. Regarding the ground axis drawn in Image 5.19 (corresponding non-equidistant ground range axis) with the converted ground-height values respectively distance values, the above described foreshortening-distortion effect and the necessity of its egalisation by the corresponding conversion is clearly recognisable.

Starting from the slant axis we now realise in slant-ground projection (following the slant pixel at Image 5.19 along the equidistant lines) a prolongation of the mountain sites tilted towards the sensor and thus we retrieve the correction of the foreshortening. This is also valid for the corresponsing heights tilting in the other direction.

Besides the correcting characteristics, the generated ground axis after the conversion appears not to be filled equidistantly with altitude values any longer. Consequently we have to determine the ground axis by interpolation routines in consistent dissolution cells. Doing so, several different methods of interpolation during the coregistration are available [KNE98].

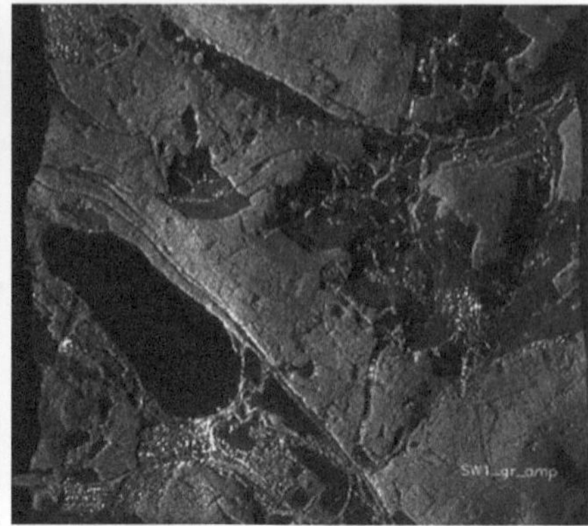

Image 5.20: Slant segment (Titisee, Germany)   **Image 5.21:** Ground segment (Titisee, Germany)

Developing the slant/ground converter, the realised algorithms had to be optimised significantly concerning the calculation time. Consequently, a 1-dimensional, linear interpolation routine in range direction was chosen for the slant/ground converter (also called ortho-projector). Images 5.20 and 5.21 illustrate a SAR scene segment of the test area "Titisee" before and after the ortho-projection. Only the amplitude image is displayed to show the obviously visible conversion distortions that originate as a result of the prevailing height differences in the current positions.

Clearly visible is a much stronger displacement of the original array in the near range. The reason is not to find in the stronger height variation in this position (see height model in Image 5.27), but rather in the steeper angle of incidence of the microwave radiation, in other words the steeper recording geometry and the related stronger distortion characteristics. Thus, the stated calculations reflect the slant geometry to the ground geometry and thus correct the so-called foreshortening effect.

5.5.2 Layover effect

Following the idea of the foreshortening effect consequently leads us to another geometric effect, the so-called layover effect. It describes such a strong shortening of the surface to be projected by the side-looking geometry and the strong height dynamic, that single pixels pass each other, so to speak, during the data acquisition.

This geometric effect (overlapping) is described in Image 5.22. Starting from the ground geometry let us have a closer look at the two pixels, pixel 3 and pixel 4. In consequence of the steep rise of the observed area, pixel 3 is closer to the sensor as the orthometrically nearer pixel 4. The obvious consequence is the permutation of the two pixels 3 and 4 whilst recording into slant coordinates. If we want to record the layover limits in the ground-coordinates system we have to project the current slant-layover limits to the ground axis. This correlation is clarified in Image 5.23 and shows in opposite to the pixel actually determining the layover a strongly enlarged overlapping range in the ground level.

However, before these layover areas can be corrected, a criterion to regognise it has to be created.

Let us again have a look at Image 5.22. Knowing that the pixel permutation is caused by the layover, we can retrieve the areas of the layover effect (layover area) by observing the ground axis respectively its gradient during its generation. If the slant altitude values strongly differ from each other in a way that two successive pixels pass each other, then we are realising the described layover effect. If this case occurs, the calculated ground-range distance, expected to increase according to the distance, will not represent

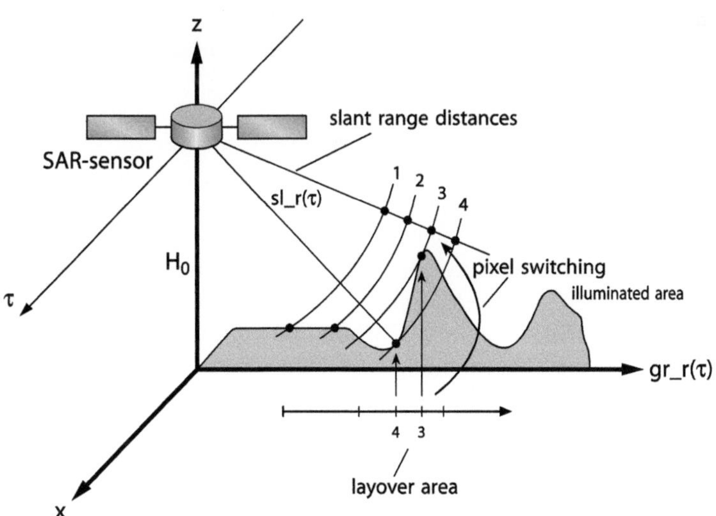

Image 5.22: Layovergeometry

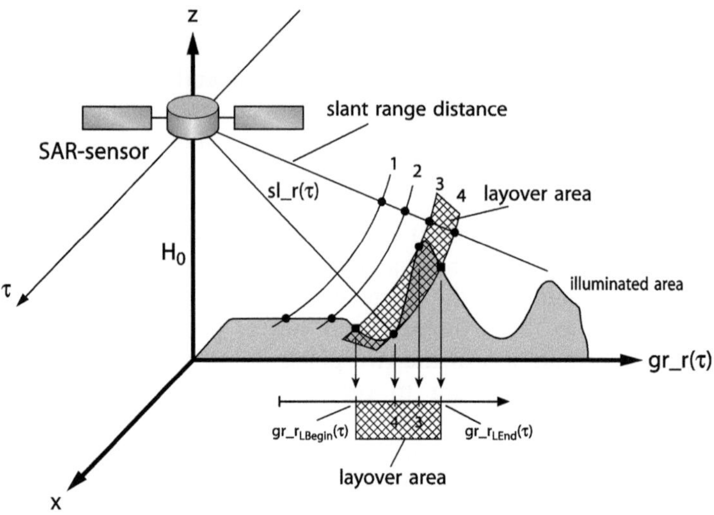

Image 5.23: Layover limits and ground geometry

a cumulative axis. The calculated distances will be shorter, in other words the ground axis will partially show a negative gradient.

The derivative of the slant axis due to the ground axis consequently becomes negative:

$$\frac{dr_L}{dgr\_r_L} \leq 0! \tag{5.17}$$

or in another description according to chapter 2.2:

$$\frac{dz_L}{dgr\_r_L} \geq \frac{gr\_r_L}{H_0 - z_L} \tag{5.18}$$

Due to these considerations, to detect the layover effect during the ortho-projection we have to consider the ground-axis gradient according to equation (5.17). The necessity of the layover control and correction is still an open question. Let us remember the derivations in chapter 2.2, saying that an altitude difference of 1.8 m per pixel is already enough to cause layover effects (for ERS-SAR results a critical altitude limit of about 10 m per pixel). Layover corrections therefore should be taken into consideration particularly for aircraft-based SAR systems with interferometric processing. A real correction in terms of correcting the wrongly recorded height values is not completely possible, though. Caused by the mentioned overlappings it is probable that different pixels are projected on one another. In this case a separation of the overlapped recorded pixels is no longer possible as only one single elevation value for them exists. Methods have to come into op-

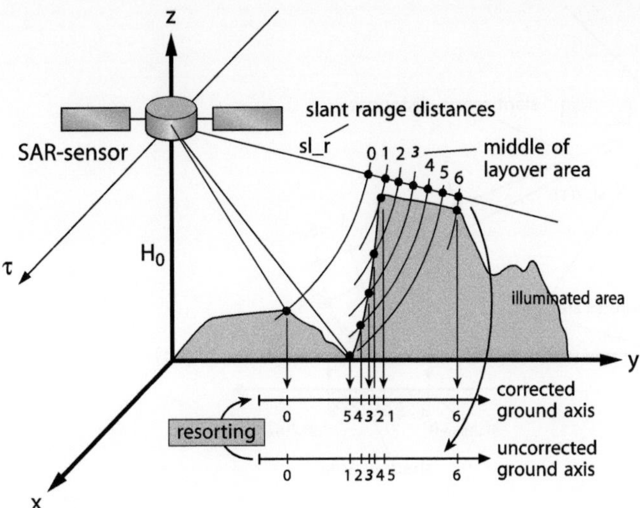

Image 5.24: Layover correction

eration that achieve the most likely restoration of the swapped pixels (see Image 5.24).

A correcting routine has to egalise the phenomenon of passing pixels due to the illuminated elevation.

For this purpose the corresponding range-axis gradients have to be observed closely. The beginning and the end of a layover area can be evaluated by the corresponding gradient. When we know the beginning and the end of an area to be corrected, the heights marked as layover are sorted in a point-symmetrical way around the central point of the layover area. The that way sorted or better resorted height values (and certainly the corresponding amplitude values) this way are treated according to their probable origin and thus fulfil the one and only optimal criterion, the geometric origins.

The this way treated now steadily increasing ground axis finally has to be interpolated because of the equidistant axis projection. Then the layover correction is finished.

5.5.3 Shadow area correction

Talking of geometric distortion and the errors resulting due to the topography, we have also to consider the shadowing effects. The illumination through a SAR antenna is comparable with the traditional illumination through sunlight. As well as with sunlight, SAR illumination creates shadows due to high topographies or other obstacles. Image 5.25 explains the mentioned effect.

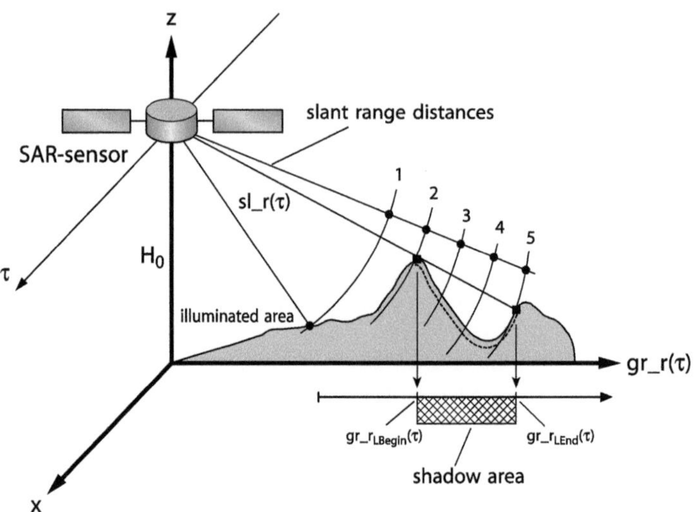

Image 5.25: Shadow geometry

The mentioned areas appear very dark respectively black due to the poor or even non-existing backscattering coefficients in the amplitude image. When backscattering coefficients are not or hardly measurable, the recorded receiving data are very similar to strongly noised data. A special problem results in generating the height values within a shaded area. Due to the strong characteristics concerning the noise of the complex received data, the height information in the shaded areas will also contain strong noisy tones which means that the height models in the shaded areas show significant errors that can be very well within a range of several hundred meters.

Image 5.26 shows a non-corrected height model of the "Titisee" area. In the north-eastern part of the height model you can recognise the height errors in form of multiple, steep peaks.

When calculating the coherence as a standardised cross-correlation coefficient of the shaded parts in the image, only very low coefficient values can be expected. The observed area appears geometrically de-correlated. The described characteristic of the amplitude and coherence in the shaded areas reveal the way that has to be taken to detect the corresponding shaded areas. With the help of a sort of data-fusion technique both in the amplitude field and in the coherence array correspondingly low values have to be looked for generally depending on the kind of the used data source. For aircraft-SAR data (Dornier), coherences of small 0.6 and 8-bit-coded amplitude values lower than 25 are ascertained. The data fusion of both arrays with the help of a boolean operator (OR) results in a marking matrix containing all detected shaded areas.

After recognising and marking of the corresponding shaded areas, especially the strongly defective height-model values have to be corrected. Thus, for the now implemented algorithm above all apply (as well as for

Image 5.26: Amplitude image

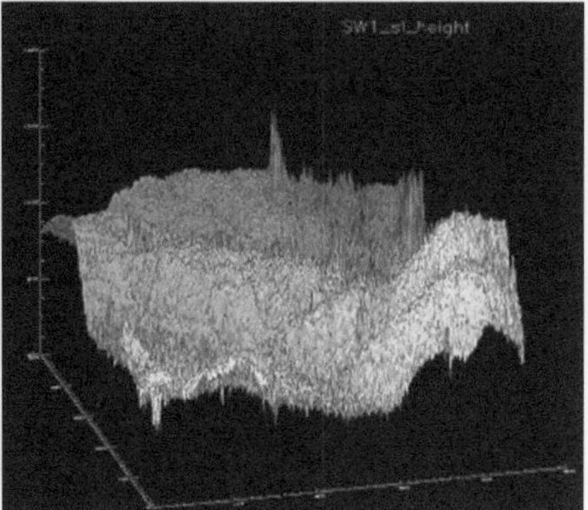

Image 5.27: Non-corrected DHM

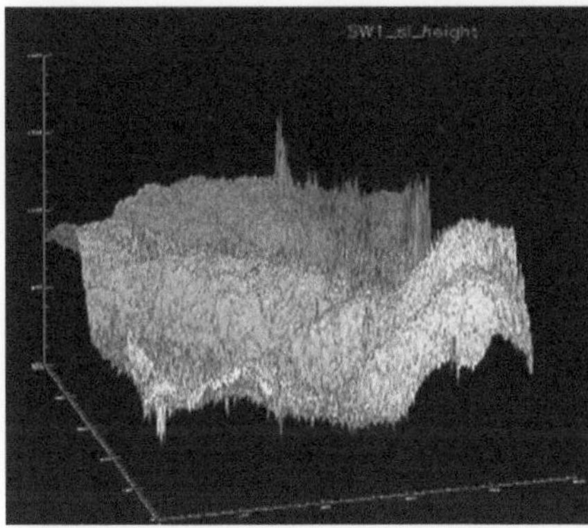

Image 5.28: Non-corrected DHM (scene Titisee), slant-elevation data

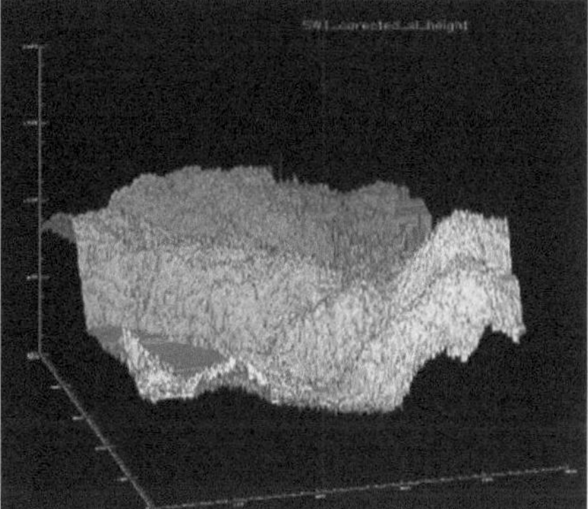

Image 5.29: Corrected DHM (scene Titisee), slant-elevation data

the foreshortening correction and the shade correction) those criteria with regard of the high procession speed. Thus, algorithms basing on iterative or model-based foundations could not be taken into consideration. In fact, it has to be referred to traditional filtering techniques.

Shaded areas mostly found behind high objects such as forests. The assumption that within the shaded areas elevations can be found whose pixel still can be measured behind the actual shaded area, seems to make sense.

To correct these height values the next "correct" (outside the shaded area) height values in increasing range direction have to be used to fill the erroneous height values within the shaded area. For this purpose, traditional techniques have to be used to retrieve the height values. Filling the shaded areas is performed with the help of a fitting routine due to the coherence. Images 5.28 to 5.34 illustrate the results of SAR overflights with DO-SAR.

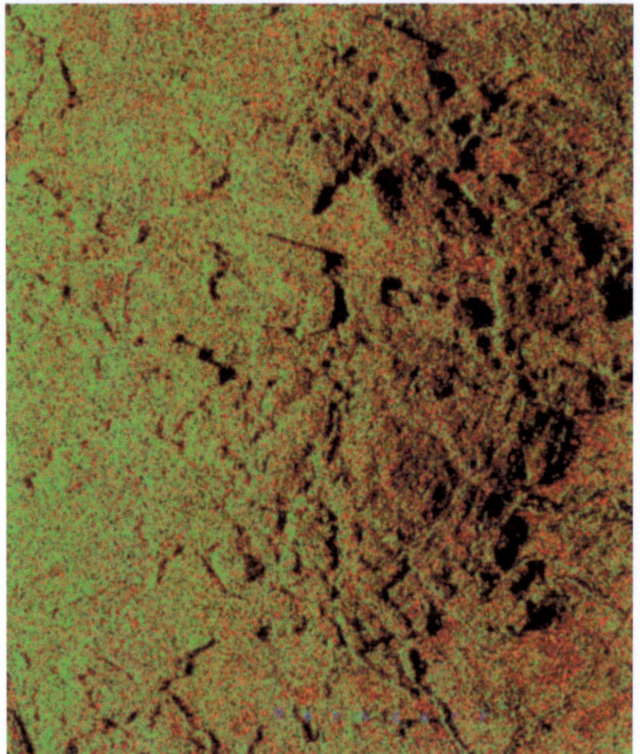

Image 5.30: Scene Markdorf, Germany

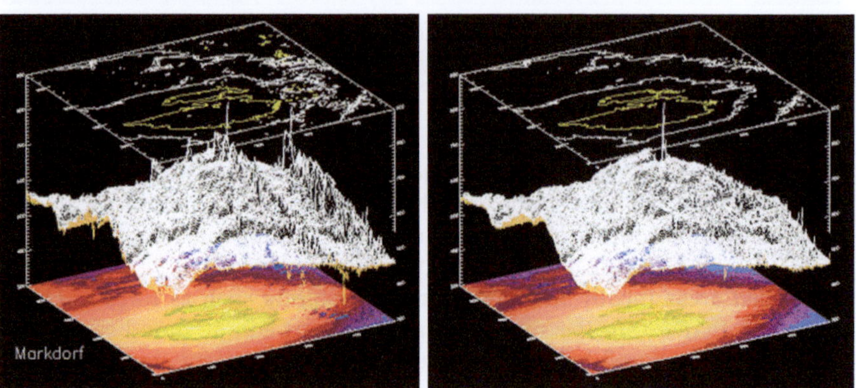

Image 5.31: Non-corrected DHM fill-out   **Image 5.32:** Corrected DHM

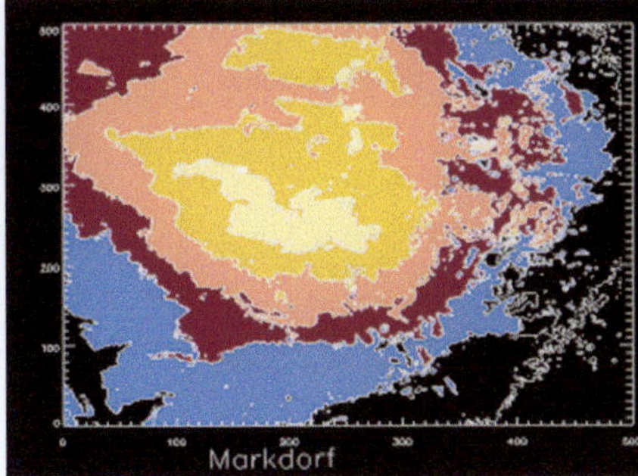

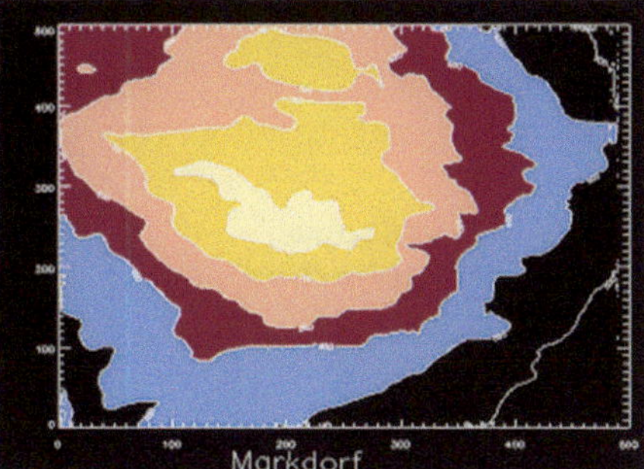

Image 5.33: Non-corrected level curves   **Image 5.34:** Corrected, averaged level lines

A final description of the algorithms is done in Image 5.35. All partial routines are discussed in [HEI97].

Image 5.30 illustrates the amplitude image used as a basis for further height models. In the indicated area "Markdorf" in the eastern half of the image, an "obscuration" of the backscattering effects in the shaded area is to see.

The following Images 5.31 to 5.34 illustrate in addition to the DHM illustrations of the test area "Markdorf, the corresponding level curves, in order to document the substantial improvements concerning the creation of height models (Image 5.34).

5.6 Height-passpoint generation

5.6.1 Introduction

After conversion of the explicit phase into the terrain height, a terrain model is available which is in the side-view (slant range) coordinates system of the SAR, though. To make the DHM usable for further applications, in particular for the geoscience, it has to be referenced onto the earth's surface.

This step is commonly referred to as geo-referencing.

The process of geo-referencing is described usually by the following four processing steps [SCHW95]:

transformation of the orbital data into the local cartesian coordinates system

determination of the cartesian coordinates as spatial position of each pixel on the earth's surface

transformation of the cartesian coordinates into geographic coordinates

transformation of the geographic coordinates into map coordinates.

Principally, in the course of this work parts of the above mentioned evaluations have been implemented. Though the purpose, the generation of geo-

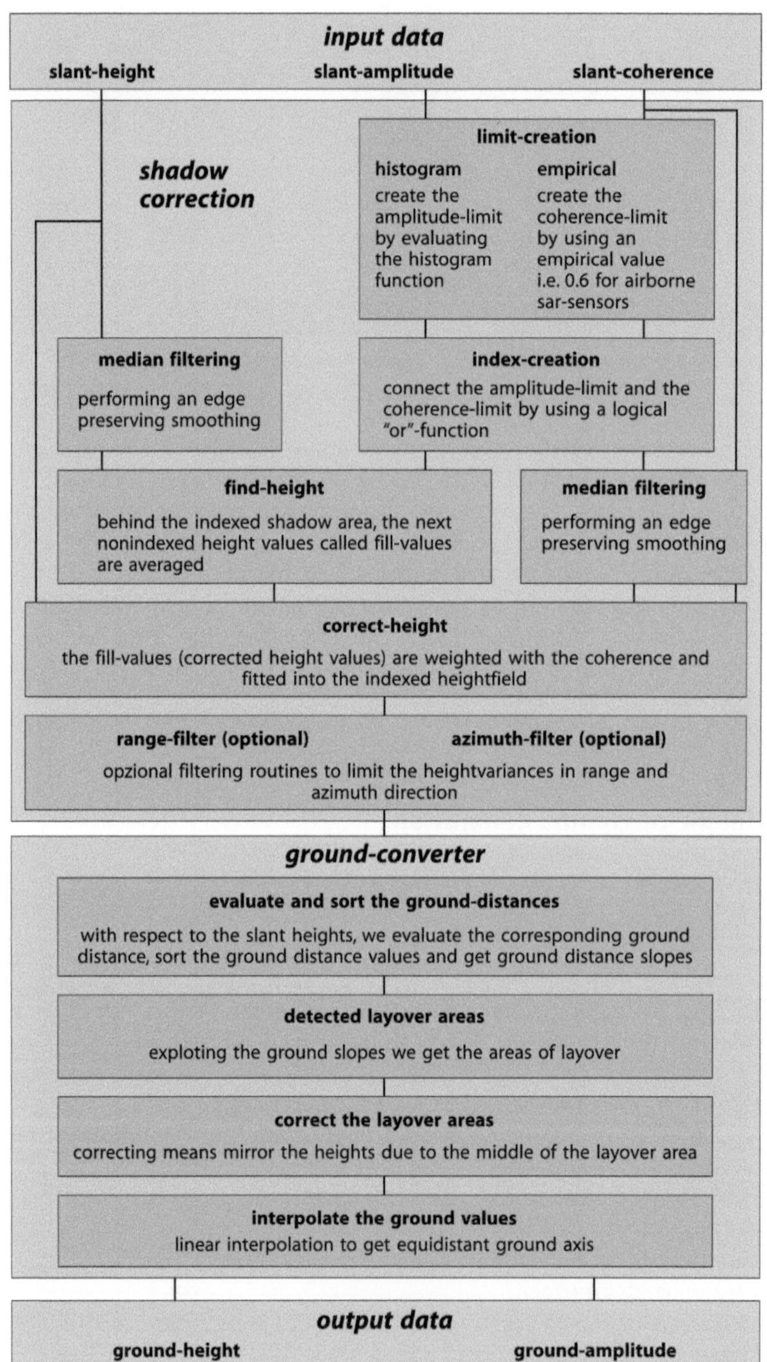

Image 5.35: Implementation of the ground conversion

referenced digital terrain model, is defined identically, still let us look for an alternative way with the aid of our novel algorithms.

The basic idea is the recording of several reference points within the interesting surface in a way that those points can be measured as reference points without additional employment of hardware and measurement techniques, and the measurement coordinates easily be transformed into the necessary geometrically relevant coordinates system.

If it is possible to generate out of the particular overflight data and the corresponding position data of the recording sensor carriers the corresponding reference points (that otherwise are positioned and measured in the area to be observed) thus, with the help of a sort of modified coregistration via the ascertained reference points, the entire DHM (consisting of ground coordinates) can be referenced on the actually prevailing conditions of the corresponding earth's surface. Pre-condition is a common measurement coordinates system.

As reference-point coordinates system the WGS 84 (World Geodetic System 1984) is the best choice as all SAR-carrier movements (required to ascertain the reference points) are monitored constantly by GPS (Global Positioning System).

Thus, no additional measurement effort using the corresponding GPS information is required and both SAR-overflight data and SAR carrier-positioning data exist in the same format. A further advantage when using this method is the reduction of the problem concerning the transformation of the coordinates to a geodesian standard problem, by name the transformation of WGS 84 data into map-coordinate data.

On the basis of the geometrically corrected data, the procedure can comprehensively be displayed as follows:

generating of reference points (will be described in the following)

modified coregistration (merely the input data for the registrating process differ from the usually used SLC data)

conversion of the WGS 84 coordinates into the correspondingly needed coordinates (e.g. Gauss Krüger).

To newly develop the mentioned requirements, there is only one method to generate appropriate pass points in the SAR scenes.

5.6.2 Principle of finding pass points

Pre-condition to realise the following pass point generation is the interferometric overflight of the same area from two different positions, a method following the stereometric principle.

Usually, the SAR overflight (aircraft SAR) is done in a way that every illuminated SAR scene overlaps with the corresponding neighbouring one by 30% (DO-SAR), both in the near range (small range distance) and in the far range (large range distance). The requirement of a double interfero-

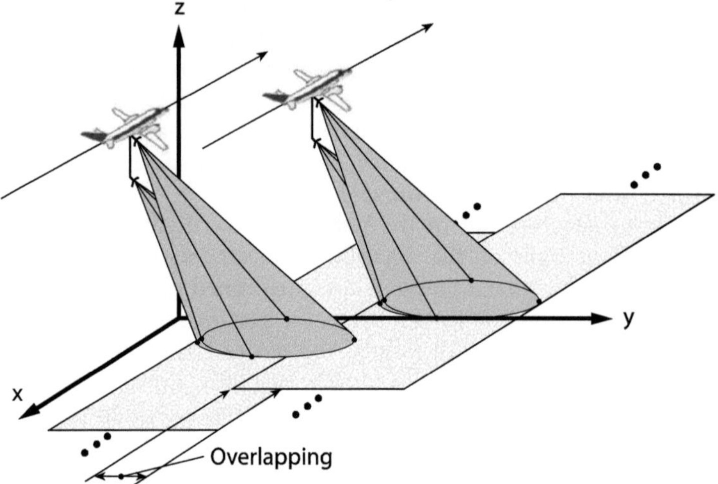

Image 5.36: Overlapping overflight for aircraft SAR, Schwarzwald (SW), Germany

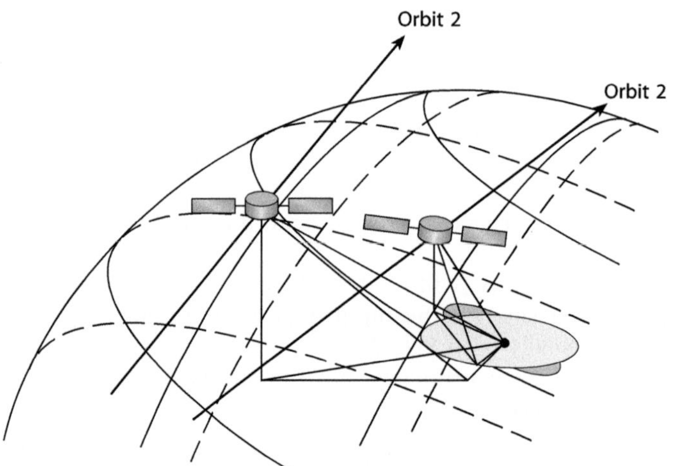

Image 5.37: Overlapping overflight for SAR satellites

metric image recording of the same area does not cause additional over-flights in the case of overlapping image recording.

In the mentioned overlapping areas a pass point geometry is defined according to Image 5.38 *3-dimensional formation for the determination of the pass point elevation h*.

Within this, all geometric parameters necessary for the pass point height calculation are explained. The known following input parameters are valid for Image 5.38:

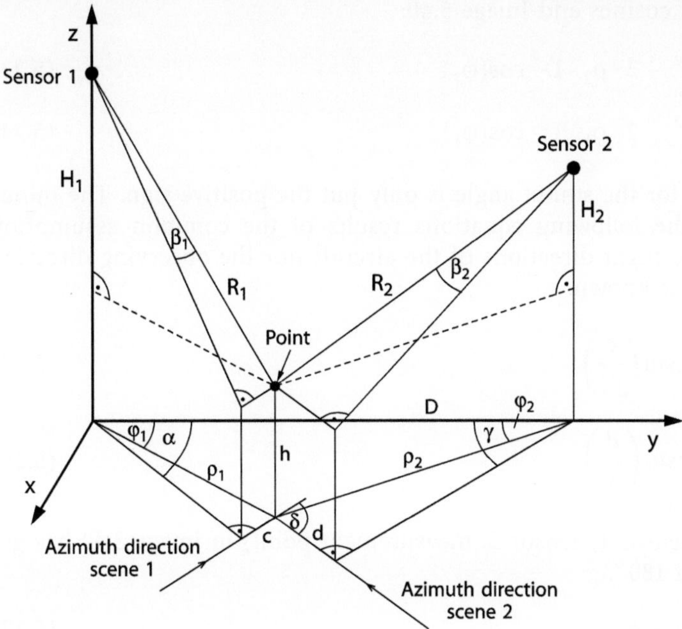

Image 5.38: 3-dimensional formation to determine the pass point elevation h

H_1, H_2: aircraft altitudes (recorded online during the overflight via GPS)

β_1, β_2: the squint angle of each antenna $0\,° < \beta_i < 90\,°$ (retrieved via doppler estimations)

R_1, R_2: slant-range distances of the antennas to the measurement point (on-board measurement)

δ: results of the coregistration and consideration according to chapter 5.6.3

D: distance of the aircraft distance projected on the earth corresponding to chapter 5.6.4

In all following images, the y-axis is placed in the on the earth's surface projected connection of the SAR antennas.

Image 5.38 offers the following equations:

$$R_1^2 = (H_1 - h)^2 + \rho_1^2 \tag{5.19}$$

$$R_2^2 = (H_2 - h)^2 + \rho_2^2 \tag{5.20}$$

$$c = R_1 \cdot \sin(\beta_1) \tag{5.21}$$

$$d = R_2 \cdot \sin(\beta_2) \tag{5.22}$$

with the law of cosines and Image 5.40:

$$\rho_1^2 = \rho_2^2 + D^2 - 2 \cdot \rho_2 \cdot D \cdot \cos(\varphi_2) \tag{5.23}$$

$$\rho_2^2 = \rho_1^2 + D^2 - 2 \cdot \rho_1 \cdot D \cdot \cos(\varphi_1) \tag{5.24}$$

In Image 5.38, for the squint angle is only put the positive sign. The minus sign in both the following equations results of the common assumption that neither the flight directions of the aircraft nor the observing direction of the antenna is known.

$$\alpha = \varphi_1 \pm \arcsin\left(\frac{c}{\rho_1}\right) \tag{5.25}$$

$$\gamma = \varphi_2 \pm \arcsin\left(\frac{d}{\rho_2}\right) \tag{5.26}$$

The triangle (sensor 1, sensor 2, measurement point) in Image 5.40 has got the angle's sum $180°$.

$$180° = \alpha + \gamma + \delta \tag{5.27}$$

The system of equations is one with nine variables (h, α, γ, φ_1, φ_2, c, d, ρ_1, ρ_2) in nine equations. The interesting parameter is h, the height in the respective measurement point. It can not be isolated but it is possible to display an implicit description due to given parameters as follows:

Equation (5.23) in equation (5.25) and equation (5.24) in equation (5.26) result in:

$$\alpha = \varphi_1 \pm \arcsin\left(\frac{R_1 \cdot \sin\beta_1}{\rho_1}\right) \tag{5.28}$$

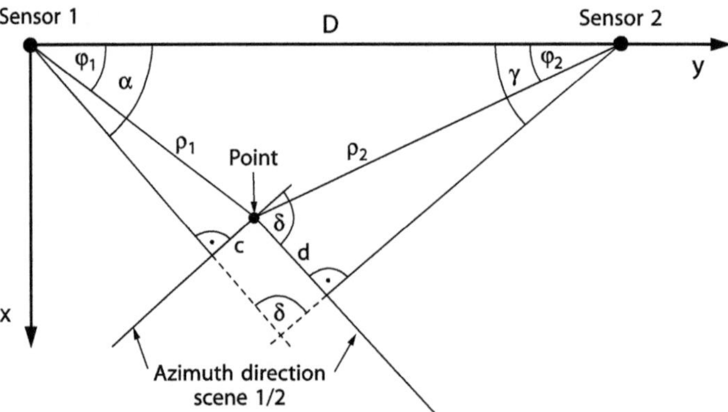

Image 5.40: Projection of the 3D-formation in the x-y level

$$\gamma = \varphi_2 \pm \arcsin\left(\frac{R_2 \cdot \sin\beta_2}{\rho_2}\right) \tag{5.29}$$

These two equations inserted in equation (5.27) result in:

$$180° - \delta = \varphi_1 \pm \arcsin\left(\frac{R_1 \cdot \sin\beta_1}{\rho_1}\right) + \varphi_2 \pm \arcsin\left(\frac{R_2 \sin\beta_2}{\rho_2}\right) \tag{5.30}$$

equation (5.19) converted to equation (5.22):

$$\rho_1^2 = R_1^2 - (H_1 - h)^2 \tag{5.31}$$

$$\rho_2^2 = R_2^2 - (H_2 - h)^2 \tag{5.32}$$

$$\varphi_1 = \arccos\left(\frac{\rho_1^2 + D^2 - \rho_2^2}{2 \cdot \rho_1 \cdot D}\right) \tag{5.33}$$

$$\varphi_2 = \arccos\left(\frac{\rho_2^2 + D^2 - \rho_1^2}{2 \cdot \rho_2 \cdot D}\right) \tag{5.34}$$

and inserted into equation (5.30):

$$\begin{aligned}
180° - \delta = \ &\arccos\left(\frac{D^2 + R_1^2 - (H_1 - h)^2 - R_2^2 + (H_2 - h)^2}{2 \cdot D \cdot \sqrt{R_1^2 - (H_1 - h)^2}}\right) \\
&\pm \arcsin\left(\frac{R_1 \sin\beta_1}{\sqrt{R_1^2 - (H_1 - h)^2}}\right) \\
&+ \arccos\left(\frac{D^2 + R_2^2 - (H_2 - h)^2 - R_1^2 + (H_1 - h)^2}{2 \cdot D \cdot \sqrt{R_2^2 - (H_2 - h)^2}}\right) \\
&\pm \arcsin\left(\frac{R_2 \sin\beta_2}{\sqrt{R_2^2 - (H_2 - h)^2}}\right)
\end{aligned} \tag{5.35}$$

To find the correct height, there are values inserted for h. The resulting absolute value for δ is compared with the absolute value of the distortion of both scenes. Only an iterative way of proceeding can deliver a result for the calibrating value as the arcus-cosine and the arcus-sine can not be transferred into one another.

5.6.2.1 Rotation angle

According to equation (5.35), the distortion of both overlapping scenes has to be known in order to ascertain the wanted height information (the distortion must only be calculated in the projection level x, y). For this purpose, according to Image 5.41 *Determination of the rotation angle* δ, the coordinates of two points in every corresponding scene have to be ascertained. Those altogether four points (respectively eight coordinates) are determined in the overlapping scene areas with the automatic search of similiar "reference points" as part of the coregistration algorithm. The further consideration is geometrically performed according to Image 5.41.

The effective distortion angle δ is a result of:

$$\delta = \delta_1 - \delta_2 \tag{5.36}$$

δ_1 and δ_2 indicate by how many degrees the connection of both coregistrated points is turned away from the negative x-axis respectively ξ-axis:

$$\delta_1 = \arctan\left(\frac{x_2 - x_1}{y_2 - y_1}\right) \tag{5.37}$$

$$\delta_2 = \arctan\left(\frac{\xi_2 - \xi_1}{\psi_2 - \psi_1}\right) \tag{5.38}$$

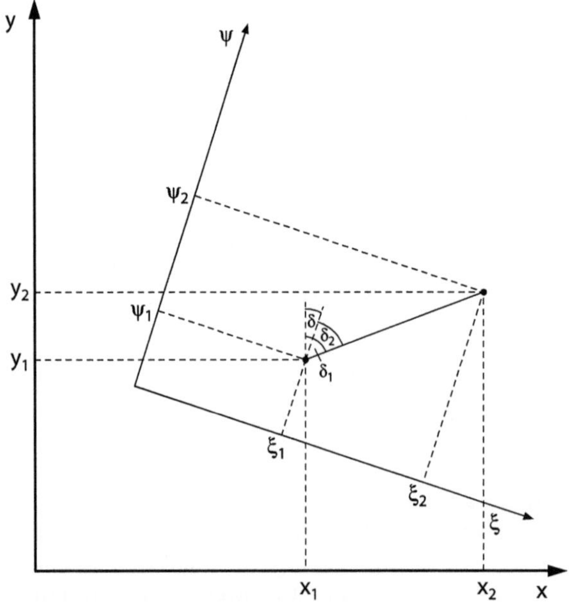

Image 5.41: Determination of the rotation angle δ

The periodicity of the tangent generally has not to be considered as the usual distortion angle can be found within a range of a few degrees (Image 5.36). If substantially wider angles are expected, the value π has to be added to the angle δ according to considerations concerning the plausibility.

5.6.2.2 Distance of the SAR carriers

In addition to the torsion angle δ, it is necessary to determine the projected distance to the earth of both SAR carrier positions to one another during the recording.

For this purpose, the spheric recording geometry of the current positions of the SAR carrier has to be taken into consideration (Image 5.42). The sensor position data recorded by GPS contain the required values for longitudinal and lateral coordinates. According to [BAR87], the following equation can be used to retrieve the shortest distance of two points along the earth's surface:

$$e = \arccos(\sin(\psi_1) \cdot \sin(\psi_2) + \cos(\psi_1) \cdot \cos(\psi_2) \cdot \cos(\lambda_1 - \lambda_2)) \qquad (5.39)$$

$$D = e \cdot R_E \qquad (5.40)$$

whereas $R_E = 6\,370\,000$ m is the earth's radius, λ_i are the longitudes and ψ_i the latitudes.

With the stated equations the necessary cartesian length D (see Image 5.40) needed to determine the pass-point height to be measured can be retrieved.

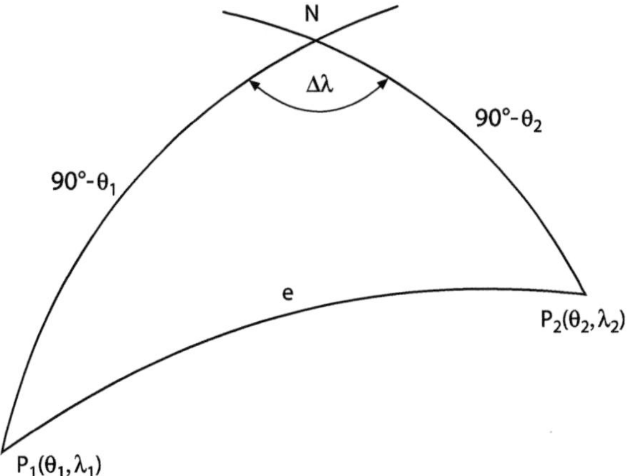

Image 5.42: Concerning the determination of the orthodrome's length "e"

5.6.2.3 Range distance of the pass points

For the calculation of the reference points height it is necessary to allocate to each interesting scene pixel its corresponding range distance. This allocation takes place due to the known parameters of the range-pixel size and the shortest range distance. For any image pixel in range direction the range distance can be determined as follows:

$$R_i = R_0 + i \cdot \Delta R \qquad (5.41)$$

R_0 indicates the near range of the SAR arrangement, i characterises the discrete array value in range direction and ΔR describes the range resolution.

5.6.2.4 Squint angle

The squint angle is kind of a 'correction' angle of the SAR antenna and it presupposes the knowledge of the doppler frequency, the radar frequency, and the flight velocity (Image 5.43).

In chapter 2.2 the theoretical basics to ascertain the missing parameters for the flight velocity and the squint angle are already discussed.

The squint angle is ascertained according to chapter 2.2:

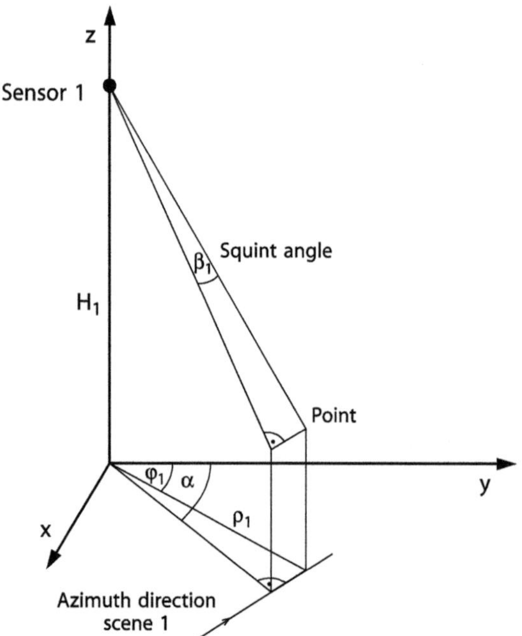

Image 5.43: Definition of the squint angle

$$\beta = \arcsin\left(\frac{f_{DC} \cdot c}{2 \cdot f_0 \cdot v}\right) \tag{5.42}$$

The required flight velocity in regard of the earth's rotation is according to chapter 3:

$$v = \sqrt{(v_x - v_a)^2 + v_y^2} \tag{5.43}$$

All parameters to solve of the equation system (equation (5.19) and following) and thus for determining the wanted information on the height are now known.

5.6.3 Implementation

As already discussed, only an iterative proceeding for the calibration's height h is able to deliver a solution. Therefore, some boundary conditions should be predetermined. To them, first of all, belongs the initial value for the estimation of the height, the possible extreme values of the heights to be calculated and the lowest height's interval to be resolved.

The lowest assumable estimated height can be determined as aircraft altitude less than the current slant-range distance.

As maximum estimation height the minor of both carrier-flight heights is used.

The height's estimation steps are predefined in intervals of 0.1 m.

With every new assumed pass-point height a theoretical rotation angle according to equation (5.35) is calculated and compared with the actual rotation angle δ. The resulting error forms the basis of a criterion for the exactness of the assumed estimated height value. A detailed description of the developed algorithms and the corresponding errors is summarised in [STO98].

A test concerning the acquired height's pass points was performed with the help of the already mentioned microwave reflectors, the so-called corner cubes. The corner cubes are positioned on different locations in the illuminated test area (DO-SAR: scene "Titisee") and measured by GPS.

Using this new method, the range of errors of the absolute elevation stretches beginning from a negligible aberration of 3 m to a maximum error of about 90 m for different corner cubes. The absolutely strong error dependency of the pass-point elevation on the longitudes and latitudes of the carrier's position must be taken into consideration. An examination of the different influencing parameters on the height's exactness resulted in a drastic dependency of the error's consequences on the orthodrome "e". That one was calculated with the help of the longitudes and latitudes of the corresponding position of the carrier which in turn had been extracted as input-data information from the overflight data.

According to [STO98] the absolute estimated height's value changes for the analysed geometry by about 200 m when the orthodrome respectively the projected carrier's distance D varies by about 0.2 m (each carrier by about 10 cm). The kind of data recording of the corresponding longitudes and latitudes of the DO-SAR system used (GPS resolution) allows only a preciseness within centimetres, in other words within those dimensions strong influences on the actual estimated value can already be detected. Thus the above mentioned extreme errors are relativised. Henceforth, it is not attributed to the algorithmic dependencies but rather requires an extremely precise registration of the respective carrier positions – a problem a solution has currently been researched for, regarding the baseline estimations necessary for the interferometric procession –, which is being currently researched, first of all in terms of the baseline estimation which is necessary for the interferometric processing.

In principle, using the presented method height calibration points in overlapping interferometric SAR scenes can be calculated (presumption is a sufficiently accurate determination of the carrier's position).

The afterwards performed modified coregistration process references with the aid of the acquired height's calibration points the whole digital height model into the pre-defined (by the calibration points) coordinate system (WGS 84), and the geo-referenced DHM is created. The probably necessary conversion into local or cartographic coordinate systems can be taken from standardised geodetic systems.

Summarising, a method for determination of height's calibration points could be introduced with the limitation of the requirements regarding the exactness. It offers the opportunity of the calibration of heights without additional hardware and measurement efforts. On the other hand it is a method to reference digital height models onto the surface of any planet without the (manually) placing of corner cubes.

Summary and perspectives

In this work examinations on the interferometric proceesing of SAR data were performed. Objective of this work was the system-theoretical description of the necessary processing steps for the SAR interferometry, beginning from receiving raw data to the generation of the interferogram. Substantial parts of the whole problem within the SAR interferometry have been solved, not only the phase-preserving SAR processing but also the correction of different geometrically caused decorrelation effects.

In order to solve the tasks concerning the SAR focussing a new method for SAR-processing was derivated: a novel precision processor was created. The geometric decorrelation effects on one hand could be egalised with the aid of novel methods for orthometric projection and correction processes of inevitable geometric effects, on the other hand, apart from improved terms concerning the registrations difficulty also novel solutions for generating height pass points.

In chapter 2 SAR basics all necessary basics for further processing steps towards the SAR signal theory were derivated and all geometric conditions for the coming approaches were created.

Chapter 3 SAR signal processing used methods of the signal theory and system theory to describe SAR systems as dynamic systems. Elementary resolution characteristics and fundamental processing steps to achieve the wanted resolution were discussed. For the derivation of the respective characteristics apart from the methods of the signal theory, especially geometric descriptive derivations were used to clarify the explanations.

After discussion of the recording problems respectively processing problems of a SAR system derivated the 2-dimensional point-target response by using the system theory. The point-target response is equivalent to the transform function of a SAR sensor and describes the system's reaction to a 2-dimensional dirac, the 2-dimensional point response.

The point-target response was calculated in multiple transformation domains, namely the range-time/azimuth-time domain, the range-frequency/ azimuth-frequency domain and the range-time/azimuth-frequency domain. This description at many levels lead to the following chapter 4 Signal-processing algorithms as a basic signal description of different processing methods and defined a common initial point for all coming processing methods.

Chapter 4 Signal-processing algorithms described different signal processing algorithms in different transformation domains. Therefore we used the derivated results according to chapter 3 SAR-signal processing. As a novel development was presented the Scaled Inverse Fourier transformation processor.

After having reflected on the basic ideas and the methods used, the SIFT processor was derivated by solving the superposition integral with the aid of all 2-dimensional target-point spectra. The illustration of the SIFT processor, beginning with the transmitting signal in chapter 3 SAR signal processing demonstrates, as far as the author knows, the first explicit illustration of SAR processing and the comparison with other processing techniques.

We discussed the following kinds of processors, the range doppler, the chirp scaling, and the SIFT processor (Scaled Inverse Fourier Transform). The named processors can be merged by assuming a standardised initial point, by name the description of the point-target response in the 2-dimensional frequency domain.

The secondary range compression (SRC) was calculated separately for all processors. As a result could be proved on the one hand the necessity of this SRC for all processors, on the other hand the similarity of the SRC to the named processors could be derivated. Afterwards the functionality of the SIFT processor was demonstrated and also its phase preservance. The phase error was about $0.5°$ and thus proved the capability of the developed algorithm that can therefore be named a precision processor (Precision SAR Processing) according to [RAN94].

The chapter concerning interferometry described the entire intergeometric processing chain after the introduction on the theory of the SAR interferometry by means of examples beginning with the raw data recording to the geometrically corrected digital terrain model. Following the interferometric processing flow there was introduced a method for the coregistration which achieves with the help of known methods a registration exactness of 1/8 pixel, and chose an interferometric approach for the automatic generation of so-called reference points.

According to the side-looking recording geometry and the thus resulting problems, a method for ortho-projection was demonstrated that recognises and corrects both the effects of foreshortening and the layover areas and shadow areas. Data-fusion techniques and geometric reflections had been used. The generation of pass points was realised with the development of an algorithm able to generate height pass points with the help of a geometric description. We found out that the exactness of the pass points depends on the input parameters and requires an absolutely accurate determination of the current SAR carrier's position in the orbit.

In addition to the generation of height models, fields of interesting activities within the SAR- and SARIF signal processing in the future are e.g. the detection of modifications on the earth's surface and the classification of areas. The possibilities of solutions revealed in this work, concerning the precise SAR processing, the automatic coregistration, the correction of recording-geometry distortions as well as the automatic generation of height pass points result in an essential presumption for the realisation of interferometric methods in the future.

All methods and algorithms developed in the course of this work were adjusted to, first of all, develop an interferometric processor. This processor had to undergo all complex processing steps, beginning with the raw data tapes of different raw data suppliers via the 2-dimensional, non-linear, distance-variant scene focussing to the geo-referenced digital height model.

A so-called generic (functionality for different SAR-input data) interferometric SAR processor had been developed over the last years by the DLR (German Research Center for Aerospace Technique) and completed in the year 1998. The center of sensor systems, project-area 2: optimal measure-values- and signal-processing, sensor data fusion, remote sensing – SAR, has already solved a lot of problems within the vast field of interferometric tasks. That means, there were problems defined in general, described in theoretical, realised using different approaches and finally implemented as solutions.

In the course of this work belong first of all reading routines for ERS-data files, reading routines for ERS tapes, reading routines for GPS-orbital data files, the development of the "Scaled Inverse Fourier Transformation Processing", the realisation of a chirp-scaling processor, the correction of the antenna tapers for ERS-1, the implementation of different doppler esti-mators, the development of a module for coregistration, the development of an ortho-projector, the correction of the geometric effects layover, sha-dows and foreshortening, the realisation of geo-referencing (this resulted in a cooperation with the specialty of practical geodesy at the University of GH-Siegen, Germany) and the development of special visualising tools.

The realised algorithms were drafted in the developing and visualising environment IDL (Interactive Date Language), additionally implemented into the program code C and partially embedded in a graphical user inter-face only developed for the interferometry processor, which can be used on different data-processing platforms. Along with the methods of phase un-wrapping already developed in [KRÄ98] and the SAR-simulation package finished in [KLA95], an extremely efficient and extensive interferometric processing system was the result that offers a great flexibility concerning its usability.

The importance of the earth's observation via SAR sensors is more and more increasing. On the one hand, the achievable preciseness of the height and the resolution limits are improved, on the other hand the kinds and numbers of recognition methods are increasing. The introduced methods offer further increasing in preciseness for the SAR signal processing and the SAR interferometry by means of further development.

Developments in the future will enlarge the field of SAR applications. Methods will become more and more important, in particular the methods of the classification and use of remote-sensing data within complex geo-in-formation systems in addition to the multi-dimensional mapping. Already now there are examinations with the help of remote-sensing data, like e.g. the forecast of risky areas concerning malaria [MON98]. New data-fusion techniques [ARN96] and the consequent use of geo-information systems will cause many interesting and useful applications for remote-sensing data.

This work defines the necessary interferometric SAR basics and provides the required steps towards signal processing.

7.1 Geometry and processing

The essentials show you the complete process of interferometrical SAR imaging. Starting with the generation of the raw data we show the development of the geo-referenced digital height model using the basic mathematical formulas and clarifying the steps by graphical descriptions and substantial results coming from airborne data and ERS data.

Side-looking geometry

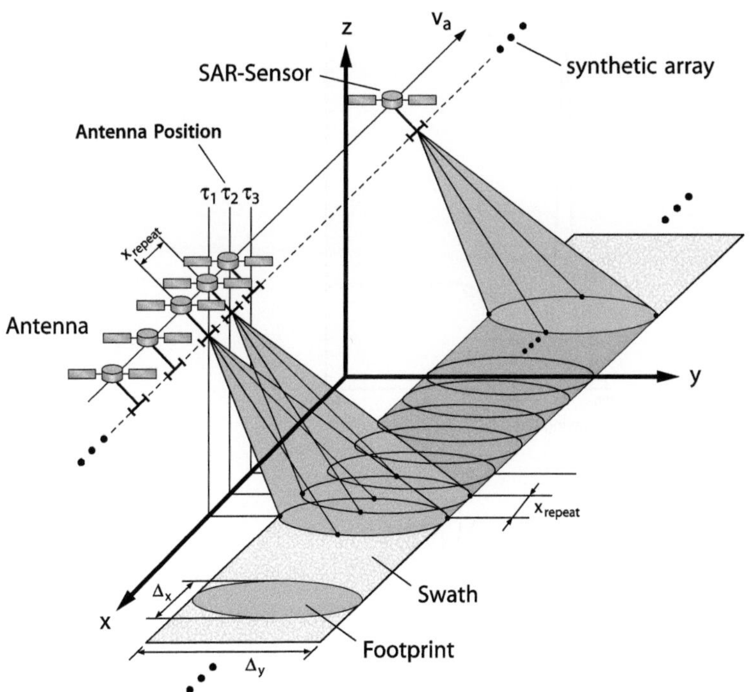

Signal theory

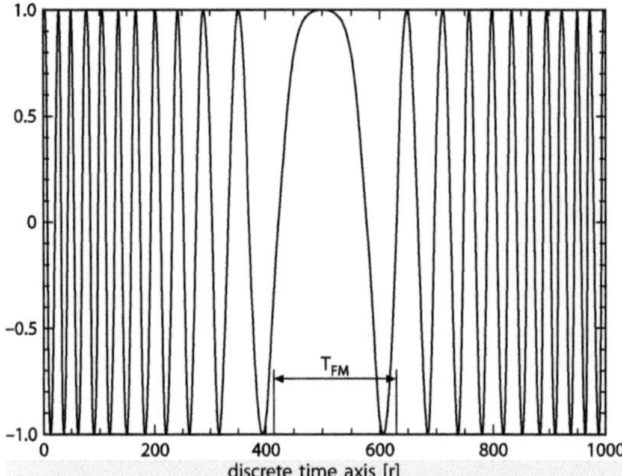

$$s(t) = \text{rect}\left(\frac{t}{T}\right) \cdot \cos(2\pi f_0 t + \pi k_r t^2)$$

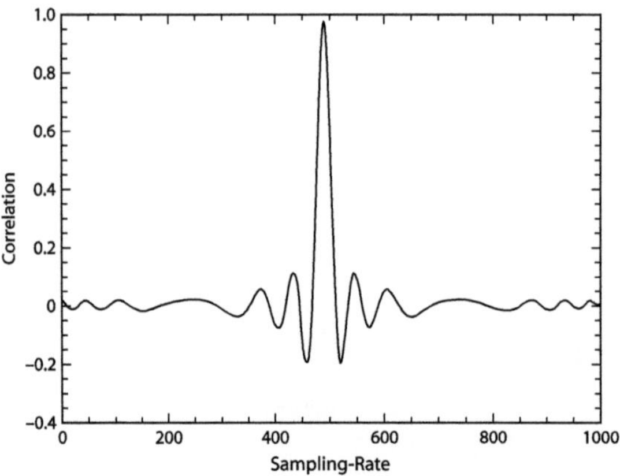

$$\varphi_{ss_T}(t) = \frac{T}{2} \cdot \Lambda\left(\frac{t}{T}\right) \cdot \text{si}\left(\pi B_r t \cdot \Lambda\left(\frac{t}{T}\right)\right)$$

Description of the point-target parabola, range migration, phase history:

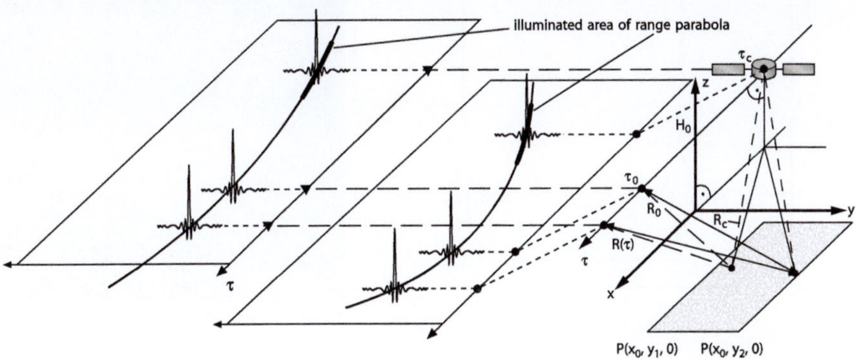

Pulse compression

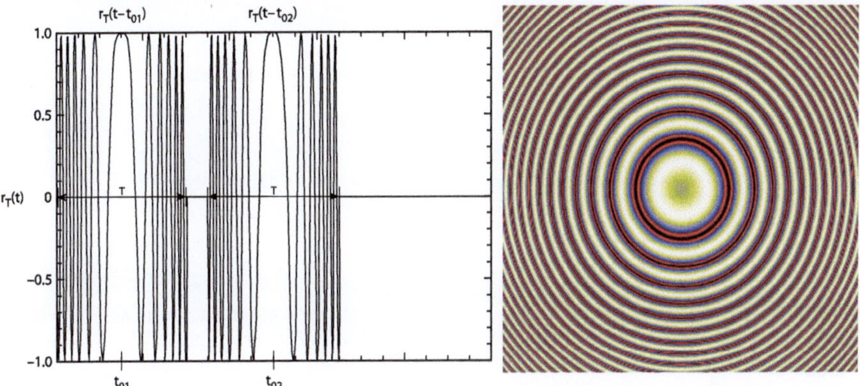

Two uncompressed chirp pulses in a one-dimensional view backscattered from two point targets.

Two-dimensional view of theoretical backscattering wavefront of point target.

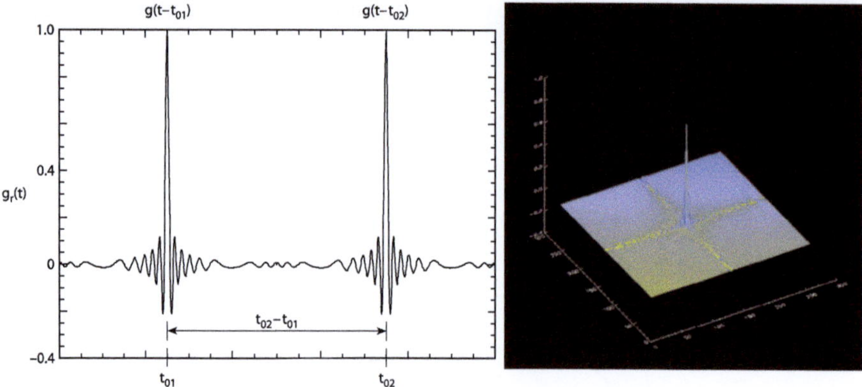

The two points targets focused using autocorrelation.

Focused two-dimensional point target.

SAR processing

System-theoretical approach: using the point target response

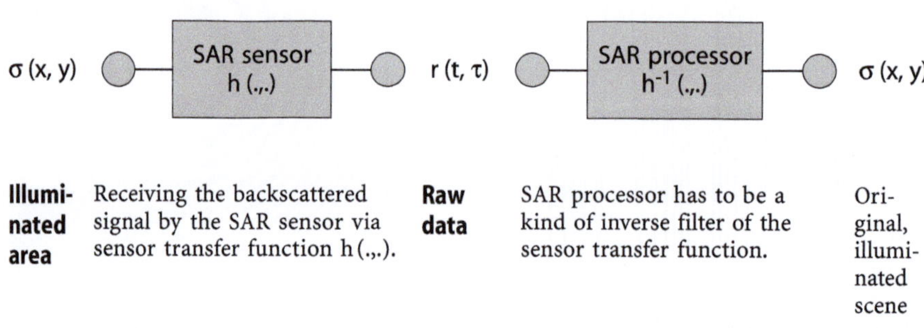

Illumi-nated area	Receiving the backscattered signal by the SAR sensor via sensor transfer function h (.,.).	**Raw data**	SAR processor has to be a kind of inverse filter of the sensor transfer function.	Ori-ginal, illumi-nated scene

Raw data spectrum

The transmitted pulse:

$$s_T(t) = \mathrm{rect}\left(\frac{t}{T}\right) \cdot \exp\{j \cdot \pi \cdot k_r \cdot t^2\}$$

The point target response in the time domain:

$$r_T(t, \tau, \tau_0, R_0) = \sigma(R_0, \tau_0) \cdot s_T(t - t_0(\tau, \tau_0, R_0)) \cdot \exp\{-j2\pi \cdot f \cdot t_0(\tau, \tau_0, R_0)\} \cdot \mathrm{rect}\left(\frac{\tau - \tau_c}{\Delta\tau}\right)$$

Range migration:

$$t_0(\tau, \tau_0, R_0) \neq \frac{2R_0}{c}$$

Phase history:

$$\phi(\tau, \tau_0, R_0) = -2\pi f_0 \cdot t_0(\tau, \tau_0, R_0)$$

The point target response in the frequency domain is evaluated by two-dimensional Fourier transformation:

$$R_T(f, f_\tau, \tau_0, R_0) = \mathscr{F}_t \mathscr{F}_\tau\{r_T(t, \tau, \tau_0, R_0)\}$$

The complete scene spectrum is obtained as the two-dimensional integral over all point target spectra:

$$R_T(f, f_\tau) = \int\limits_{-\infty}^{\infty} \int\limits_{-\infty}^{\infty} R_T(f, f_\tau, \tau_0, R_0) \cdot d\tau_0 \cdot dR_0$$

Solving the integral and substituting $\sqrt{R_0} = \sqrt{R_m + r} \approx \sqrt{R_m}$, we get the *raw data spectrum*:

$$R_T(f, f_\tau) = S_T(f) \cdot \mathrm{rect}\left[\frac{f_\tau - f_c}{B_{az}}\right] \cdot \frac{\sqrt{\frac{c}{2} \cdot \frac{1}{v} \cdot (f + f_0) \cdot e^{-j \cdot \pi/4}}}{\left[(f + f_0)^2 - \frac{f_\tau^2 \cdot c^2}{4v^2}\right]^{3/4}} \cdot \sqrt{R_m}$$
$$\cdot \sigma\left(\frac{2}{c} \cdot \sqrt{(f + f_0)^2 - \frac{f_\tau^2 \cdot c^2}{4v^2}}, f_\tau\right)$$

The *'deformed' scene spectrum* contains the complete information depending on the range frequency and the azimuth frequency and describes the range migration by coupling between the frequencies.

$$\sigma\left(\frac{2}{c} \cdot \sqrt{(f + f_0)^2 - \frac{f_\tau^2 \cdot c^2}{4v^2}}, f_\tau\right) = \sigma(f_r(f, f_\tau), f_\tau)$$

with

$$f_r(f, f_\tau) = \frac{2}{c} \cdot \sqrt{(f + f_0)^2 - \frac{f_\tau^2 \cdot c^2}{4v^2}}$$

After the amplitude compensation and range compression, the basic focusing problem is to transform this non-linear, range-invariant two-dimensional, scaled spectrum into the range- and azimuth - time domain by re-scaling the deformation.

Two basic problems

Describe the spectrum in a re-scaleable form (approximation):

$$\sigma(f_r(f, f_\tau), f_\tau) = \sigma\left(\frac{2}{c} \cdot \sqrt{(f + f_0)^2 - \frac{f_\tau^2 \cdot c^2}{4v^2}}, f_\tau\right) \overset{!}{=} \sigma(a(f_\tau) \cdot f + b(f_\tau) \cdot f_0, f_\tau)$$

After approximating the spectrum, we additionally have to transform this spectrum back into the time domain by re-scaling it (scale transformation).

$$\sigma(f, \tau) \overset{!}{=} \mathscr{F}_{\text{scale·t}}^{-1} \mathscr{F}_\tau^{-1} \{\sigma(a(f_\tau) \cdot f + b(f_\tau) \cdot f_0, f_\tau)\}$$

Approximation

The deformed scene spectrum $\sigma(f_r(f, f_\tau), f_\tau)$ contains a non-linear frequency term:

$$f_r(f, f_\tau) = \frac{2}{c} \cdot \sqrt{(f + f_0)^2 - \frac{f_\tau^2 \cdot c^2}{4v^2}}$$

This term can be modified to the expected form resulting in:

$$f_r(f, f_\tau) \cong a(f_\tau) \cdot f + b(f_\tau) \cdot f_0$$

with the scaling factors:

$$a(f_\tau) = \frac{2}{c} \cdot \frac{1}{\sqrt{1 - \frac{f_\tau^2 \cdot \lambda^2}{4v^2}}} \qquad b(f_\tau) = \frac{2}{c} \cdot \sqrt{1 - \frac{f_\tau^2 \cdot \lambda^2}{4v^2}} \; .$$

Scaled inverse fourier transformation

The scaled inverse Fourier transformation is given by:

$$S(a \cdot f)$$

$$\updownarrow \quad \text{ISFT}$$

$$s(t) = |a| \cdot \int_{\infty} S(a \cdot f) \cdot e^{j2\pi \cdot a \cdot f \cdot t} \cdot df$$

The following sequence of chirp multiplication, chirp convolution and chirp multiplication achieves the desired scaled inverse Fourier transformation:

$$S(a \cdot f) \longrightarrow \otimes \longrightarrow \boxed{\exp(-j \cdot \pi \cdot a \cdot f^2)} \longrightarrow \otimes \longrightarrow \frac{1}{|a|} \cdot s(f)$$

$$\exp(j \cdot \pi \cdot a \cdot f^2) \qquad \qquad \exp(j \cdot \pi \cdot a \cdot f^2)$$

Scaled inverse Fourier transformation realised with chirps.

Scaled inverse fourier transformation

$$\frac{1}{|a|} \cdot s(t) = \int_{\infty} S(a \cdot f) \cdot e^{j2\pi aft} \cdot df$$

with:

$$e^{j2\pi aft} = e^{j\pi at^2} \cdot e^{j\pi af^2} \cdot e^{-j\pi a(t-f)^2}$$

$$\frac{1}{|a|} \cdot s(t) = e^{j\pi at^2} \cdot \int_{\infty} S(a \cdot f) \cdot e^{j\pi af^2} \cdot e^{-j\pi a(t-f)^2} \cdot df$$

$$\frac{1}{|a|} \cdot s(t) = e^{j\pi at^2} \cdot \left[\left\{ S(a \cdot f) \cdot e^{j\pi af^2} \right\} \cdot e^{-j\pi af^2} \right]$$

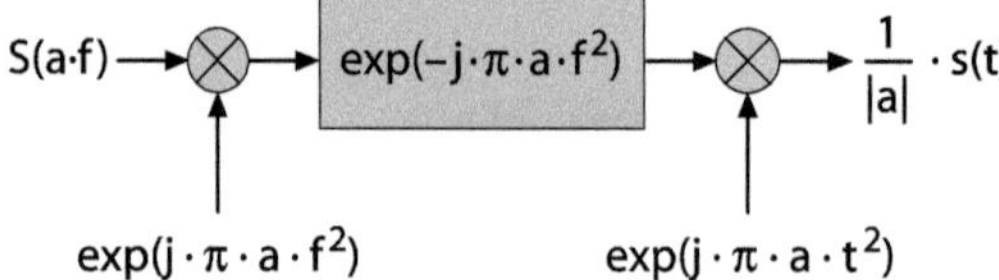

Implementation

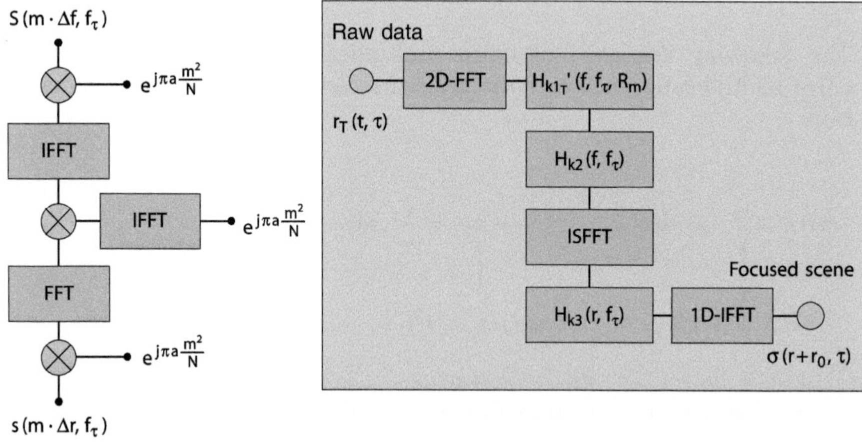

Block diagram of ISFFT

Complete processing diagram of the SIFT processor

$H_{k2T}(f, f_\tau)$ and $H_{k3T}(r, f_\tau)$ are linear phase functions to focus the scene into the slant range, azimuth-doppler domain.

Processing result: scene Flevoland

The exact processing algorithm generates point-target responses coming e.g. from corner cubes quite similar to the theoretical view on page 246 'Focused two-dimensional point target'.

7.2 Interferometry

Single-pass interferometry

DO-SAR: airborne SAR system with side-looking antennas

Multi-pass interferometry

ERS1/2 tandem mission

Shuttle radar topography mission (SRTM)

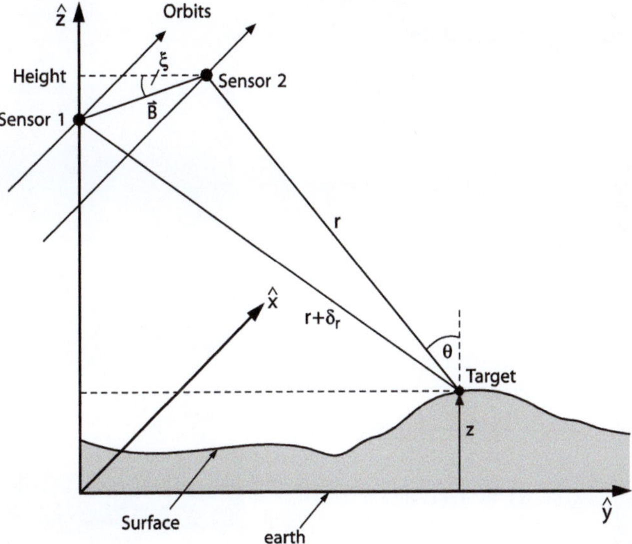

Basic geometry

$$z = H_1 - R_1 \cdot \left(\sqrt{1 - a^2} \cdot \cos\xi - a \cdot \sin\xi \right)$$

$$a = \sin\left(\frac{(R_1 + \delta_r)^2 - R_1^2 - B^2}{2 \cdot B \cdot R_1} \right)$$

$$\delta_r = \frac{\lambda}{4\pi} \cdot \delta_\varphi$$

B: baseline
ξ: baseline angle
δ_φ: phase information of interferogram (phase unwrapped)

Generating DHM starting with raw data

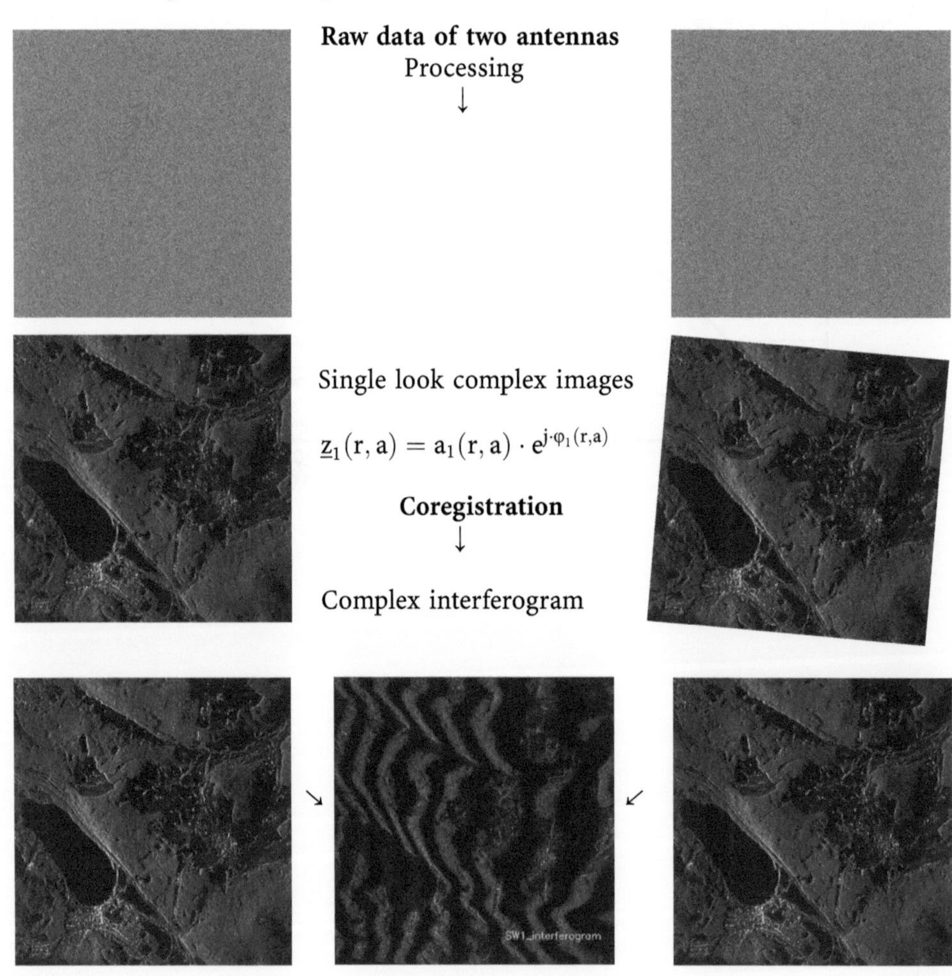

$$\underline{z}_1(r, a) = a_1(r, a) \cdot e^{j \cdot \varphi_1(r,a)}$$

$$\underline{z}(r,a) = \underline{z}_1(r,a)$$
$$\cdot\, \underline{z}_2(r,a)^*$$
$$= a(r,a)$$
$$\cdot\, e^{j\varphi(r,a)}$$

$$\varphi(r,a) = \varphi_1(r,a)$$
$$- \varphi_2(r,a)$$

$$\varphi_m(r,a) \Rightarrow \varphi_h(r,a)$$

The phase

$$\varphi_m(r,a) = \mathrm{atan}_2$$
$$\cdot\left(\frac{\mathrm{Im}\{\underline{z}(r,a)\}}{\mathrm{Re}\{\underline{z}(r,a)\}}\right)$$

Phase unwrapping

Generating height model

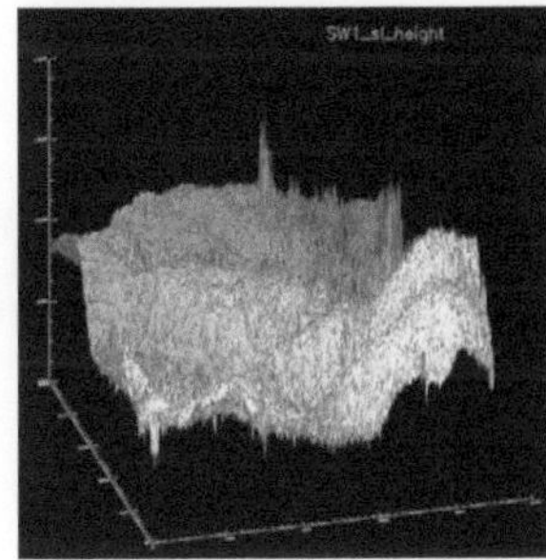

Orthoprojection

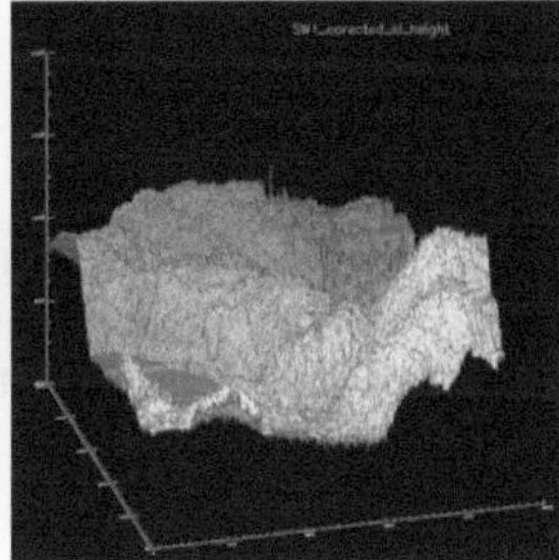

Digital height model

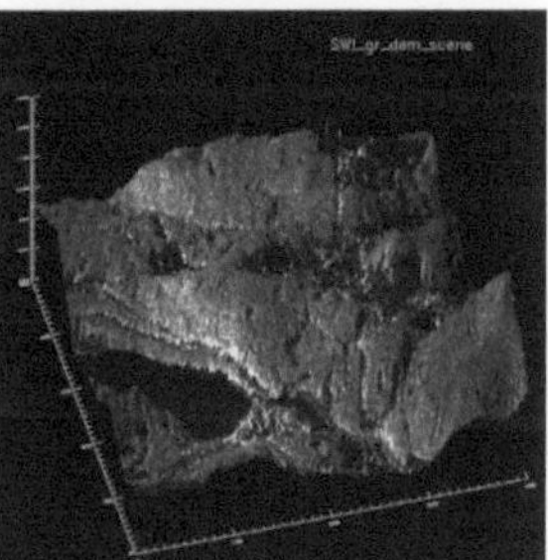

Coregistration

Rough registration
■ Reference points in slave image
■ Registration of corresponding reference points in master image

Precise registration
■ Oversampling by zero-padding
■ Calculate the transform functions
■ Transformation and re-sampling

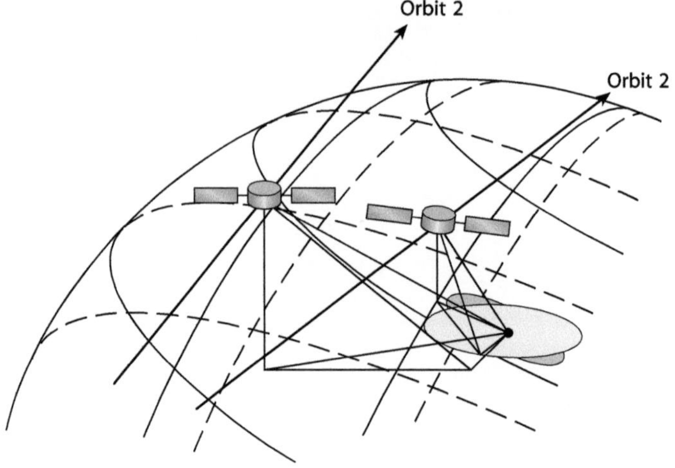

Phase unwrapping

Coming from the SLC images:

$$\underline{z}(r, a) = \underline{z}_1(r, a) \cdot \underline{z}_2(r, a)^* = a(r, a) \cdot e^{j\varphi(r,a)}$$

with

$$\varphi(r, a) = \varphi_1(r, a) - \varphi_2(r, a)$$

we have to evaluate the phase of the interferogram to get the height model of the surface. While evaluating the measured phase we got a so-called wrap-around effect coming from the arctan function:

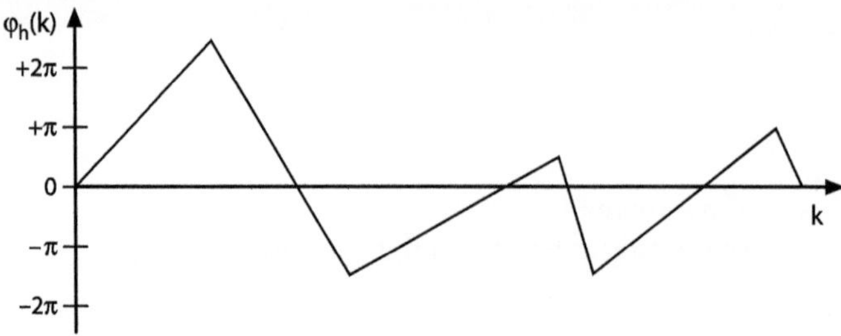

Phase corresponding to the original height

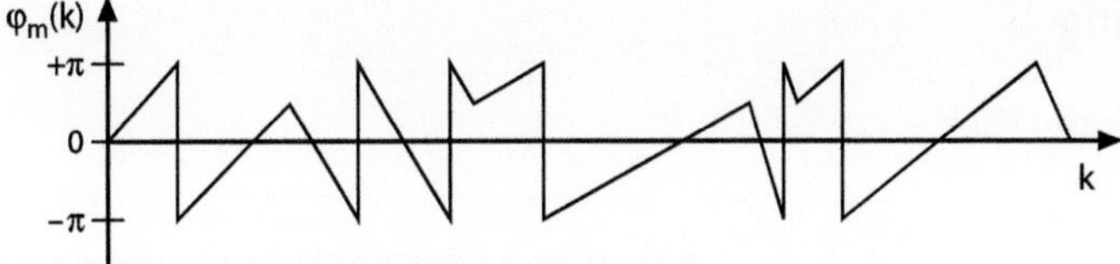

Wrap-around effect of the interferometrical phase

$$\varphi_m(r, a) = \arctan_2 \left(\frac{\mathrm{Im}\{\underline{z}(r, a)\}}{\mathrm{Re}\{\underline{z}(r, a)\}} \right)$$

To get the original height model we have to rewrap or unwrap the interferometrical phase.

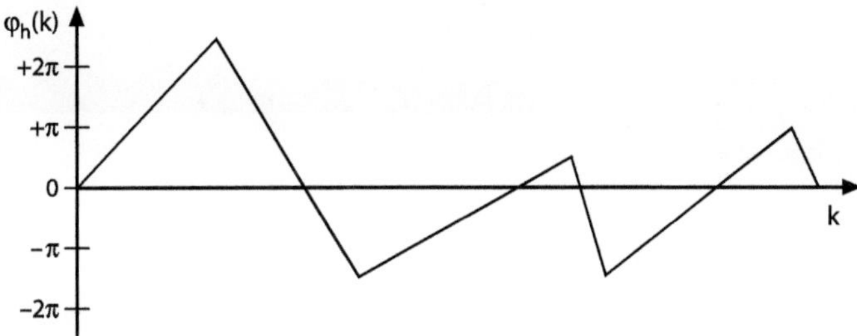

→ Phase unwrapping

$$\varphi_m(r, a) \Rightarrow \varphi_h(r, a)$$

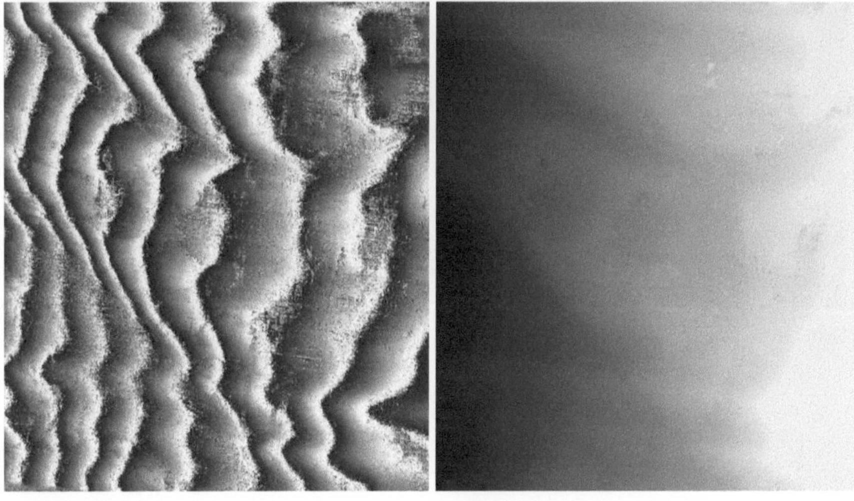

Wrapped phase   Unwrapped phase

7.3 Geocoding

Orthometrical projection

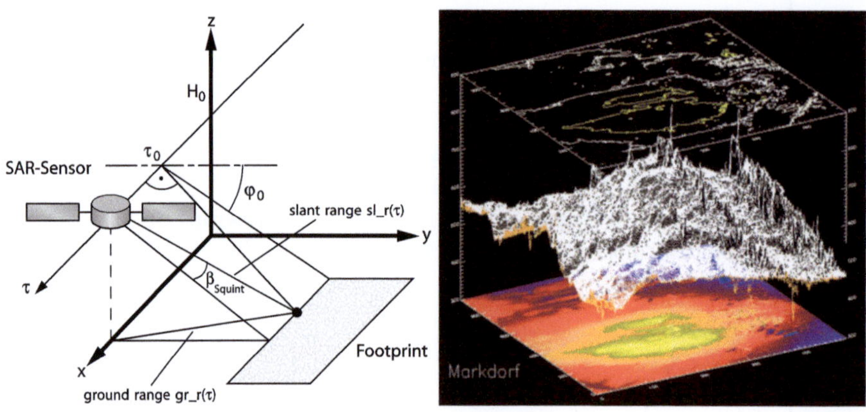

Side looking geometry (slant range)

Resulting effects: foreshortening, layover, shadowing

Foreshortening/orthoprojection

Slant range image Titisee

Ground range image Titisee

Results shadow correction

Height model without shadow effects

Corrected height model of Titisee

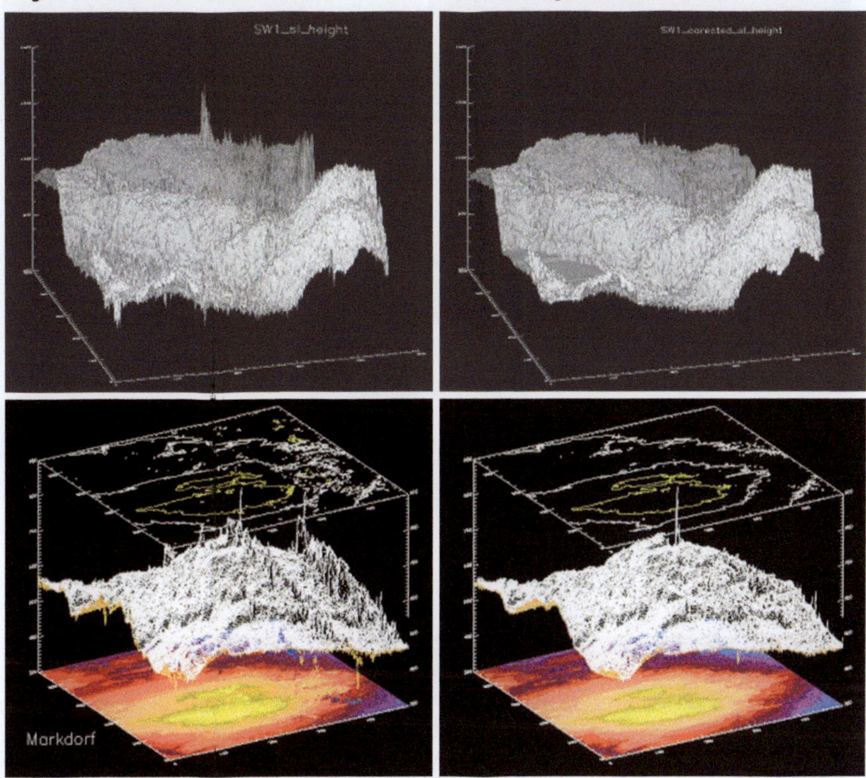

Height model without shadowing effects

Corrected height model of Markdorf

Referencing due to height

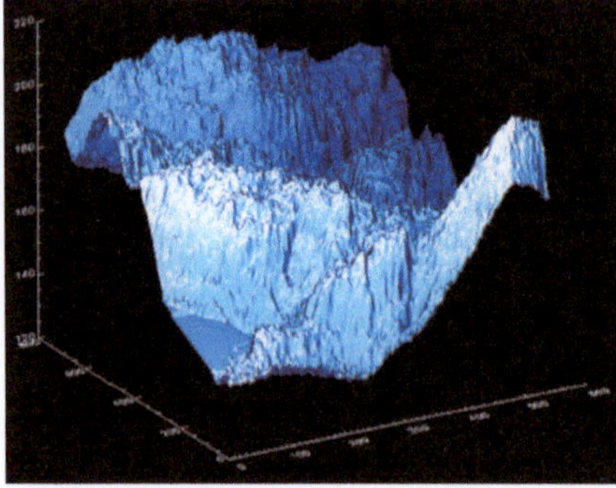

Digital height model with height errors

↓

Referencing in height i.e. using corner cubes

Corresponding amplitude image

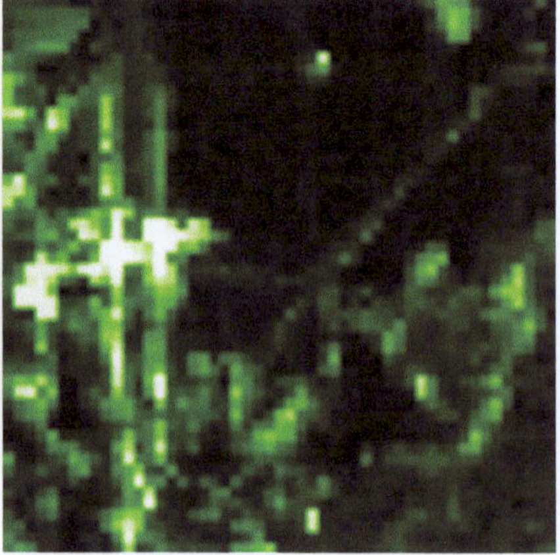

Corner cubes

Corrected digital height model

Referencing due to map

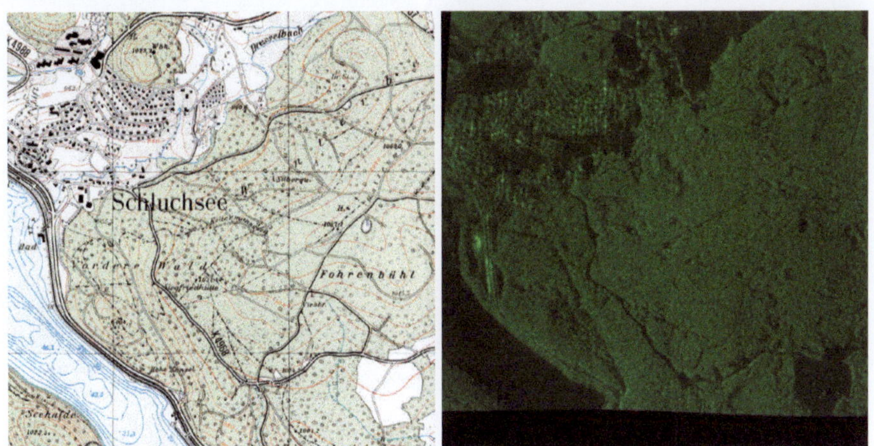

Map of Schluchsee   Part of scene Schluchsee

Overlapping map and SAR image

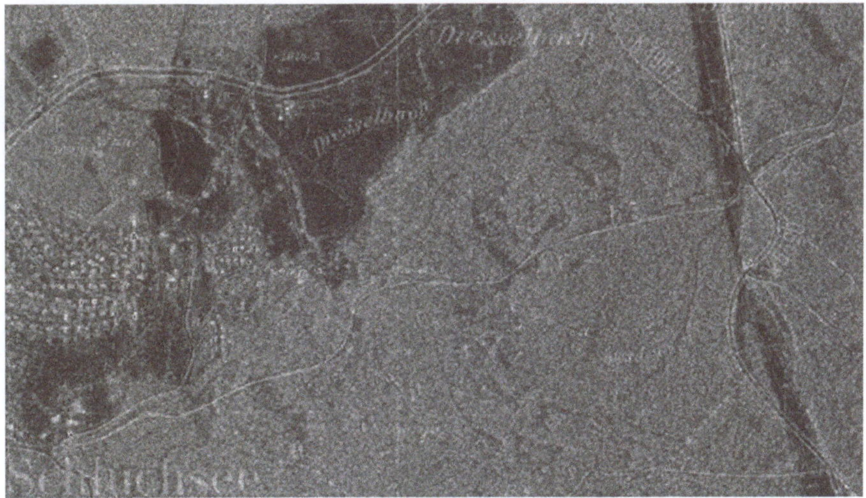

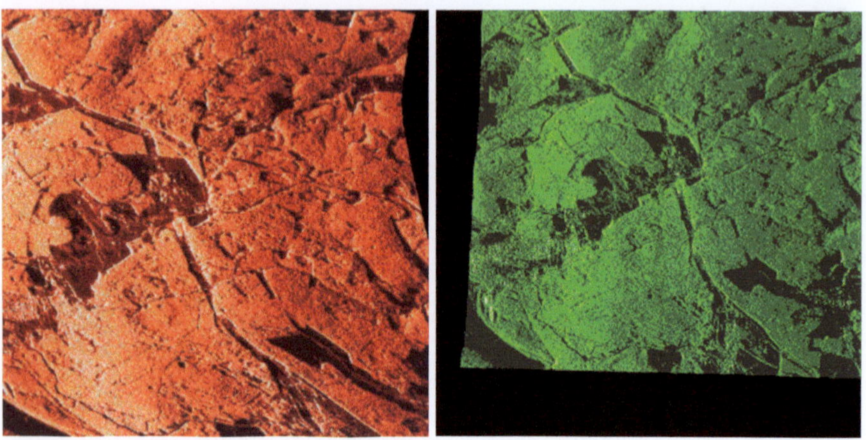

Scene   Referenced scene

Appendix

A-3.4.1.9 Determination of the rectangular in the frequency domain

As shown in chapter 3, to solve the integral in the frequency domain completely we need to calculate the rectangular in equation (3.139) as a function of the time. The complete scene spectrum has to be discussed as a general matter due to the frequency, especially with respect to the azimuth frequency.

The stationary point τ^* in equation (3.121) is a function of the azimuth frequency. We will show how to specify the frequency position in the frequency domain by inserting the stationary point into the rectangular function in the time domain:

$$\tau^* = \tau_0 - \frac{f_\tau \cdot R_0 \cdot \dfrac{c}{2v^2}}{\left[(f + f_0)^2 - f_\tau^2 \cdot \dfrac{c^2}{4v^2}\right]^{\frac{1}{2}}} \tag{3.140}$$

$$\Rightarrow \mathrm{rect}\left(\frac{\tau^* - \tau_c}{\Delta\tau}\right) = \mathrm{rect}\left(\frac{\tau_0 - \dfrac{f_\tau \cdot R_0 \cdot \dfrac{c}{2v^2}}{\left[(f + f_0)^2 - f_\tau^2 \cdot \dfrac{c^2}{4v^2}\right]^{\frac{1}{2}}} - \tau_c}{\Delta\tau}\right) \tag{3.141}$$

The rectangular function is an even function:

$$\mathrm{rect}\left(\frac{\tau_0 - \tau_c}{\Delta\tau}\right) = \mathrm{rect}\left(\frac{\tau_c - \tau_0}{\Delta\tau}\right) \tag{3.142}$$

$$\Rightarrow \mathrm{rect}\left(\frac{\tau^* - \tau_c}{\Delta\tau}\right)$$

$$= \mathrm{rect}\left(\frac{f_\tau \cdot R_0 \cdot c - (\tau_0 - \tau_c) \cdot 2v^2 \cdot \left[(f + f_0)^2 - f_\tau^2 \cdot \dfrac{c^2}{4v^2}\right]^{\frac{1}{2}}}{\Delta\tau \cdot 2v^2 \cdot \left[(f + f_0)^2 - f_\tau^2 \cdot \dfrac{c^2}{4v^2}\right]^{\frac{1}{2}}}\right) \tag{3.143}$$

In general we formulate:

$$\mathrm{rect}\left(\frac{\tau^* - \tau_c}{\Delta\tau}\right) = \mathrm{rect}\left(\frac{g(f_\tau, f)}{h(f_\tau, f)}\right) = \begin{cases} 1 & \text{for } |g(f_\tau, f)| \leq \frac{1}{2}|h(f_\tau, f)| \\ 0 & \text{other!} \end{cases}$$

$$(3.144)$$

View the numerator:

$$f_\tau \cdot R_0 \cdot c - (\tau_0 - \tau_c) \cdot 2v^2 \cdot \left[(f + f_0)^2 - f_\tau^2 \cdot \frac{c^2}{4v^2}\right]^{\frac{1}{2}} \begin{cases} > 0 \\ < 0 \end{cases} \qquad (3.145)$$

Excluding the root function, every term of the numerator can acquire any sign or can be equal to zero e.g.:

$$f_\tau \cdot R_0 \cdot c \begin{cases} \geq 0 \\ < 0 \end{cases} (\tau_0 - \tau_c) \cdot 2v^2 \begin{cases} \geq 0 \\ < 0 \end{cases} \left[(f + f_0)^2 - f_\tau^2 \cdot \frac{c^2}{4v^2}\right]^{\frac{1}{2}} \begin{cases} > 0 \end{cases}$$

View the denominator:

$$\Delta\tau \cdot 2v^2 \cdot \left[(f + f_0)^2 - f_\tau^2 \cdot \frac{c^2}{4v^2}\right]^{\frac{1}{2}} > 0 \qquad (3.146)$$

The denominator function is positive at any time.

As shown above, the denominator is always positive. In the following considerations we still have to consider the sign of the nominator!

Using equation (3.144) and the positive denominator, we receive the argument of the rectangular function:

$$\left|\frac{f_\tau \cdot R_0 \cdot c - (\tau_0 - \tau_c) \cdot 2v^2 \cdot \left[(f + f_0)^2 - f_\tau^2 \cdot \frac{c^2}{4v^2}\right]^{\frac{1}{2}}}{\Delta\tau \cdot 2v^2 \cdot \left[(f + f_0)^2 - f_\tau^2 \cdot \frac{c^2}{4v^2}\right]^{\frac{1}{2}}}\right| \leq \frac{1}{2} \qquad (3.147)$$

$$\left|\frac{f_\tau \cdot R_0 \cdot c \cdot \left[(f + f_0)^2 - f_\tau^2 \cdot \frac{c^2}{4v^2}\right]^{-\frac{1}{2}} - (\tau_0 - \tau_c) \cdot 2v^2}{\Delta\tau \cdot 2v^2}\right| \leq \frac{1}{2} \qquad (3.148)$$

The sign of $(\tau_0 - \tau_c) \cdot 2v^2$ defines the position of the rectangular function!

A-3.4.1.9.1 Case discrimination

It is obvious that:

$$\frac{f_\tau \cdot R_0 \cdot c}{\left[(f + f_0)^2 - f_\tau^2 \cdot \frac{c^2}{4v^2} \right]^{\frac{1}{2}}} - (\tau_0 - \tau_c) \cdot 2v^2 \begin{cases} > 0 \\ < 0 \end{cases} \tag{3.149}$$

A-3.4.1.9.1.1. First main case

$$\frac{f_\tau \cdot R_0 \cdot c}{\left[(f + f_0)^2 - f_\tau^2 \cdot \frac{c^2}{4v^2} \right]^{\frac{1}{2}}} - (\tau_0 - \tau_c) \cdot 2v^2 \geq 0 \tag{3.150}$$

This means:

$$\frac{f_\tau \cdot R_0 \cdot c}{\left[(f + f_0)^2 - f_\tau^2 \cdot \frac{c^2}{4v^2} \right]^{\frac{1}{2}}} \geq (\tau_0 - \tau_c) \cdot 2v^2 \tag{3.151}$$

In this case:

$$\left| \frac{\dfrac{f_\tau \cdot R_0 \cdot c}{\left[(f + f_0)^2 - f_\tau^2 \cdot \frac{c^2}{4v^2} \right]^{\frac{1}{2}}} - (\tau_0 - \tau_c) \cdot 2v^2}{\Delta\tau \cdot 2v^2} \right|$$

$$= \frac{\dfrac{f_\tau \cdot R_0 \cdot c}{\left[(f + f_0)^2 - f_\tau^2 \cdot \frac{c^2}{4v^2} \right]^{\frac{1}{2}}} - (\tau_0 - \tau_c) \cdot 2v^2}{\Delta\tau \cdot 2v^2} \leq \frac{1}{2} \tag{3.152}$$

It follows:

$$\frac{f_\tau \cdot R_0 \cdot c}{\left[(f + f_0)^2 - f_\tau^2 \cdot \frac{c^2}{4v^2} \right]^{\frac{1}{2}}} \leq (\tau_0 - \tau_c) \cdot 2v^2 + \Delta\tau \cdot v^2 \tag{3.153}$$

A-3.4.1.9.1.2 Second main case

$$\frac{f_\tau \cdot R_0 \cdot c}{\left[(f + f_0)^2 - f_\tau^2 \cdot \frac{c^2}{4v^2} \right]^{\frac{1}{2}}} - (\tau_0 - \tau_c) \cdot 2v^2 < 0 \tag{3.154}$$

This applies if:

$$\frac{f_\tau \cdot R_0 \cdot c}{\left[(f + f_0)^2 - f_\tau^2 \cdot \frac{c^2}{4v^2} \right]^{\frac{1}{2}}} < (\tau_0 - \tau_c) \cdot 2v^2 \tag{3.155}$$

In this case equation (3.148) means:

$$\left| \frac{\dfrac{f_\tau \cdot R_0 \cdot c}{\left[(f + f_0)^2 - f_\tau^2 \cdot \frac{c^2}{4v^2} \right]^{\frac{1}{2}}} - (\tau_0 - \tau_c) \cdot 2v^2}{\Delta\tau \cdot 2v^2} \right| =$$

$$- \frac{\dfrac{f_\tau \cdot R_0 \cdot c}{\left[(f + f_0)^2 - f_\tau^2 \cdot \frac{c^2}{4v^2} \right]^{\frac{1}{2}}} - (\tau_0 - \tau_c) \cdot 2v^2}{\Delta\tau \cdot 2v^2} \leq \frac{1}{2} \tag{3.156}$$

It follows that:

$$\frac{f_\tau \cdot R_0 \cdot c}{\left[(f + f_0)^2 - f_\tau^2 \cdot \frac{c^2}{4v^2} \right]^{\frac{1}{2}}} \geq (\tau_0 - \tau_c) \cdot 2v^2 - \Delta\tau \cdot v^2 \tag{3.157}$$

A-3.4.1.9.1.3 Summary of the two main cases:

Case 1:

$$(\tau_0 - \tau_c) \cdot 2v^2 \leq \frac{f_\tau \cdot R_0 \cdot c}{\left[(f + f_0)^2 - f_\tau^2 \cdot \frac{c^2}{4v^2} \right]^{\frac{1}{2}}} \leq (\tau_0 - \tau_c) \cdot 2v^2 + \Delta\tau \cdot v^2 \tag{3.158}$$

Case 2: (adjoin to the lower limit of case one)

$$(\tau_0 - \tau_c) \cdot 2v^2 - \Delta\tau \cdot v^2 \leq \frac{f_\tau \cdot R_0 \cdot c}{\left[(f + f_0)^2 - f_\tau^2 \cdot \dfrac{c^2}{4v^2}\right]^{\frac{1}{2}}} < (\tau_0 - \tau_c) \cdot 2v^2 \quad (3.159)$$

$$(\tau_0 - \tau_c) \cdot 2v^2 - \Delta\tau \cdot v^2 \leq \frac{f_\tau \cdot R_0 \cdot c}{\left[(f + f_0)^2 - f_\tau^2 \cdot \dfrac{c^2}{4v^2}\right]^{\frac{1}{2}}} \leq (\tau_0 - \tau_c) \cdot 2v^2 + \Delta\tau \cdot v^2$$

$$(3.160)$$

A-3.4.1.9.2 Solution of the inequality

We solve the inequality by squaring. Therefore we need to have a look for the signs. We differentiate three cases.

See additionally $\text{rect}\left(\dfrac{\tau_0 - \tau}{\Delta\tau}\right)$.

1) lower limit > 0 $f_\tau > 0$
   upper limit > 0

2) lower limit ≤ 0 $f_\tau \begin{cases} \geq 0 \\ \leq 0 \end{cases}$
   upper limit ≥ 0

3) lower limit < 0 $f_\tau < 0$
   upper limit < 0

A-3.4.1.9.2.1 Case 1

The direction of the inequality will not be changed!

$$[(\tau_0 - \tau_c) \cdot 2v^2 - \Delta\tau \cdot v^2]^2 \leq \frac{(f_\tau \cdot R_0 \cdot c)^2}{(f + f_0)^2 - f_\tau^2 \cdot \dfrac{c^2}{4v^2}}$$

$$\leq [(\tau_0 - \tau_c) \cdot 2v^2 + \Delta\tau \cdot v^2]^2 \quad (3.161)$$

$$4v^4 \cdot \left[(\tau_0 - \tau_c)^2 + \frac{\Delta\tau^2}{4} - \Delta\tau \cdot (\tau_0 - \tau_c)\right] \leq \frac{f_\tau^2 \cdot R_0^2 \cdot c^2}{(f + f_0)^2 - f_\tau^2 \cdot \dfrac{c^2}{4v^2}}$$

$$\leq 4v^4 \cdot \left[(\tau_0 - \tau_c)^2 + \frac{\Delta\tau^2}{4} + \Delta\tau \cdot (\tau_0 - \tau_c)\right] \quad (3.162)$$

As the lower limit is higher than zero, we get:

$$(\tau_0 - \tau_c) \cdot 2v^2 - \Delta\tau \cdot v^2 > 0 \tag{3.163}$$

$$(\tau_0 - \tau_c) - \frac{\Delta\tau}{2} > 0 \tag{3.164}$$

$$(\tau_0 - \tau_c) > \frac{\Delta\tau}{2} \tag{3.165}$$

$$v^2 \cdot \left[(\tau_0 - \tau_c)^2 + \frac{\Delta\tau^2}{4} - \Delta\tau \cdot (\tau_0 - \tau_c) \right] \leq \frac{f_\tau^2 \cdot R_0^2 \cdot \dfrac{c^2}{4v^2}}{(f + f_0)^2 - f_\tau^2 \cdot \dfrac{c^2}{4v^2}}$$
$$\leq v^2 \cdot \left[(\tau_0 - \tau_c)^2 + \frac{\Delta\tau^2}{4} + \Delta\tau \cdot (\tau_0 - \tau_c) \right] \tag{3.166}$$

$$\left[(\tau_0 - \tau_c)^2 + \frac{\Delta\tau^2}{4} - \Delta\tau \cdot (\tau_0 - \tau_c) \right] \cdot \left[(f + f_0)^2 - f_\tau^2 \cdot \frac{c^2}{4v^2} \right] \cdot \frac{v^2}{R_0^2}$$
$$\leq f_\tau^2 \cdot \frac{c^2}{4v^2} \leq \left[(\tau_0 - \tau_c)^2 + \frac{\Delta\tau^2}{4} + \Delta\tau \cdot (\tau_0 - \tau_c) \right]$$
$$\cdot \left[(f + f_0)^2 - f_\tau^2 \cdot \frac{c^2}{4v^2} \right] \cdot \frac{v^2}{R_0^2} \tag{3.167}$$

Look for the lower limit:

$$\left[(\tau_0 - \tau_c)^2 + \frac{\Delta\tau^2}{4} - \Delta\tau \cdot (\tau_0 - \tau_c) \right] \cdot (f + f_0)^2 \cdot \frac{v^2}{R_0^2}$$
$$\leq f_\tau^2 \cdot \frac{c^2}{4v^2} \cdot \left[1 + \frac{v^2}{R_0^2} \cdot \left[(\tau_0 - \tau_c)^2 + \frac{\Delta\tau^2}{4} - \Delta\tau \cdot (\tau_0 - \tau_c) \right] \right] \tag{3.168}$$

$$f_\tau^2 \cdot \frac{c^2}{4v^2} \geq \frac{\left[(\tau_0 - \tau_c)^2 + \dfrac{\Delta\tau^2}{4} - \Delta\tau \cdot (\tau_0 - \tau_c) \right] \cdot (f + f_0)^2 \cdot \dfrac{v^2}{R_0^2}}{\left[1 + \dfrac{v^2}{R_0^2} \cdot \left[(\tau_0 - \tau_c)^2 + \dfrac{\Delta\tau^2}{4} - \Delta\tau \cdot (\tau_0 - \tau_c) \right] \right]} \tag{3.169}$$

$$f_\tau \geq \frac{2v}{c} \cdot (f + f_0) \cdot \frac{v}{R_0} \cdot \sqrt{\frac{(\tau_0 - \tau_c)^2 + \dfrac{\Delta\tau^2}{4} - \Delta\tau \cdot (\tau_0 - \tau_c)}{1 + \dfrac{v^2}{R_0^2} \cdot \left[(\tau_0 - \tau_c)^2 + \dfrac{\Delta\tau^2}{4} - \Delta\tau \cdot (\tau_0 - \tau_c)\right]}}$$

$$(3.170)$$

We have to regard the positive root, because of the positive lower limit:

$$f_\tau \geq \frac{2v^2}{c} \cdot (f + f_0) \cdot \left|(\tau_0 - \tau_c) - \frac{\Delta\tau}{2}\right| \cdot \frac{1}{\sqrt{R_0^2 + v^2 \cdot \left[(\tau_0 - \tau_c) - \frac{\Delta\tau}{2}\right]^2}} \qquad (3.171)$$

$$f_\tau \geq \frac{2v^2}{c} \cdot (f + f_0) \cdot \left|(\tau_0 - \tau_c) - \frac{\Delta\tau}{2}\right|$$
$$\cdot \frac{1}{\sqrt{R_0^2 + v^2 \cdot (\tau_0 - \tau_c)^2 + v^2 \cdot \left[\dfrac{\Delta\tau^2}{4} - \Delta\tau \cdot (\tau_0 - \tau_c)\right]}} \qquad (3.172)$$

with

$$\sqrt{R_0^2 + v^2 \cdot (\tau_0 - \tau_c)^2} = R(\tau_c)$$

$$f_\tau \geq \frac{\dfrac{2v^2}{c} \cdot (f + f_0)}{R(\tau_c)} \cdot \left|(\tau_0 - \tau_c) - \frac{\Delta\tau}{2}\right| \cdot \frac{1}{\sqrt{1 + \dfrac{v^2 \cdot \Delta\tau^2}{R^2(\tau_c)} \cdot \left[\dfrac{1}{4} - \dfrac{(\tau_0 - \tau_c)}{\Delta\tau}\right]}} \qquad (3.173)$$

The upper limit in case one is positive as well:

$$f_\tau^2 \cdot \frac{c^2}{4v^2} \left[1 + \left[(\tau_0 - \tau_c)^2 + \frac{\Delta\tau^2}{4} + \Delta\tau \cdot (\tau_0 - \tau_c)\right] \cdot \frac{v^2}{R_0^2}\right]$$
$$\leq \left[(\tau_0 - \tau_c)^2 + \frac{\Delta\tau^2}{4} + \Delta\tau \cdot (\tau_0 - \tau_c)\right] \cdot \frac{v^2}{R_0^2} \cdot (f + f_0)^2 \qquad (3.174)$$

$$f_\tau^2 \cdot \frac{c^2}{4v^2} \leq \frac{\left[(\tau_0 - \tau_c)^2 + \frac{\Delta\tau^2}{4} + \Delta\tau \cdot (\tau_0 - \tau_c)\right] \cdot \frac{v^2}{R_0^2} \cdot (f + f_0)^2}{1 + \left[(\tau_0 - \tau_c)^2 + \frac{\Delta\tau^2}{4} + \Delta\tau \cdot (\tau_0 - \tau_c)\right] \cdot \frac{v^2}{R_0^2}} \tag{3.175}$$

$$f_\tau \leq \frac{2v^2}{c \cdot R_0} \cdot (f + f_0) \cdot \sqrt{\frac{(\tau_0 - \tau_c)^2 + \frac{\Delta\tau^2}{4} + \Delta\tau \cdot (\tau_0 - \tau_c)}{1 + \left[(\tau_0 - \tau_c)^2 + \frac{\Delta\tau^2}{4} + \Delta\tau \cdot (\tau_0 - \tau_c)\right] \cdot \frac{v^2}{R_0^2}}} \tag{3.176}$$

Positive sign of upper limit results in:

$$f_\tau \leq \frac{2v^2}{c} \cdot (f + f_0) \cdot \frac{\left|(\tau_0 - \tau_c) + \frac{\Delta\tau}{2}\right|}{\sqrt{R_0^2 + v^2 \cdot \left((\tau_0 - \tau_c) + \frac{\Delta\tau}{2}\right)^2}} \tag{3.177}$$

$$f_\tau \leq \frac{2v^2}{c} \cdot (f + f_0) \cdot \frac{\left|(\tau_0 - \tau_c) + \frac{\Delta\tau}{2}\right|}{\sqrt{R_0^2 + (\tau_0 - \tau_c)^2 \cdot v^2} \cdot \sqrt{1 + \frac{v^2 \cdot \left[\frac{\Delta\tau^2}{4} + (\tau_0 - \tau_c) \cdot \Delta\tau\right]}{R_0^2 + (\tau_0 - \tau_c)^2 \cdot v^2}}} \tag{3.178}$$

with

$$R_0^2 + v^2 \cdot (\tau_0 + \tau_c)^2 = R^2(\tau_c)$$

follows:

$$f_\tau \leq \frac{\frac{2v^2}{c} \cdot (f + f_0) \cdot \left|(\tau_0 - \tau_c) + \frac{\Delta\tau}{2}\right|}{R(\tau_c) \cdot \sqrt{1 + \frac{v^2 \cdot \Delta\tau^2}{R^2(\tau_c)} \cdot \left[\frac{1}{4} + \frac{(\tau_0 - \tau_c)}{\Delta\tau}\right]}} \tag{3.179}$$

$$f_\tau \le \frac{\dfrac{2v^2}{c} \cdot (f + f_0)}{R(\tau_c)} \cdot \left|(\tau_0 - \tau_c) + \frac{\Delta\tau}{2}\right| \cdot \frac{1}{\sqrt{1 + \dfrac{v^2 \cdot \Delta\tau^2}{R^2(\tau_c)} \cdot \left[\dfrac{1}{4} + \dfrac{(\tau_0 - \tau_c)}{\Delta\tau}\right]}}$$

(3.180)

A-3.4.1.9.2.2 Summary case 1

$$f_\tau > 0\,;\ \tau_0 - \tau_c > \frac{\Delta\tau}{2}$$

$$\frac{\dfrac{2v^2}{c} \cdot (f + f_0)}{R(\tau_c)} \cdot \left|(\tau_0 - \tau_c) - \frac{\Delta\tau}{2}\right| \cdot \frac{1}{\sqrt{1 + \dfrac{v^2 \cdot \Delta\tau^2}{R^2(\tau_c)} \cdot \left[\dfrac{1}{4} - \dfrac{(\tau_0 - \tau_c)}{\Delta\tau}\right]}} \le f_\tau$$

$$\le \frac{\dfrac{2v^2}{c} \cdot (f + f_0)}{R(\tau_c)} \cdot \left|(\tau_0 - \tau_c) + \frac{\Delta\tau}{2}\right|$$

$$\cdot \frac{1}{\sqrt{1 + \dfrac{v^2 \cdot \Delta\tau^2}{R^2(\tau_c)} \cdot \left[\dfrac{1}{4} + \dfrac{(\tau_0 - \tau_c)}{\Delta\tau}\right]}}$$

(3.181)

(Centroid frequency):

$$f_c = \frac{1}{2}(f_u + f_0)$$

(3.182)

$$f_c = \frac{1}{2} \cdot \frac{1}{R(\tau_c)} \cdot \frac{2v^2}{c} \left[\left|(\tau_0 - \tau_c) - \frac{\Delta\tau}{2}\right| \cdot \frac{1}{\sqrt{1 + \dfrac{v^2 \cdot \Delta\tau^2}{R^2(\tau_c)} \cdot \left[\dfrac{1}{4} - \dfrac{(\tau_0 - \tau_c)}{\Delta\tau}\right]}}\right.$$

$$\left. + \left|(\tau_0 - \tau_c) + \frac{\Delta\tau}{2}\right| \cdot \frac{1}{\sqrt{1 + \dfrac{v^2 \cdot \Delta\tau^2}{R^2(\tau_c)} \cdot \left[\dfrac{1}{4} + \dfrac{(\tau_0 - \tau_c)}{\Delta\tau}\right]}}\right] \cdot (f + f_0) \quad (3.183)$$

with the general approximation $\dfrac{1}{\sqrt{1 \pm x}} \cong 1 \mp \dfrac{x}{2}$ and $v^2 \cdot \Delta\tau^2 \ll R^2(\tau_c)$ and without the absolute value[1] follows:

$$f_c \approx \frac{1}{2} \cdot \frac{2v^2}{c} \cdot \frac{1}{R(\tau_c)} \cdot 2 \cdot (\tau_0 - \tau_c) \cdot (f + f_0) \tag{3.184}$$

$$f_c \approx \frac{2v^2}{c} \cdot \frac{1}{R(\tau_c)} \cdot (\tau_0 - \tau_c) \cdot (f + f_0) \tag{3.185}$$

Azimuth bandwidth B_{az}

$$B_{az} = f_0 - f_u \tag{3.186}$$

$$B_{az} \cong - \frac{2v^2}{c} \cdot \frac{1}{R(\tau_c)} \cdot (f + f_0) \cdot \left[(\tau_0 - \tau_c) - \frac{\Delta\tau}{2} \right] + \frac{2v^2}{c} \cdot \frac{1}{R(\tau_c)}$$
$$\cdot (f + f_0) \cdot \left[(\tau_0 - \tau_c) + \frac{\Delta\tau}{2} \right] \tag{3.187}$$

$$B_{az} \cong \frac{2v^2}{c} \cdot \frac{1}{R(\tau_c)} \cdot (f + f_0) \cdot \Delta\tau \tag{3.188}$$

A-3.4.1.9.2.3 Case 2

We look for the case when both limits are less than zero.

lower limit <0
upper limit <0
$\qquad f_\tau < 0 \quad \Rightarrow \tau_0 - \tau_c < -\dfrac{\Delta\tau}{2}$

$$(\tau_0 - \tau_c) \cdot 2v^2 - \Delta\tau \cdot v^2 \leq \frac{f_\tau \cdot R_0 \cdot c}{\left[(f + f_0)^2 - f_\tau^2 \cdot \dfrac{c^2}{4v^2} \right]^{\frac{1}{2}}} \leq (\tau_0 - \tau_c) \cdot 2v^2 + \Delta\tau \cdot v^2$$

$$\tag{3.189}$$

---

[1] Case 1 defines for both limits positive sign

Squaring means multiplying with inverted sign. Therefore we invert the inequality:

$$4v^4 \cdot \left[(\tau_0 - \tau_c) - \frac{\Delta\tau}{2}\right]^2 \geq \frac{(f_\tau \cdot R_0 \cdot c)^2}{(f + f_0)^2 - f_\tau^2 \cdot \dfrac{c^2}{4v^2}} \geq 4v^4 \cdot \left[(\tau_0 - \tau_c) + \frac{\Delta\tau}{2}\right]^2$$

$$(3.190)$$

Lower limit:

$$f_\tau^2 \cdot R_0^2 \cdot \frac{c^2}{4v^2} \geq v^2 \cdot \left[(\tau_0 - \tau_c) + \frac{\Delta\tau}{2}\right]^2 \cdot \left[(f + f_0)^2 - f_\tau^2 \cdot \frac{c^2}{4v^2}\right] \qquad (3.191)$$

$$f_\tau^2 \cdot \frac{c^2}{4v^2} \cdot \left[1 + \frac{v^2 \cdot \left[(\tau_0 - \tau_c) + \dfrac{\Delta\tau}{2}\right]^2}{R_0^2}\right] \geq \frac{v^2}{R_0^2} \cdot \left[(\tau_0 - \tau_c) + \frac{\Delta\tau}{2}\right]^2 \cdot (f + f_0)^2$$

$$(3.192)$$

$$f_\tau^2 \cdot \frac{c^2}{4v^2} \geq \frac{v^2 \cdot \left[(\tau_0 - \tau_c) + \dfrac{\Delta\tau}{2}\right]^2 \cdot (f + f_0)^2}{R_0^2 + v^2 \cdot (\tau_0 - \tau_c)^2 + v^2 \cdot \left[\Delta\tau \cdot (\tau_0 - \tau_c) + \dfrac{\Delta\tau^2}{4}\right]} \qquad (3.193)$$

with

$$R_0^2 + v^2 \cdot (\tau_0 + \tau_c)^2 = R^2(\tau_c)$$

$$f_\tau^2 \geq \frac{4v^4}{c^2} \cdot \frac{1}{R^2(\tau_c)} \cdot (f + f_0)^2 \cdot \left[(\tau_0 - \tau_c) + \frac{\Delta\tau}{2}\right]^2 \cdot \frac{1}{1 + \dfrac{v^2 \cdot \Delta\tau^2}{R^2(\tau_c)} \cdot \left[\dfrac{(\tau_0 - \tau_c)}{\Delta\tau} + \dfrac{1}{4}\right]}$$

$$(3.194)$$

$$|f_\tau| \geq \frac{2v^2}{c} \cdot \frac{1}{R(\tau_c)} \cdot |f + f_0|$$

$$\cdot \left|(\tau_0 - \tau_c) + \frac{\Delta\tau}{2}\right| \cdot \frac{1}{\left[1 + \dfrac{v^2 \cdot \Delta\tau^2}{R^2(\tau_c)} \cdot \left[\dfrac{(\tau_0 - \tau_c)}{\Delta\tau} + \dfrac{1}{4}\right]\right]^{\frac{1}{2}}} \qquad (3.195)$$

with $f_\tau < 0$ follows:

$$f_\tau \le -\frac{2v^2}{c} \cdot \frac{1}{R(\tau_c)} \cdot (f + f_0) \cdot \left|(\tau_0 - \tau_c) + \frac{\Delta\tau}{2}\right|$$
$$\cdot \frac{1}{\left[1 + \frac{v^2 \cdot \Delta\tau^2}{R^2(\tau_c)} \cdot \left[\frac{(\tau_0 - \tau_c)}{\Delta\tau} + \frac{1}{4}\right]\right]^{\frac{1}{2}}} \tag{3.196}$$

The upper limit has to be less than zero:

$$\frac{f_\tau^2 \cdot R_0^2 \cdot c^2}{(f + f_0)^2 - f_\tau^2 \cdot \frac{c^2}{4v^2}} \le 4v^4 \cdot \left[(\tau_0 - \tau_c) - \frac{\Delta\tau}{2}\right]^2 \tag{3.197}$$

$$f_\tau^2 \cdot R_0^2 \cdot \frac{c^2}{4v^2} \le v^2 \cdot \left[(\tau_0 - \tau_c) - \frac{\Delta\tau}{2}\right]^2 \cdot \left[(f + f_0)^2 - f_\tau^2 \cdot \frac{c^2}{4v^2}\right] \tag{3.198}$$

$$f_\tau^2 \cdot \frac{c^2}{4v^2} \left[1 + \frac{v^2 \cdot \left[(\tau_0 - \tau_c) - \frac{\Delta\tau}{2}\right]^2}{R_0^2}\right] \le \frac{v^2}{R_0^2} \cdot \left[(\tau_0 - \tau_c) - \frac{\Delta\tau}{2}\right]^2 \cdot (f + f_0)^2$$

$$\tag{3.199}$$

Comparing with the lower limit in equation (3.192) and analogous usage of equation (3.194) results in:

$$f_\tau^2 \le \frac{4v^4}{c^2} \cdot \frac{1}{R^2(\tau_c)} \cdot (f + f_0)^2 \cdot \left[(\tau_0 - \tau_c) - \frac{\Delta\tau}{2}\right]^2 \cdot \frac{1}{1 + \frac{v^2 \cdot \Delta\tau^2}{R^2(\tau_c)} \cdot \left[\frac{1}{4} - \frac{(\tau_0 - \tau_c)}{\Delta\tau}\right]} \tag{3.200}$$

$$|f_\tau| \le \frac{2v^2}{c} \cdot \frac{1}{R(\tau_c)} \cdot |f + f_0| \cdot \left|(\tau_0 - \tau_c) - \frac{\Delta\tau}{2}\right| \cdot \frac{1}{\left[1 + \frac{v^2 \cdot \Delta\tau^2}{R^2(\tau_c)} \cdot \left[\frac{1}{4} - \frac{(\tau_0 - \tau_c)}{\Delta\tau}\right]\right]^{\frac{1}{2}}} \tag{3.201}$$

with $f_\tau < 0$ follows:

$$f_\tau \geq -\frac{2v^2}{c} \cdot \frac{1}{R(\tau_c)} \cdot (f + f_0) \cdot \left|(\tau_0 - \tau_c) - \frac{\Delta\tau}{2}\right|$$

$$\cdot \frac{1}{\left[1 + \frac{v^2 \cdot \Delta\tau^2}{R^2(\tau_c)} \cdot \left[\frac{1}{4} - \frac{(\tau_0 - \tau_c)}{\Delta\tau}\right]\right]^{\frac{1}{2}}} \tag{3.202}$$

A-3.4.1.9.2.4 Summary case 2

$$\tau_0 - \tau_c + \frac{\Delta\tau}{2} < 0 \quad \Rightarrow \quad \tau_0 - \tau_c < -\frac{\Delta\tau}{2}$$

$$\frac{2v^2}{c} \cdot \frac{1}{R(\tau_c)} \cdot (f + f_0) \cdot \left[(\tau_0 - \tau_c) - \frac{\Delta\tau}{2}\right] \cdot \frac{1}{\left[1 + \frac{v^2 \cdot \Delta\tau^2}{R^2(\tau_c)} \cdot \left[\frac{1}{4} - \frac{(\tau_0 - \tau_c)}{\Delta\tau}\right]\right]^{\frac{1}{2}}}$$

$$\leq f_\tau \leq \frac{2v^2}{c} \frac{1}{R(\tau_c)} \cdot (f + f_0) \cdot \left[(\tau_0 - \tau_c) + \frac{\Delta\tau}{2}\right]$$

$$\cdot \frac{1}{\left[1 + \frac{v^2 \cdot \Delta\tau^2}{R^2(\tau_c)} \cdot \left[\frac{1}{4} - \frac{(\tau_0 - \tau_c)}{\Delta\tau}\right]\right]^{\frac{1}{2}}} \tag{3.203}$$

Now we can calculate the centroid frequency the way we did in case 1:

$$f_c \cong \frac{2v^2}{c} \cdot \frac{1}{R(\tau_c)} \cdot (\tau_0 - \tau_c) \cdot (f + f_0) \tag{3.204}$$

as well as the bandwidth:

$$B_{az} \cong \frac{2v^2}{c} \cdot \frac{1}{R(\tau_c)} \cdot (f + f_0) \cdot \Delta\tau \tag{3.205}$$

A-3.4.1.9.2.5 Case 3

Lower limit: $-\Delta\tau \cdot v^2 + (\tau_0 - \tau_c) \cdot 2v^2 \leq 0$
Upper limit: $+\Delta\tau \cdot v^2 + (\tau_0 - \tau_c) \cdot 2v^2 \geq 0$

$$
\left.\begin{aligned}
\tau_0 - \tau_c &\leq \dfrac{\Delta\tau}{2} \\[2mm]
\tau_0 - \tau_c &\geq -\dfrac{\Delta\tau}{2}
\end{aligned}\right\} \quad \Rightarrow \quad -\dfrac{\Delta\tau}{2} \leq \tau_0 - \tau_c \leq \dfrac{\Delta\tau}{2}
$$

Look for the upper limit in equation (3.160):

$$
\frac{f_\tau \cdot R_0 \cdot c}{\left[(f + f_0)^2 - f_\tau^2 \cdot \dfrac{c^2}{4v^2}\right]^{\frac{1}{2}}} \leq (\tau_0 - \tau_c) \cdot 2v^2 + \Delta\tau \cdot v^2 \tag{3.206}
$$

To evaluate the upper critical frequency, we remember case 3. The upper limit was defined higher than zero. Thus, the azimuth frequency has to be less than the critical frequency $f_\tau < f_{\text{limit}}$.

$$
\frac{f_\tau^2 \cdot R_0^2 \cdot c^2}{(f + f_0)^2 - f_\tau^2 \cdot \dfrac{c^2}{4v^2}} \leq 4v^4 \cdot \left[(\tau_0 - \tau_c) + \frac{\Delta\tau}{2}\right]^2 \tag{3.207}
$$

$$
\frac{f_\tau^2 \cdot c^2}{4v^2} \leq \frac{v^2}{R_0^2} \cdot \left[(\tau_0 - \tau_c) + \frac{\Delta\tau}{2}\right]^2 \cdot \left[(f + f_0)^2 - f_\tau^2 \cdot \frac{c^2}{4v^2}\right] \tag{3.208}
$$

$$
f_\tau^2 \cdot \frac{c^2}{4v^2} \cdot \left[1 + \frac{v^2 \cdot \left[(\tau_0 - \tau_c) + \dfrac{\Delta\tau}{2}\right]^2}{R_0^2}\right] \leq \frac{v^2}{R_0^2} \cdot \left[(\tau_0 - \tau_c) + \frac{\Delta\tau}{2}\right]^2 \cdot (f + f_0)^2
$$

$$
\tag{3.209}
$$

$$
f_\tau^2 \cdot \frac{c^2}{4v^2} \leq \frac{v^2 \cdot \left[(\tau_0 - \tau_c) + \dfrac{\Delta\tau}{2}\right]^2 \cdot (f + f_0)^2}{R_0^2 + v^2 \cdot (\tau_0 - \tau_c)^2 + v^2 \cdot \left[\Delta\tau \cdot (\tau_0 - \tau_c) + \dfrac{\Delta\tau^2}{4}\right]} \tag{3.210}
$$

Following the steps of equation (3.193) to equation (3.194) results in:

$$f_\tau^2 \leq \frac{4v^4}{c^2} \cdot \frac{1}{R^2(\tau_c)} \cdot (f + f_0)^2 \cdot \left[(\tau_0 - \tau_c) + \frac{\Delta\tau}{2}\right]^2 \cdot \frac{1}{1 + \dfrac{v^2 \cdot \Delta\tau^2}{R^2(\tau_c)} \cdot \left[\dfrac{(\tau_0 - \tau_c)}{\Delta\tau} + \dfrac{1}{4}\right]}$$

$$(3.211)$$

$$|f_\tau| \leq \frac{2v^2}{c} \cdot \frac{1}{R(\tau_c)} \cdot |f + f_0| \cdot \left|(\tau_0 - \tau_c) + \frac{\Delta\tau}{2}\right| \cdot \frac{1}{\left[1 + \dfrac{v^2 \cdot \Delta\tau^2}{R^2(\tau_c)} \cdot \left[\dfrac{(\tau_0 - \tau_c)}{\Delta\tau} + \dfrac{1}{4}\right]\right]^{\frac{1}{2}}}$$

$$(3.212)$$

We have to take the positive solution as $f < f_{\text{limit}}$ is a result of the inequality!

$$f_\tau \leq \frac{2v^2}{c} \cdot \frac{1}{R(\tau_c)} \cdot (f + f_0) \cdot \left[(\tau_0 - \tau_c) + \frac{\Delta\tau}{2}\right] \cdot \frac{1}{\left[1 + \dfrac{v^2 \cdot \Delta\tau^2}{R^2(\tau_c)} \cdot \left[\dfrac{(\tau_0 - \tau_c)}{\Delta\tau} + \dfrac{1}{4}\right]\right]^{\frac{1}{2}}}$$

$$(3.213)$$

After the evaluation of the upper critical frequency, we will have a look for the lower limit:

$$\frac{f_\tau \cdot R_0 \cdot c}{\left[(f + f_0)^2 - f_\tau^2 \cdot \frac{c^2}{4v^2}\right]^{\frac{1}{2}}} \geq (\tau_0 - \tau_c) \cdot 2v^2 - \Delta\tau \cdot v^2 \tag{3.214}$$

In equation (3.214) we see $f_\tau > f_{\text{limit}}$ because the lower limit has to be less than zero and the right side of the equation has to be lower than zero.

$$\frac{f_\tau^2 \cdot R_0^2 \cdot c^2}{(f + f_0)^2 - f_\tau^2 \cdot \dfrac{c^2}{4v^2}} \leq 4v^4 \cdot \left[(\tau_0 - \tau_c) - \frac{\Delta\tau}{2}\right]^2 \tag{3.215}$$

$$f_\tau^2 \cdot \frac{c^2}{4v^2} \cdot \left[1 + \frac{v^2 \cdot \left[(\tau_0 - \tau_c) - \dfrac{\Delta\tau}{2}\right]^2}{R_0^2}\right] \leq \frac{v^2}{R_0^2} \cdot \left[(\tau_0 - \tau_c) - \frac{\Delta\tau}{2}\right]^2 \cdot (f + f_0)^2$$

$$(3.216)$$

$$f_\tau^2 \leq \frac{4v^4}{c^2} \cdot \frac{1}{R^2(\tau_c)} \cdot (f + f_0)^2 \cdot \left[(\tau_0 - \tau_c) - \frac{\Delta\tau}{2}\right]^2 \cdot \frac{1}{1 + \dfrac{v^2 \cdot \Delta\tau^2}{R^2(\tau_c)} \cdot \left[\dfrac{1}{4} - \dfrac{(\tau_0 - \tau_c)}{\Delta\tau}\right]}$$

(3.217)

$$|f_\tau| \leq \frac{2v^2}{c} \cdot \frac{1}{R(\tau_c)} \cdot |f + f_0| \cdot \left|(\tau_0 - \tau_c) - \frac{\Delta\tau}{2}\right| \cdot \frac{1}{\left[1 + \dfrac{v^2 \cdot \Delta\tau^2}{R^2(\tau_c)} \cdot \left[\dfrac{1}{4} - \dfrac{(\tau_0 - \tau_c)}{\Delta\tau}\right]\right]^{\frac{1}{2}}}$$

(3.218)

Due to the inequality we have to take the negative solution because $f_\tau > f_{\text{limit}}$:

$$f_\tau \geq \frac{2v^2}{c} \cdot \frac{1}{R(\tau_c)} \cdot (f + f_0) \cdot \left[(\tau_0 - \tau_c) - \frac{\Delta\tau}{2}\right] \cdot \frac{1}{\left[1 + \dfrac{v^2 \cdot \Delta\tau^2}{R^2(\tau_c)} \cdot \left[\dfrac{1}{4} - \dfrac{(\tau_0 - \tau_c)}{\Delta\tau}\right]\right]^{\frac{1}{2}}}$$

(3.219)

A-3.4.1.9.2.6 Summary case 2

$$\frac{2v^2}{c} \cdot \frac{1}{R(\tau_c)} \cdot (f + f_0) \cdot \left[(\tau_0 - \tau_c) - \frac{\Delta\tau}{2}\right] \cdot \frac{1}{\left[1 + \dfrac{v^2 \cdot \Delta\tau^2}{R^2(\tau_c)} \cdot \left[\dfrac{1}{4} - \dfrac{(\tau_0 - \tau_c)}{\Delta\tau}\right]\right]^{\frac{1}{2}}}$$

$$\leq f_\tau \leq \frac{2v^2}{c} \cdot \frac{1}{R(\tau_c)} \cdot (f + f_0) \cdot \left[(\tau_0 - \tau_c) + \frac{\Delta\tau}{2}\right]$$

$$\cdot \frac{1}{\left[1 + \dfrac{v^2 \cdot \Delta\tau^2}{R^2(\tau_c)} \cdot \left[\dfrac{1}{4} + \dfrac{(\tau_0 - \tau_c)}{\Delta\tau}\right]\right]^{\frac{1}{2}}}$$

(3.220)

It follows for the centroid frequency:

$$f_c \cong \frac{2v^2}{c} \cdot \frac{1}{R(\tau_c)} \cdot (\tau_0 - \tau_c) \cdot (f + f_0)$$

(3.221)

and the bandwidth:

$$B_{az} \cong \frac{2v^2}{c} \cdot \frac{1}{R(\tau_c)} \cdot (f + f_0) \cdot \Delta\tau$$

(3.222)

A-3.4.1.10 Summary of the position of the rectangular function

Centroid frequency:

$$f_c = \frac{1}{2}(f_0 + f_u) \cong \frac{2v^2}{c} \cdot \frac{1}{R(\tau_c)} \cdot (\tau_0 - \tau_c) \cdot (f + f_0) \tag{3.223}$$

Bandwidth:

$$B_{az} = f_0 - f_u \cong \frac{2v^2}{c} \cdot \frac{1}{R(\tau_c)} \cdot (f + f_0) \cdot \Delta\tau \tag{3.224}$$

Look for the centroid frequency:

$$f_c \cong \frac{2v^2}{c} \cdot \frac{1}{R(\tau_c)} \cdot (\tau_0 - \tau_c) \cdot (f + f_0)$$

with $\lambda = \dfrac{c}{f_0}$ follows:

$$f_c \cong \frac{2v^2}{\lambda} \cdot \frac{1}{R(\tau_c)} \cdot (\tau_0 - \tau_c) \cdot \left(1 + \frac{f}{f_0}\right) \tag{3.225}$$

In equation (3.225) we see the following effect with consequences for the doppler frequency. The doppler frequency does not run constantly into the azimuth direction. The doppler frequency arises due to the term $1 + f/f_0$. We neglect this term because in the first approximation $f \ll f_0$.

A-3.4.2 Point target spectrum in the two-dimensional frequency domain

As final result we get for the two-dimensional point-target spectrum the following term:

$$R_T(f, f_\tau, R_0, \tau_0) = S_T(f) \cdot \sigma(R_0, \tau_0) \cdot \mathrm{rect}\left[\frac{f_\tau - f_c}{B_{az}}\right] \cdot \sqrt{\frac{c \cdot R_0}{2}} \cdot \frac{1}{v}$$

$$\cdot \frac{f + f_0}{\left[(f + f_0)^2 - \dfrac{f_\tau^2 \cdot c^2}{4v^2}\right]^{3/4}} \cdot e^{-j \cdot \pi/4} \cdot e^{-j \cdot 2\pi f_\tau \tau_0}$$

$$\cdot e^{-j4\pi \frac{R_0}{c} \cdot \sqrt{(f+f_0)^2 - \frac{f_\tau^2 c^2}{4v^2}}} \tag{3.227}$$

This term contains all SAR sensor parameters necessary for the processing. We use this result to return to the derivation of the point-target spectrum and different processing algorithms chapter 3.4 page 120.

Bibliography

[ADA95] Adam, N., "Phase Unwrapping in SAR-Interferogrammen", Diplomarbeit Rostock/Oberpfaffenhofen, 1995

[AND79] Anderson, B. D. O.; Moore J. B., "Optimal filtering", New Jersey: Prentice Hall 1979

[ARN96] Arndt, C., "Informationsgewinnung und -verarbeitung in nichtlinearen dynamischen Systemen", Dissertation 1996, Uni-GH-Siegen, INV

[ARN98] Arndt, C.; Hein, A.; Loffeld, O., "Information Theory in Data Fusion", Proceedings 'Fusion of Earth data', 28–30 January 98, Sophia Antipolis, Frankreich

[BAM89] Bamler, R., "Das $\pi \times$ Daumen X-SAR Formula Package", Arbeitsversion PDXSFP; WT-DA-MV 13.04.89

[BAM91/1] Bamler, R.; Runge, H., "A Novel PRF-Ambiguity Resolver", IEEE Transactions on Geoscience and Remote Sensing, pp. 1035–1038, IGARSS 1991

[BAM91/2] Bamler, R.; Runge, H., "PRF-Ambiguity Resolving by Wavelength Diversity", IEEE Transactions on Geoscience and Remote Sensing, Vol. 29, No. 6, November 1991

[BAM92/1] Bamler, R., "A Comparison of Range-Doppler and Wave-Domain SAR Focussing Algorithms", IEEE Trans. Geosci. Remote Sensing 30, 706–713, 1992

[BAM92/2] Bamler, R., "A Novel High Precision SAR Focussing Algorithm Based on Chirp Scaling", IEEE Trans. Geosci. Remote Sensing 1992, 91-72810/92$03.00

[BAM93] Bamler, R.; Just, D., "Phase Statistics and Decorrelation in SAR Interferograms", IGARRS '93, Tokyo, pp. 980–984, 1993

[BAR83] Barber, B.C., "Theory of digital Imaging from orbital Synthetic Aperture Radar", Technical Report 83079, Royal Aircraft Establishment, November 1983

[BAR87] Bartsch, H.J., "Taschenbuch mathematischer Formeln", 10. Auflage, Verlag Harri Deutsch, VEB Leipzig 1987

[BEN79] Bennet, J.R.; Cumming, I.G., "A digital processor for the production of Seasat synthetic aperture radar imaginary", Proc. SURGE Workshop Frascati, ESA-SP-154, 1997

[BOL98] Bolter, R.; Gelautz, M.; Leberl, F., "Simulation Based SAR-Stereo Analysis in Layover Areas", 0-7803-4403-0/98$10.00, Proceeding IGARSS '98

[BUC94] Buckreus, S., "Bewegungskompensation für flugzeuggetragene SAR-Systeme", DLR-Forschungsbericht 94-17

[BRE98] Breit, H.; Bamler, R., "An InSAR Processor for On-Board Performance Monitoring of the SRTM/X-SAR Interferometer", 0-7803-4403-0/98$10.00, Proceeding IGARSS '98

[BRO91] Bronstein, I.N., "Taschenbuch der Mathematik", 25. Auflage, Teubner, 1991

[CAF91]   **Cafforio, C.; Prati, C.; Rocca, F.,** "SAR Data Focussing Using Seismic Migration Techniques", IEEE Trans. Aerosp. Electr. Syst., AES-27, 194–207, 1991

[CAN89]   **Cantafio, L. J.,** "Space-Based Radar Handbook", Artech House 1989

[CAR87]   **Carver, K.; et al.,** "Earth Observing System Reports Vol. 2f", Synthetic Apertur Radar, Instrument Panel Report, NASA, Washington 1987

[COO67]   **Cook, C. E.; Bernfeld, M.,** "Radar Signals", Academic Press, New York, San Francisco, London 1967

[DAV94]   **Davidson, G. W.,** "Image Formation From Squint Mode Synthetic Aperture Radar Data", Dissertation, Department of Electrical Engineering, Faculty of Applied Science, University of British Columbia, 1994

[DET89]   **Detlefsen, J.,** "Radartechnik", Springer 1989

[ELA88]   **Elachi, C.,** "Spaceborne Radar Remote Sensing Applications and Techniques", IEEE Press, New York 1988

[ERS92]   **Vass, P.; Battrick, B.,** "ERS-1-System", ESA Publications, ESRIN, ESTEC, ESA SP-1146, 1992

[FIT88]   **Fitch, J. P.,** "Synthetic Aperture Radar", Springer 1988

[FLI91]   **Fliege, N.,** "Systemtheorie", B. G. Teubner, Stuttgart, 1991

[FOR95]   **Fornaro, G.; Franceschetti, G.,** "Image registration in interferometric SAR processing", IEE Proceedings Radar, Sonar, Navigation, Vol. 142, No. 6, 1995

[FOR96]   **Fornaro, G.; Franceschetti, G.,** "Interferometric SAR Phase Unwrapping Using Green's Formulation", IEEE Transactions on Geoscience and Remote Sensing, Vol. 34, No. 3, pp. 720–727, 1996

[FRA90]   **Franceschetti, G.; Schirinzi, G.,** "A SAR Processor Based on Two-Dimensional FFT Codes", IEEE Trans. Aerosp. Electr. Syst.; AES-26, pp. 356–365, 1990

[FRA92]   **Franceschetti, G.; Lanari, R.; Pascazio; V. Schirinzi, G.,** "WASAR A Wide-Angle SAR Processor", Proceedings IEE, 139-F, 107–114, 1992

[GAB89]   **Gabrial, A. K.; Goldstein, R. M.; Zebker, H. A.,** "Mapping Small Elevation Changes Over Large Areas Differential Radar Interferometry", Journal of Geophysical Research, Vol. 94, No. B7, July 1989

[GAT94]   **Gatelli, F.; Guarnieri, M.; Parizzi, F.; Pasquali, P.; Prati, C.; Rocca, F.,** "The Wavenumber Shift in SAR Interferometry", IEEE Transactions on Geoscience and Remote Sensing, Vol. 32, No. 4, pp. 855–865, 1994

[GEL96]   **Gelautz, M.; Leberl, F.; Kellerer-Pirklbauer, W.,** "Image Enhancement and Evaluation: SAR Layover and Shadows", Proceeding EUSAR '96, 71–76, 1996

[GEU94]   **Geudtner, D.; Schwäbisch, M.; Winter, R.,** "SAR Interferometrie with ERS-1 Data", Proceedings Piers '94, Noordwijk 1994

[GEU95]   **Geudtner, D.,** "Die interferometrische Verarbeitung von SAR-Daten der ERS-1", Dissertation, Universität Stuttgart, Oberpfaffenhofen, 1995

[GOO]   **Goodman, J. W.,** "Introduction to Fourier Optics", McGraw-Hill Book Company

[GRA74]   **Graham, L. C.,** "Synthetic Interferometer Radar for Topographic Mapping", Proceedings of the IEEE, Vol. 62, No. 6, pp. 763–768, 1974

[HAF94]   **Hafner, H.; et al.,** "Geometrische und radiometrische Vorverarbeitung von SAR-Aufnahmen für geographische Anwendungen", ZPF 4/94, 123–128, (Wichman) Heidelberg, 1994

[HAR70]   **Harger, R. O.,** "Synthetic Aperture Radar Systems", Academic Press, New York 1970

[HAR93]   Hartl, P.; Wu, X., "SAR Interferometry: Experiences with various Phase Unwrapping methods", Proceedings Second ERS-1 Symposium, Hamburg, Germany, 1993

[HEI95]   Hein, A., "Zweidimensionale ortsvariante SAR-Bildverarbeitung mit Hilfe skalierter Fouriertransformation", Diplomarbeit, INV, Uni-GH-Siegen, 1995

[HEI97]   Hein, A.; Loffeld, O., "The Slant To Ground Converter Module – including Layover and Shadow Correction", Reference Guide, Zentrum für Sensorsysteme, Uni-GH-Siegen, 1997

[HEI98]   Hein, A.; Loffeld, O.; Arndt, Ch., "Correcting Shadowing, Layover and Foreshortening Height Errors in interferometric SAR-images", International Association of Science and Technology for Development – IASTED '98, pp. 529–532, 11–14 February 98, Canary Islands, Spain (http://www.gpds.ulpgc.es/icspc98)

[JUS94]   Just, D.; Bamler, R., "Phase statistics of interferograms with applications to synthetic aperture radar", Applied Optics, Vol. 33, No. 20, pp. 4361–4368, 1994

[KAU92]   Kaudewitz, M. J., "Untersuchung und Analyse der wesentlichen X-SAR-Verarbeitungsparameter und ihrer Abhängigkeit von den Orbitalparametern", Studienarbeit, INV, Uni-GH-Siegen, 1992

[KER83]   Kernigham, B. W., Ritchie, D. M., "Programmieren in C", Carl Hanser München Wien 1983

[KLA90]   Klaus, F., "Entwicklung, Adaption und Analyse von „Fixed Interval Kalman-Smoothern" für die Doppler-Centroid-Estimation beim Synthetic Aperture Radar", Diplomarbeit am INV/Uni-GH-Siegen, 1990

[KLA91]   Klaus, F.; Loffeld, O., "A Concept for the Simulation of SAR Data", Proc. IGARSS '91, International Geoscience and Remote Sensing Symposium, 2391–2394, Helsinki 1991

[KLA92]   Klaus, F.; Loffeld, O., "Simulation of Planetary Surfaces for a SAR Data Simulator", Proc. IGARSS '92, International Geoscience and Remote Sensing Symposium, Houston, 1992

[KLA93]   Klaus, F.; Blewonska, U.; Loffeld, O., "Simulation of Shuttle Trajectory and Attitude for a SAR-Simulator", Proc. IGARSS '93, International Geoscience and Remote Sensing Symposium, Tokyo 1993

[KLA94]   Klaus, F., "SAR-Simulation", Dissertation, Universität-GH-Siegen, August 1994

[KLA95]   Klaus, F., "Simulation von Radar-Sensoren mit synthetischer Apertur", Dissertation Universität Siegen, INV, Fortschritt-Bericht VDI Reihe 8, Nr. 447, VDI-Verlag, Düsseldorf 1995

[KLA96]   Klaus, F.; Hein, A.; Loffeld, O., "Generating Interferometric SAR Raw Data Using SISAR", Proceedings EUSAR '96, 195–208, Königswinter 1996

[KLA89]   Klausing, H., "Realisierbarkeit eines Radars mit synthetischer Apertur durch rotierende Antennen", MBB-Publikation, MBB-UA-1150-89-Pub, 1989

[KNE98]   Knedlik, S., "Co-Registrierung von SAR-Bildern", Studienarbeit am Institut für Nachrichtenverarbeitung der Uni-GH-Siegen, 1998

[KOV]   Kovaly, J. J., "Synthetic Aperture Radar", Artech

[KRÄ96/1] Krämer, R.; Loffeld, O., "A Novel Procedure For Cutline Detection", Proceedings EUSAR '96, Königswinter, March 26–28, pp. 253–256, 1996

[KRÄ96/2] Krämer, R.; Loffeld, O., "Phase Unwrapping for SAR Interferometry with Kalman Filters", Proceedings EUSAR '96, Königswinter; March 26–28, pp. 165–169, 1996

[KRÄ98]   **Krämer, R.,** "Kalman-Filter basierende Phasen- und Parameterestimation zur Lösung der Phasenvieldeutigkeitsproblematik bei der Höhenmodellerstellung aus SAR-Interferogrammen", Dissertation Universität Siegen, INV, 1998

[LAN]   **Lanari, R.,** "A New Method for the Compensation of the SAR Range Cell Migration based on the Z-Chirp Transform", in Vorbereitung zur Veröffentlichung IEEE Transactions

[LI90]   **Li, F. K.; Goldstein, R. M.,** "Studies of Multibaseline Spaceborne Interferometric Synthetic Aperture Radars", IEEE Trans. Geo. Rem. Sens., Vol. 28, Nr. 1, 88–97, 1990

[LI91]   **Li, A.; Loffeld, O.,** "Two Dimensional SAR Processing in the Frequency Domain", Proc. IGARSS '91, International Geoscience and Remote Sensing Symposium, Helsinki, 1065–1068, 1991

[LI93]   **Li, A.,** "Systemtheoretische Untersuchung und Entwicklung neuartiger SAR Signalverarbeitungsalgorithmen", Dissertation Universität Siegen, INV, 1993

[LIN92/1]   **Lin, Q.; Vesecky, J. F.; Zebker, H. A.,** "Registration of Interferometric SAR Images", Proc. IGARSS '92, 1579–1581, 1992

[LIN92/2]   **Lin, Q.; Vesecky, J. F.; Zebker, H. A.,** "New Approaches in Interferometric SAR Data Processing", IEEE Transactions on Geoscience and Remote Sensing, Vol. 30, No. 3, pp. 560–567, 1992

[LIN92/3]   **Lin, Q.; Vesecky, J. F.; Zebker, H. A.,** "Registration of interferometric SAR images", Proceedings of the International Geoscience and Remote Sensing Symposium IGARSS '92, pp. 1579–1581, 1992

[LIN94]   **Lin, Q.; Vesecky, J. F.; Zebker, H. A.,** "Comparison of elevation derived from INSAR data with DEM over large relief terrain", International Journal Remote Sensing, Vol. 15, No. 9, pp. 1775–1790, 1994

[LOF]   **Loffeld, O.,** Vorlesungsskript "SAR-Signale und SAR-Signalverarbeitung", INV, Universität Siegen

[LOF89/1]   **Loffeld, O.,** "Kalman Filtering for X-SAR Doppler Centroid Improvement, Part I, Kalman Filter Design", DLR Contract Report, Contr. No. 5-565-4408, München, 1989

[LOF89/2]   **Loffeld, O.,** "Kalman Filtering for X-SAR Doppler Centroid Improvement, Part II, Kalman Filter Implementation and Tests", DLR Contract Report, Contr. No. 5-565-4408, München, 1989

[LOF89/3]   **Loffeld, O.,** "Mathematical Model of the Shuttle's Attitude", DLR Study Report, Contr. No. 5-565-4395, München, 1989

[LOF90/1]   **Loffeld, O.,** "Estimationstheorie", Band 1+2, R. Oldenbourg, München Wien 1990

[LOF90/2]   **Loffeld, O.,** "Doppler-Centroid-Estimation für Synthetic Aperture Radar im X-Band mit Kalman-Filtern", Beitrag zur Broschüre ‚Forschungsland NRW – Thema Raumfahrt'; im Auftrag des Ministers für Wissenschaft und Forschung des Landes NRW, 1990

[LOF90/3]   **Loffeld, O., Sack, A.,** "Kalman Filters for the Demodulation of Angular Modulated Signals", Proc. ICSP 1990, pp. 631–634, Beijing, China 1990

[LOF90/4]   **Loffeld, O.,** "Doppler Centroid Estimation With Kalman Filters", Conf. Proceedings IGARSS '90, International Geoscience and Remote Sensing Symposium, Washington, May 1990

[LOF91/1]   **Loffeld, O.,** "Estimating Range And Time Varying Doppler Centroids With Kalman Filters", Proceedings IGARSS '91, Espoo, Finland, 1043–1045, 1991

[LOF91/2]   **Loffeld, O.,** "Demodulation of noisy phase or frequency modulated signals with Kalman filters", Proceeding IGARSS 1991

[LOF93]   **Loffeld, O.,** "Estimating Range and Time Varying Doppler Centroids with Kalman Filters", Proc. IGARSS '93, International Geoscience and Remote Sensing Symposium, Tokyo, 1993

[LOF94/1]   **Loffeld, O.,** "Phase Unwrapping for SAR Interferometry with Kalman Filters", Proc. IGARSS '94, International Geoscience and Remote Sensing Symposium, Pasadena, Aug. 1994

[LOF94/2]   **Loffeld, O.,** "Demodulation of noisy phase or frequency modulated signals with Kalman filters", ICASSP94, IV, pp. 177–190, Adelaide, Australia 1994

[LOF96/1]   **Loffeld, O.; Hein, A.,** "SAR Processing by 'Inverse Scaled Fourier Transformation', Proceedings EUSAR '96, Königswinter, 143–146, 1996

[LOF96/2]   **Loffeld, O.; Hein, A.,** "A Phase Preserving SAR-Processor by Inverse Scaled Fourier Transformation", Invited paper at Piers '96, Innsbruck, 541–544, 1996

[LOF96/4]   **Loffeld, O.; Arndt, C.; Hein, A.,** "Estimating the derivative of modulo-mapped phases", Proceeding FRINGE '96, ESA Workshop ETH Zürich, 1–3 October 96 (http://www.geo.unizh.ch/rsl/fringe96/)

[LOF98/1]   **Loffeld, O.; Hein, A.,** "Focussing SAR Images By Inverse Scaled Fourier Transformation", Proceeding IASTED '98, 537–541, 11–14 February 98, Canary Islands, Spain (http://www.gpds.ulpgc.es/icspc98)

[LOF98/2]   **Loffeld, O.; Hein, A.,** "SAR Focusing: Scaled Inverse Fourier Transformation and Chirp Scaling", IEEE International Geoscience and Remote Sensing Symposium IGARSS '98, 6–10 July 1998, Seattle WA, USA

[LÜK92]   **Lüke, H. D.,** "Signalübertragung", 5. Auflage, Springer, 1992

[MAD93]   **Madsen, S. N.; Zebker, H. A.; Martin, J. M.,** "Topographic Mapping Using Radar Interferometry: Processing Techniques", IEEE Transactions on Geoscience and Remote Sensing, Vol. 32, No. 1, 1993

[MAS93/1]   **Massonnet, D.,** "Validation of ERS-1 interferometry at CNES", Proceedings Second ERS-1 Symposium, Hamburg, Germany, 1993

[MAS93/2]   **Massonnet, D.; Rabaute, T.,** "Radar Interferometry: Limits and Potential", IEEE Transactions on Geoscience and Remote Sensing, Vol. 31, No. 2, 1993

[MAS93/3]   **Massonet, D.; et al.,** "The Displacement Field of the Landers Earthquake Mapped by Radar Interferometrie", Nature, Vol. 364, July 1993

[MAY83]   **Maybeck, P. S.,** "Stochastic Models, Estimation and Control", Vol. 1–3, Academic Press, New York, 1979, 1982, 1983

[MCD]   **McDonough, R. N.; Raff, B. E.; Kerr, J. L.,** "Image Formation from Spaceborne Synthetic Aperture Radar Signals", Johns Hopkins APL Technical Digest, Vol. 6, No. 4

[MCG74]   **McGarty, T. P.,** "Stochastic Systems and State Estimation", Wiley, New York 1974

[MCG82]   **McGowan, Kuc,** "A Direct Relation Between a Signal Time Series and Its Unwrapped Phase", IEEE Transactions on Acoustics, Speech, and Signal Processing, Vol. ASSP-30, No. 5, pp. 719–726, 1982

[MEI89]   **Meier, E.,** "Geometrische Korrektur von Bildern orbitgestützter SAR-Systeme", Dissertation, Universität Zürich, 1989

[MIT82]   **Mittra, R.; Imbriale, W. A.; Maanders, E. J.,** "Satellite Communication Antenna Technology", North-Holland 1982

[MON98]   **Montgomery, B. C.,** "A remote sensing analysis of vector abundance and malaria risk associated with selected villages in southern Belize", 0-7803-4403-0/98$10.00, Proceedings IGARSS '98

[MOR90]   Moreira, J. R., "A new Method of Aircraft Motion Error Extraction from Radar Raw Data for Real Time Motion Compensation", IEEE Trans. Geosci. Remote Sensing 28, 620–626, 1990

[MOR92]   Moreira, J. R., "Bewegungsextraktionsverfahren für Radar mit synthetischer Apertur", Dissertation TU München 1992

[MOR95]   Moreira, J.; Schwäbisch, M.; Fornaro, G.; Lanari, R.; Bamler, R.; Just, D.; Steinbrecher, U.; Breit, H.; Eineder, M.; Franceschetti, G.; Geudtner, D.; Rinkel, H., "X-SAR Interferometry: First Results", IEEE Transactions on Geoscience and Remote Sensing, Vol. 33, No. 4, pp. 950–955, 1995

[NEU88]   Neufang, O., Rühl, H., "Elektronik Wörterbuch", 3. Auflage, Zimmermann-Neufang, Ulmen, 1988

[OPP89]   Oppenheim, A. V., Schafer, R. W., "Discrete Signal Processing", 1989 Prentice Hall, Englewood Cliffs, New Jersey 07632

[ÖTT85]   Öttl, H.; Valdoni, F.; et al., "DFVLR-Mitteilungen Nr. 85-17", Deutsche Forschungs- und Versuchsanstalt für Luft- und Raumfahrt, Oberpfaffenhofen 1985

[PAP68]   Papoulis, A., "Systems and Transforms with Applications in Optics", McGraw-Hill Book Company 1968

[PRA89/1]   Prati, C.; Rocca, F., "Limits of the Resolution of Elevation MaPs from Sereo SAR Images", IEEE Trans. Geo. Rem. Sens., Vol. 11, No. 12, 2215–2235, 1990

[PRA89/2]   Prati, C.; Rocca, F.; Monti Guarnieri, A., "Effects of Speckle and Additive Noise on the Altimetric Resolution of Interferometric SAR (ISAR) Surveys", Proc. IGARSS '89, 2469–2472, 1989

[PRA90/1]   Prati, C.; Rocca, F., "Limits to the resolution of elevation maPs from stereo SAR images", Int. J. remote sensing, Vol. 11, No. 12, pp. 2215–2235, 1990

[PRA90/2]   Prati, C.; Rocca, F.; Monti Guarnieri, A., "Seismic Migration For SAR Focusing: Interferometrical Applications", IEEE Transactions on Geoscience and Remote Sensing, Vol. 28, No. 4, pp. 627–640, 1990

[PRA93/1]   Prati, C.; Rocca, F., "Improving Slant-Range Resolution With Multiple SAR Surveys", IEEE Transactions on Aerospace and Electronic Systems, Vol. 29, No. 1, pp. 135–144, 1993

[PRA93/2]   Prati, C.; Rocca, F., "Use of the spectra shift in SAR interferometry", Proceedings Second ERS-1 Symposium, Hamburg, Germany, 1993

[RAN89]   Raney, R. K.; Vachon, P. W., "A Phase Preserving SAR Processor", Proc. IGARSS '89, Vancouver, 2588–2591, 1989

[RAN92/1]   Raney, R. K., "An exact wide Field digital Imaging Algorithm", Int. J. of Remote Sensing 13, 991–998, 1992

[RAN92/2]   Raney, R. K., "A New And Fundamental Fourier Transform Pair", IEEE 1992, 91-72810/92$03.00

[RAN94]   Raney, K. R.; Runge, H.; Bamler, R.; Cumming, I. G.; Wong, F. H., "Precision SAR Processing Using Chirp Scaling", IEEE Trans. Geosci. Remote Sensing, Vol. 32, No. 4, July 1994

[ROC89]   Rocca, F.; Prati, C.; Guarnieri, A. M., "New Algorithms for Processing SAR Data", European Space Agency Contract Report, Politecnico di Milano Dipartimento di Electronica, July 1989

[ROD92]   Rodriguez, E.; Martin, J. M., "Theory and design of interferometric synthetic apereture radars", IEEE Proceedings F, Vol. 139, No. 2, 1992

[ROS76]   Rosenfeld, A.; Kahn, A. C., "Digital Picture Processing", Academic Press, New York 1976

[RUN92]   Runge, H.; Bamler, R., "A novel high Precision SAR focussing Algorithm Based on Chirp Scaling", IGARSS '92, 372–375, Texas, 1992

[SCHA86] Schanda, E., "Radar mit synthetischer Apertur (SAR)", Mikrowellen Magazin 12, 534–543, 6/1986

[SCHN81] Schneider, M., "Himmelsmechanik", B.I.-Wissenschaftsverlag 1981

[SCHR98] Schreiber, R., "A New Method for Accurate Co-Registration of SAR Images", 0-7803-4403-0/98$10.00, Proceedings IGARSS '98

[SCHW95/1] Schwäbisch, M.; Geudtner, D., "Improvement of Phase and Coherence Map Quality Using Azimuth Prefiltering: Examples from ERS-1 and X-SAR", Proceeding IGARSS '95

[SCHW95/2] Schwäbisch, M., "Erzeugung digitaler Geländemodelle mit Methoden der SAR-Interferometrie", DLR-Nachrichten, Heft 79, pp. 20–24, 1995

[SCHW95/3] Schwäbisch, M., "Die SAR-Interferometrie zur Erzeugung digitaler Geländemodelle", Dissertation, Universität Stuttgart, Oberpfaffenhofen; 1995

[SCL993] Sclabassi, R. J.; Sun, M., "Diskrete Augenblicksfrequenz und ihre Berechnung", IEEE Transaction on Signal Processing Vol. 41, No. 5, May 1993

[SEI93] Seifert, P.; Zink, M., "Synthetik-Apertur-Radar-Technik und Anwendung", Physik in unserer Zeit 24, 24 (1/1993)

[SKI91] Skifstad, K. B., "High Speed range estimation based on intensity grade analysis", Springer 1991

[SKO70] Skolnik, M. I., "Radar Handbook", Mc-Graw-Hill Book Company 1970

[SMA93] Small, D.; Werner, C.; Nüesch, D., "Baseline Modelling for ERS-1 SAR Interferometry", Proc. IGARSS '93, 1024–1209, 1993

[SMI91] Smith, A. M., "A New Approach to Range-Doppler SAR-Processing", Int. J. of Remote Sensing 12, 235–251, 1991

[SOM95] Sommer, D., "Simultane Zustands- und Parameterestimation für Remote Sensing Anwendungnen mit Extended Linearized Kalman-Filtern", Diplomarbeit INV/Universität GH-Siegen 1995

[STE91] Steinbrecher, U.; Bamler, R.; Runge, H.; Loffeld, O., "Performance Analysis of Kalman Filters For Doppler Centroid Estimation", Proc. IGARSS '91; International Geoscience and Remote Sensing Symposium, Helsinki, 1059–1063, 1991

[STO98] Stoever, I.; Kunz, A., "Automatische Höhenpaßpunktgeneration in interferometrischen SAR-Bildern", Studienarbeit, INV, Universität GH-Siegen, 1998

[STO78] Stolt, R. H., "Migration by Fourier Transform", Geophysics, Vol. 43, 23–48, 1978

[ULAB82] Ulaby, F. T.; Moore, R. K.; Funk, A. K., "Microwave Remote Sensing", Vol. 2, Addison-Wesley, Reading MA 1982

[WAH84] Wahl, F. M., "Digitale Bildverarbeitung", Springer 1984

[WEG93] Wegmüller, U.; Werner, C. H.; Rosen, P., "Derivation of terrain slope from SAR interferometric phase gradient", Proceedings Second ERS-1 Symposium, Hamburg, Germany, 1993

[WER93] Werner, C. H.; Wegmüller, U.; Small, D., "Applications of Interferometrically derived Terrain Slopes: Normalization of SAR Backscatter and the Interferometric Correlation Coefficient", Proceedings Second ERS-1 Symposium, Hamburg, Germany, 1993

[WIE49] Wiener, N., "Extrapolation, Interpolation and Smoothing of Stationary Time Series, with Engineering Applications", Wiley, New York 1949

[WU76] Wu, C., "A Digital System to Produce Imaginary from SAR-Data", AIAA Systems Design Driven by Sensors, Jet Propulsion Lab, Pasadena, Calif., October 18–20, 1976

[ZEB86] Zebker, H. A.; Goldstein, R. M., "Topographic Mapping from Interferometric Synthetic Aperture Radar Observations", J. f. Geophys. Res., Vol. 91, Nr. B5, 4993–4999, 1986

[ZEB92]   Zebker, H. A.; Villasenor, J., "Decorrelation in Interferometric Radar Echoes", IEEE Transactions on Geoscience and Remote Sensing, Vol. 30, No. 5, 1992

[ZEB92]   Zebker, H. A., "Decorrelation in Interferometric Radar Echoes", IEEE Transactions on Geosci. and Remote Sensing, Vol. 30, No. 5, pp. 950–959, 1992

[ZEB93]   Zebker, H. A.; Werner, C. L.; Rosen, P. A.; Hensley, S., "Accuracy of Topographic Maps Derived from ERS-1 Interferometric Radar", IEEE Transactions on Geoscience and Remote Sensing, Vol. 32, No. 4, pp. 823–836, 1993

[ZIN93]   Zink, M., "Kalibration von SAR-Systemen", Dissertation, DLR, 05.1993

List of symbols

α_{3dB}	3-dB aperture of antenna
ε_{3dB}	3-dB aperture of synthetic array
A_w	effective antenna
$A(x)$	antenna function
B_{az}	bandwidth in azimuth
B_r	bandwidth of transmitted pulse
β	squint angle
c	velocity of light
$C(\phi)$	antenna-aperture function
$D_s(r)$	density of radiation
D_x	length of antenna in azimuth
D_y	length of antenna in elevation
δ	pitch angle
$\delta(t)$	delta function
Δx	resolution in azimuth
Δx_e	single 3-dB aperture of antenna's footprint in azimuth
Δx_z	double 3-dB aperture of antenna's footprint in azimuth
Δy	3-dB aperture of antenna's footprints in ground-range direction
Δf_τ	signal bandwidth in azimuth
ΔR_u	range resolution uncompressed
ΔR_k	range resolution compressed
ΔR_y	ground-range resolution
$\Delta \tau$	puls-length of azimuth signal
$\Delta \tau_b$	time of pulse response of azimuth bandpass
$E(\phi)$	field force
f	range frequency
f_a	sample frequency
f_0	carrier frequency of transmitted pulse
$f_0(t)$	current frequency
f_c	centroid frequency of azimuth spectrum
f_{dc}	doppler-centroid frequency
f_R	frequency in range
f_τ	azimuth frequency
φ_0	incidence angle
ϕ	aperture angle
ϕ_{HB}	peak width at half-height
ϕ_{HW}	angle at half-height

$\varphi_{ss_T}(t)$	autocorrelation of transmitted signal in equivalent lowpass domain
$\psi_{ss_T}(f)$	autocorrelation- respectively power density spectrum of transmitted signal in equivalent lowpass domain
$g_T(\tau, t)$	range-compressed signal of point target in equivalent lowpass domain
G	antenna factor
$G_T(f_\tau, f)$	spectrum of range-compressed signal of point target in equivalent lowpass domain
gr_r	ground-range distance
H_0	height of SAR sensor
$h_T(t)$	pulse response of transmitted pulse in equivalent lowpass domain
$H_k(f_\tau, f)$	transfer function
k	Boltzmann constant
$K(x)$	fresnel integral
k_{az}	doppler rate
k_r	frequency rate of signal
L_{ges}	loss factor
λ	wave length
$\Lambda(t/T)$	triangle function
$N_{transmit}$	number of echoes
ω	circular frequency
δ	roll angle
P_e	received power
P_R	backscattering power
P_r	power of noise
P_s	transmitted power
PRF	pulse repetition frequency
ψ	yaw angle
ψ_p	projection of squint angle
$\psi_{sg}(f)$	cross-power density of signals s(t) and g(t)
$r(\tau)$	projection of slant-range distance
$r(t)$	received signal
$\mathbf{R}$	distance vector of SAR sensor
$R(\tau)$	azimuth time-dependent distance
R_0	minimum distance of SAR sensor to point target
R_{0max}	maximum range for CW radar
$R_{0maxsyn}$	maximum range for synthetic aperture radar
R_m	mean of minimum distance
R_c	distance between center of footprint and SAR sensor
$rect(t/T)$	rectangular function
ρ	phase shift of received signal
S, x_{repeat}	range between transmitted pulses
SAR	synthetic aperture radar
$s(t)$	transmitted pulse

$s_T(t)$	equivalent lowpass signal of $s(t)$
$S_{T_\cap}(f)$	time-limited chirp signal
$S_{T_\infty}(f)$	time-unlimited chirp signal
$si(x)$	$\sin(x)/x$-function
$Si(x)$	integral sine
sl_r	slant-range distance
SNR	signal-to-noise ratio
$\sigma(x_0,y_0,0)$	scattering coefficient of point target at position $(x_0,y_0,0)$
$T, t_{transmit}$	puls-length of transmitted signal
T_{sep}	time to separate two signals
t_0	time delay $(=2R_0/c)$
t_c	time delay $(=2R_c/c)$
t_v	time delay between transmitted and received signal $(=2R(\tau)/c)$
ϑ_0	effective noise temperature
τ	azimuth time
τ^*, t^*	stationary points
τ_0	azimuth time when target perpendicular
τ_c	azimuth time when target in epicenter of azimuth signal
v_a	velocity of SAR sensor
v_{res}	resulting azimuth components of velocity of SAR sensor
v	velocity of SAR sensor with respect to earth's rotation
x	azimuth-distance coordinate
x_0	x-coordinate of point target on earth
x_{repeat}	puls-repetition distance
x_{pulse}	azimuth puls-length
y	range-distance coordinate
y_0	y-coordinate of point target on earth
$\mathscr{F}$	fourier transformation
Φ	off-nadir angle

MIX
Papier aus verantwortungsvollen Quellen
Paper from responsible sources
FSC® C105338

If you have any concerns about our products,
you can contact us on
ProductSafety@springernature.com

In case Publisher is established outside the EU,
the EU authorized representative is:
Springer Nature Customer Service Center GmbH
Europaplatz 3, 69115 Heidelberg, Germany

Printed by Libri Plureos GmbH
in Hamburg, Germany